Glaciers

Glaciers

THE POLITICS OF ICE

Jorge Daniel Taillant

OXFORD
UNIVERSITY PRESS

Oxford University Press is a department of the University of
Oxford. It furthers the University's objective of excellence in research,
scholarship, and education by publishing worldwide.

Oxford   New York
Auckland   Cape Town   Dar es Salaam   Hong Kong   Karachi
Kuala Lumpur   Madrid   Melbourne   Mexico City   Nairobi
New Delhi   Shanghai   Taipei   Toronto

With offices in
Argentina   Austria   Brazil   Chile   Czech Republic   France   Greece
Guatemala   Hungary   Italy   Japan   Poland   Portugal   Singapore
South Korea   Switzerland   Thailand   Turkey   Ukraine   Vietnam

Oxford is a registered trademark of Oxford University Press
in the UK and certain other countries.

Published in the United States of America by
Oxford University Press
198 Madison Avenue, New York, NY 10016

Library of Congress Cataloging-in-Publication Data
Taillant, Jorge Daniel, 1968-
Glaciers : the politics of ice / Jorge Daniel Taillant.
pages cm
Includes bibliographical references and index.
ISBN 978–0–19–936725–2 (hardback)
1. Glaciers.   2. Glacial erosion.   3. Glaciers—Environmental aspects.
4. Glaciers—Social aspects.   I. Title.
GB2405.T35 2015
551.31'2—dc23
2014043381

1 3 5 7 9 8 6 4 2
Printed in the United States of America
on acid-free paper

To Romina Picolotti, a friend and partner in life for her inspiration and example, for her environmental stewardship and for her unwavering commitment to create a better and more sustainable world

{ CONTENTS }

{ ACKNOWLEDGMENTS }

Many people have opened my eyes to the fantastic world of glaciers. Thanks to them, I am getting to know these fabulous water towers before they vanish.

Perhaps the first thanks that not only I owe, *but that many of us owe*, is to the people who introduced the idea that we needed glacier protection legislation. In Chile, Leopoldo Sanchez Grunert, Rosana Bórquez, Sara Larrain, Rodrigo Polanco, Juan Carlos Urquidi, and Antonio Horvath were the spearheads of the glacier protection movement, laying out the first proposal ever to come up with a glacier protection law. *Thank you!*

Special mention goes to Marta Maffei, the mother of Argentina's national glacier protection law, along with her legal advisor Andrea Burucua of ECOSUR. Their contribution to glaciology and to the protection of glaciers around the world has been paramount and will survive for many generations to come. Without their visionary legislative work, we would not be so far along in our societal effort to protect this very crucial and vulnerable natural resource. I've interviewed both Marta and Andrea during my research in the preparation of this book *Thank you!*

I personally am indebted to those who have helped me learn about glaciers and periglacial environments. My inspiring professors have devoted far more than their mere academic and professional responsibilities to the protection and conservation of ice. Cedomir Marangunic (Geoestudios, of Chile), Juan Carlos Leiva (IANIGLA), Benjamín Morales Arnao (Patronato de las Montañas Andinas, Peru), and Bernard Francou (IRD), were my instructors in the first course I ever took on glaciology organized by the United Nation's Environmental Program (UNEP) in Chile. Cedomir, Juan Carlos, Benjamín, and Bernard have been very patient in answering my innumerable questions on the recognition of glaciers through satellite imagery and on the function and vulnerability of glaciers and geoforms of the periglacial environment. They have donated much of their personal time to long technical conversations and e-mail exchanges concerning my research and publications. Cedo has provided me with unique insights into the world of rock glaciers and has voluntarily contributed his time and energy to our work, coming to conferences on his own financing, offering materials, and being always ready to lend a hand. Juan Carlos has inundated me with academic material, research papers, and studies, and his work has been an anchor in evolving discussions on glacier protection in Argentina. Benjamín has provided inspiration from

Peru, where, after decades of work to avoid glacier tsunamis in the Cordillera Blanca Mountains, he is showing us that we can still do much to protect our glaciers and our communities. He has also invested much energy and work on glacier education. In 2011, Bernard dubbed the sort of work we were doing at the Center for Human Rights and Environment (CEDHA) "cryoactivism." It was the first time I ever heard that concept, and we adopted it instantly because it summarizes exactly what we do! Over long coffee breaks and dinner hours during various glacier courses and on our visits to several glaciers in Ecuador and Chile, Bernard has shared his deep and first-hand knowledge of the receding glaciers of the Tropical Andes. *Thank you!*

I am especially indebted to three expert glaciologists whose direct contribution to our work has been paramount: Juan Pablo Milana, Alexander Brenning, and Mateo Martini; they are always available to deepen my inquiries about the particularities of the cryosphere. They have spent valuable unpaid time reviewing my reports and other materials. Juan Pablo is a constant point of reference and source of information. Be it an SMS message, an e-mail, or a phone call, many times monthly, he patiently reviews, criticizes (sometimes too much), and edits much of the cryoactivism I engage on. I have come to respect and admire his career-long contribution to glaciology. Juan Pablo is perhaps one of the most knowledgeable specialists of the periglacial environment, having spent the better portion of his career amongst the rock glaciers and frozen grounds of the Central Andes. Alexander Brenning deserves a fair share of gratitude as he was the first brave soul in the academic glacier world to lend a hand, to reach out to nonacademic actors concerned with glacier protection to help them learn, explore, and understand these sensitive environments and help guide my (and CEDHA's) learning process in the task of inventorying glaciers and understanding the impacts caused by mining operations. Alex's unselfish commitment to his work and to the protection of glaciers has been an inspiration to us all. Mateo Martini, Doctor in Geological Studies of the University of Cordoba, Argentina, has been our unwavering trainer on rock-glacier recognition, not only reviewing all of our inventories but also sitting with the CEDHA team to show us the nuances among periglacial forms. Without him, we would have never been able to generate much of the material that we have produced over the years. *Thank you!*

I must recognize the folks at the IANIGLA, Argentina's Snow and Glacier Institute, particularly the contribution of Dario Trombotto, one of the world's most knowledgeable geocryologists and periglacial environment specialists. Dario has patiently answered many questions about the dynamics of rock glaciers and permafrost and has openly shared much of his work, publications, pictures, and other materials that have been critical to my work and research. Gabriel Cabrera, Juan Carlos Leiva (again), Lydia Espizua, Ricardo Villalba, and Mariano Castro of the IANIGLA have contributed to my knowledge of glacier areas and periglacial environments in the Central Andes, as well as materials that have been used in this book. Juan Carlos, Gabriel, and Lydia particularly have helped with my research on the impacts to glacier and periglacial environments of the Pascua Lama and Veladero mining projects. Ricardo has been instrumental in understanding the politics and evolution of the glacier law from the viewpoint of the IANIGLA. He also took a strong public stance after the presidential veto of the first glacier law by speaking at numerous public events, which undoubtedly influenced the comeback of the glacier law two years later. *Thank you!*

Thanks goes also to Stephen Gruber of the University of Zurich who shared his permafrost model with us and with whom I've maintained several communications to discuss the particularities of his global permafrost mapping tool. *Thank you!*

In San Juan, I must thank Silvia Villalonga of the Fundación Ciudadanos Independientes (FuCI), and Diego Seguí, both of whom offered me valuable information on the history of the early stages of legal debates and filings in the courts to protect glaciers. Through our various social networks, we've also met many individuals who have helped with the research for our various reports and for this book, offering information, pictures, videos, and other material that we've been able to utilize. Gustavo Manrique, Marcelo Scanu, Federico Barberis, Fernando Berdugo, Fredys Espejo, and Rudolf Posch are a few of the contributors of material that is published in this book. Thanks also to Ana Paula Forte who volunteered time to help with our glacier inventories. I'd also like to thank many individuals (many of whom are working in the mining sector and in the scientific community) who, for personal reasons, have decided to keep their contributions anonymous but whom have been instrumental to my research and publications. *Thank you!*

In La Rioja Province, I am indebted to Paulo Dalessandro who provided visual material and logistical support to visit the Famatina Mountains and to the communities of Famatina and Chilecito, which have shown unwavering commitment to protect their environment and glaciers. Particular thanks goes to Marcela Crabbe and Carina Diaz Moreno who have welcomed me to their communities and shared their vision and advocacy to protect water in the region. I also must thank Raul Vidable of the Fundación Ambiente y Desarrollo who helped us analyze the province's legal protection framework for glaciers. *Thank you!*

In Catamarca Province, Sergio Martinez has provided much help in logistics, critical research information, and assistance with our visit to show communities the first images they saw of rock glaciers in the Aconquija Mountains. *Thank you!*

In Chile, I am indebted to a number of people. I owe special gratitude to the president of the Diaguita Indigenous Community, Sergio Campusano, who opened his community's doors so that I could visit and learn first-hand about their lands, their glaciers, and their wonderful mountain environment.

Gustavo Freixas of the Dirección General de Aguas of the Coquimbo Region shared his vast experience working in high altitude glacier regions, and Javier Narbona Naranjo, Chief of the Hydrological Department has been unwavering in his support for the yearly UNEP glaciology course I've taken (and taught) each year. Also, a thank-you goes to José Luis Rodriguez of the Fundación Huilo Huilo who has helped with my research and provided critical technical input. Thank you also to Constanza San Juan and Carolina Pérez Soto of the Valle del Huasco and Jhon Melendez of the indigenous community Comunidad Diaguita-Patay, who opened their doors and invited me to the Huasco Valley to share my research with the community. *Thank you!*

This book has lots of pictures and without the generous contributions of the photographers to this book, you would not have been able to see these magnificent images. Thanks to Federico Barberis, Marcelo Scanu, Juan Pablo Milana, Mateo Martini, Mariano Castro, Rudolph Posch, Klaus Thymann, Jorge Garcia Dihinx, Robert Cassady, James Yungel, Constanza Taboada, Devon Libby, Ricardo Accieto, Raul Contreras, Cedomir Marangunic, Benjamin Morales Arnao, Gerhard Huedepohl, and to Google Earth, and also to many anonymous photo donors, all of whom have contributed images to this and other works. *Thank you!*

I also must thank those who have contributed financially to my (and CEDHA's) cryoactivism, including the Wallace Global Fund (at an earlier stage): Richard Mott; the Unitarian Universalist Service Committee (UUSC): Patricia Jones; and Patagonia: Raul and Cristobal Costa. Global Green Grants contributed to one of my trips to the Huasco Valley in Chile (indirectly by supporting the community to invite me), and the United Nations contributed to my participation first as student and later as trainer at its yearly glaciology course. I must also recognize the contribution in kind made by the Digital Globe Foundation: thanks to Devon Libby who provided us with free access to satellite images of the Pascua Lama area, without which much of our analysis would have been impossible, given that Argentina's own public satellite image agency remains closed to civil society consultation. *Thank you!*

Thanks to Kalia Moldogazieva of Kyrgyzstan. Kalia is an environmental leader in Kyrgyzstan and has been advocating for the protection of glaciers for many years following the severe impacts to glaciers by the Kumtor mining project. She joined us in Rio for the Earth Summit in 2012 for the session we held on "Glaciers and Sustainability in the Anthropocene." Kalia is now working toward the adoption of yet another glacier protection law. *Thank you!*

I'd like to recognize John Bonine and the late Svitlana Kravchenko, at the University of Oregon, who have been inspirational life-long human rights and environmental advocates—*and friends*. After Svitlana's passing, John invited me to a symposium in her honor and challenged me to come up with something very new and cutting edge for the human rights and environment community—something that would be the type of seemingly impossible policy objectives that Svitlana would have come up with in her quest to advance environmental protection and international environmental law. It was from this challenge that I came up with Chapter 10 of this book on "The Human Right . . . to Glaciers?" *Thank you!*

I must also thank CEDHA's staff for the many hours of work, often unpaid, spent putting together materials, publishing reports, drafting press releases, and collecting and managing our communications lists to get the information out where it needs to be! Special thanks goes to Fernanda Baissi who has been a communication pillar in my work. Also to Rafael Huber who devoted volunteer time to reading and editing several sections of this book. Just as we were going to print, Rafael (who lives in Switzerland), inspired by the text, was taking his parents Adriana and Cyril to visit glaciers up in the Swiss Alps! *Thank you!*

Much appreciation to Jeremy Lewis of Oxford University Press and to Mark Carey of the University of Oregon (who introduced us) and their belief in my work on glacier protection. *Thank you!*

Thanks to my children, Angelina and Ulises, who have had to miss out on dad's free evening time in the run up to the presentation of the manuscript of this book—they gave up endless hours of downtime on the Internet as I hoarded up our very slow Internet line to do much of the research that went into this publication—and we all know how much kids value Internet time these days! *Thank you!*

And finally, a special and personal thanks to Romina Picolotti, a friend, a partner, and an inspiration to this and many other environmental causes.

—Jorge Daniel Taillant

{ HOW TO READ THIS BOOK }

If you're wondering whether you should read this book, go to the Introduction and read the first page. My guess is you'll find it at least interesting enough to leaf through on a flight, on your way home from work, over coffee on a weekend, or during a cold winter evening—or even to share some of the cool glacier stuff (like the anecdotes in Chapter 8) with your kids. Additionally, you'll learn something about glaciers, which we don't get a chance to do very often. Given that glaciers are melting into oblivion and will soon disappear, it's now or never!

Because modern times and technology call for modern approaches, I reach back to 1963 and emulate the introduction to a book by Julio Cortazar, an Argentine who changed up the order of things. Cortazar invited his readers to choose one of two ways to read his revolutionary novel, *Hopscotch*: either read start to finish in sequential order as you would read any book, *but only through Chapter 56*. Or, begin at Chapter 73 and then follow a random irrational sequence which he laid out, even suggesting that the reader *not read* one of the chapters. I'll do something similar.

This book is several books in one. The odd-numbered chapters are a sequential story about the politics behind the passage of the world's first glacier protection law. I had a very first-hand experience with this unique process, which is partially why I decided to tell this story, and so I've reproduced quite a bit of dialogue between several of the individuals involved—*I was one of them*, but there were several others. Although I add a little bit of a novelistic tone to the prose to make it more interesting for the reader, it's quite a revealing tale and very close to what actually happened. It's interesting not only because of the glacier story behind it, but because it reveals the behind-the-curtain scenes to real politics in action in a country where everyone is obsessed with politics. I have had a very close and personal relationship with many of the people I am quoting, so I feel confident enough to put words in their mouths. Some of the conversations I was present to hear or engage in, and others were recounted to me directly by those who had them. On a very select few occasions, I took the liberty to guess at what people might be thinking . . . and, of course, maybe I got it wrong, so I am happy to entertain from readers questions, comments, or even debate on any part of the story I am telling.

One of the chapters, Chapter 9, is about a lot of the personal work I have done to promote glacier protection. I tried at first to keep it in the third person, but it was very odd to speak about myself at a distance. So I switched the prose to the first person. Sorry if that switch bothers anyone. I also inadvertently switch back and forth between "I" and "we," and, just so the reader is clear, when I say "we" it refers to the organization I founded and have worked with for more than a decade, the Center for Human Rights and Environment (CEDHA). It's hard sometimes to divorce yourself from something so dear. Those in the nonprofit world will be familiar with the type of activities and events described in this chapter because they are typical of the advocacy actions of nonprofits working to influence public policy. Those from other sectors may find this section interesting precisely because it offers an introspection on how nongovernmental organizations think and operate.

The even-numbered chapters are about glaciers, *generally*. You can read the even-numbered chapters in sequential order to focus on glacier learning, or you can read the entire book in sequential order—and get a break every other chapter from *the politics of ice* that I recount in the odd-numbered chapters. All the chapters, for the most part, can also be read independently from one another—which leads to some (although not much) repetition of some text. If you're a schoolteacher, particularly for primary or secondary school-aged kids and are interested in using any of this material with your students, please contact me because we are developing educational tools about glaciers, with much of this material especially prepared for the classroom.

Of the even-numbered chapters, Chapter 2 offers some basic definitions about glaciers. Chapter 4 focuses on the enigmatic world of the *periglacial environment*. If you've never heard of the periglacial environment, I guarantee you'll be impressed with this one, and you'll wonder how you could have not known about this part of our ecosystem: incredibly, about a fourth of the land mass of the planet is in periglacial areas! If this is new to you, and you become interested in the topic, I suggest that you read both Chapters 2 and 4 sequentially.

Chapter 6 is about climate change. No book about glaciers written in this anthropogenic era could avoid this topic. The idea behind Chapter 6 is to bring the climate change issue into practical perspective with regards to glaciers and show that it is possible to take action and revert the impacts that our changing climate is having on us. It is a story about the real dangers that people confront when glaciers melt, but it is also a story about how some people are taking action to deal with this melt.

Chapter 8 was a tough one. Down to the final weeks before submitting the final manuscript of the book, I hadn't yet written Chapter 8. To be honest, the intermixing of political story and glacier description didn't quite add up evenly! I didn't have a clear topic for Chapter 8, but I had a lot of material that was left over that I didn't want to leave out: so I came up with a hodge-podge of information, and that's why it probably seems a bit out of place, a bit unbalanced, and maybe unrelated to the rest. That's because it is! Nonetheless, Chapter 8 is a fun chapter, great for the kids and for bedside reading!

Chapter 10 is a reprint of an academic article I wrote recently positing a "Human Right ... to Glaciers?" which will surely bother many lawyers reading this book, but I include it anyway. I wrote this article on the occasion of a symposium in honor of the late Svitlana Kravchenko, wife and life partner of a good friend, John Bonine. Both Svitlana and John, dedicated environmental layers whom I met when they crashed an experts meeting of the United Nations in Geneva many years ago to discuss trying to link human rights and environment policies, have helped change the international environmental world. Svitlana was invited as an expert to the meeting, John was not—but he snuck in anyway, during a coffee break. I admired his intrepidness, as I myself had leveraged entry into the experts meeting at the very last minute by convincing the organizations that our viewpoint from the Global South was absolutely essential to the conversations that would take place. Together, the "uninvited bunch" gave some pretty incisive input that day, helping steer the UN into new territory. It just goes to show that what Woody Allen and Steve Jobs said, respectively, seems to be very true, "99% of success is simply being there" and "the people who are crazy enough to think they can change the world probably will." We've been working on the right to water for years, starting back when we got terrible smirks from lawyers for even suggesting it. Well, today, they've pretty much accepted that one, so why not challenge the world anew with the human right to glaciers? If it really bothers you, I'll say what Cortazar said in the Introduction to Hopscotch: *don't read it!*

So now pull out the smartphones and tablets and let's see some pictures and videos about glaciers.

My kids don't go anywhere without their phones and tablets. New computers don't even have very big hard drives anymore, and that's because people nowadays see things live on their mobile device. Phones are to modern society what computers were just a few years ago and what encyclopedias were just a few decades ago. This book has lots of footnotes and references to places, people, and things, *and to lots of glaciers*. Most of us just ignore the footnotes when we read, but I encourage you to look more closely, read this book with a live Wi-Fi connection, and visit some of the places I mention in the text and in the notes. They're quite amazing!

When have you ever seen a glacier on your phone? Well, now you will. Here's a first one, give it a shot. So, you thought glaciers were near the poles in extremely cold countries? Open Google Maps or whatever map program you use, put it in satellite mode (this is important!), and type the following address in the search box, exactly as it is here (comma and all):

0 9 27 S, 37 18 49 E

Give the phone a moment to gyrate the planet. Moving the Earth around can take a few seconds. Zoom in and give your phone a moment to load the satellite image. You'll see a few white splotches. Zoom in further. This is Lewis Glacier on Mt. Kenya in Africa. Yes, there are glaciers in Africa! The few remaining glaciers on this mountain are some of the few remaining perennial ice bodies on the African continent, most of which are slated to disappear in the next few years. They tower at nearly 5,000 meters above sea level (about 16,400 feet), and they're practically on the Equator. Go figure! That's about as far as you can get from the polar ice caps. Now you've seen African glaciers, just before they disappear!

So, each time you see a GPS coordinate cited in this book, and you're interested in seeing a cool glacier, enter the coordinate as it appears into Google Maps or Google Earth and go!

One final note on the GPS locations: sometimes you'll see the location sited with degrees, minutes, and seconds, like this:

27°22′10.56″ S 66°16′52.60″ W

In this case, there is no comma in the address. This address can be pasted *as is* into Google Earth to visit the site. Alternatively, you can remove the degrees, minutes, and seconds and add a comma to the address as follows:

27 22 10.56 S, 66 16 52.60 W

You can also take only the first three sets of digits of the directional coordinates (in this case S and W) of the address, and you will generally arrive at the very same spot, or very near to it. In this case, the address could be copied as:

27 22 10 S, 66 16 52 W

These are simply a few tips to make your Google Earth or Google Maps navigation a bit simpler!

Oh, and I forgot one last important thing!

Enjoy the pictures!

{ INTRODUCTION }

In this book, you will read about a fascinating world of ice critical to human existence, but one that you have probably never heard of. If you keep your smartphone or tablet handy or have Google Earth running on your computer while you read, you will be able to see this frozen world in a mere few keystrokes.

You will read about invisible glaciers, unimaginable glacier ice that can cause tsunami-like waves, and even *ice that can catch on fire*! You will read about colossal amounts of ice that many millions of people depend on but take for granted even as it becomes one of the world's most important, most vulnerable, and yet most ignored and unprotected natural resources. You will learn about ice and glaciers located in places you likely never imagined might be glaciated, including regions of Mexico, Armenia, Iran, Colombia, Australia, Indonesia, Venezuela, Turkey, Kenya, Uganda, Nevada, and Southern California. You will learn why glaciers are important to your daily existence and about the chilling consequences that could occur if the entire polar icecaps, holding most of the world's freshwater, melt away. *It's already happening!*

You will read incredible stories about people and companies that have drilled into glaciers in search of gold and others that have sent missiles into glaciers in an attempt to destroy them. You will read crazy accounts, some true, some fiction, of ancient pyramids and millenary mummies being unearthed from the ice to tell us incredibly detailed information about the past. You will read about surfers who daringly surf glacier waves and glaciers that have swallowed up entire commercial airplanes and many years later spit out bits and pieces of perfectly frozen human history and tragedy. You will hear about people stranded on ice for months on end and even about soccer games played on floating ice. You will read about hidden glacier-buried treasures and people who stashed messages for humanity in ice decades ago only to have their words appear one day to a sole trekker walking by. You will learn about what trees and glaciers have in common and why they are sometimes enemies, and how we can use glaciers to study our prehistoric atmosphere. You will read about horrific creeping ice that lurks on beaches and invades homes unexpectedly. You will learn about the trembling and roaring sounds glaciers make and the dangers of their almost imperceptible but often deadly advance.

You will read about the seemingly impossible (ridiculous?) attempts to *move* glaciers from one location to another. You will read about people who have lived in sight of glaciers and never knew that they were there and why they were oblivious to their existence. You will read about so-called *rock glaciers* and even *fossil* glaciers that teach us about our past environment and about climate change. You will read about industries knowingly plowing bulldozers into glaciers and dynamiting glaciers to seek for and get at mineral deposits hidden beneath ice.

You will learn about people with the apparently crazy but very real idea of actually fabricating glaciers to store water for a dry season, grafting glaciers as you would a plant to grow a new one, or painting them with spray paint to intentionally provoke melting, or covering them up with tarps to keep them cool. You'll even learn how to make your own glacier right in your refrigerator! You'll read about one experiment to use sawdust to make glacier surfaces grow and about another idea to use chicken wire and fences to change wind patterns and create *glaciosystems* that are conducive to glacier formation. Yes, there are really people out there making glaciers!

You'll hear about a guy who ended up in jail because he was illegally stealing away chunks of glacier ice so that local bar-goers could have the ultimate "whiskey on the *glacier* rocks." You'll read about people who bare it to the buff to make art on glaciers. You'll even hear about the creation of a "glacier republic" that issues citizenship and passports. You'll hear about a woman who writes poems and rock songs about glaciers and about one community that treks up to glaciers each year and carries away pieces of the glacier tied to their backs because the water contained in glacier ice is deemed by the community to be pure and magical. You'll read about emerging "cryoactivism" as environmental groups around the world from Chile to Argentina to Kyrgyzstan to Bhutan and Nepal are teaming up to protect glacier resources. And you will read about one society's quest (in the absence of glacier protection policies) to establish the world's first national glacier protection law, an effort that is now spilling over to other societies for emulation.

This book is about *ice*—specifically, *glacier ice*—and not necessarily about the glaciers we are most likely to recognize in far away and remote places like Alaska, Patagonia, Greenland, Norway, or Antarctica where you might take a luxurious cruise ship and see the glaciers from afar. Rather, this book is about the *other* glaciers, ones that are closer to home in places like the Sierra Nevada of California, in Montana, France, Austria, Italy, and Switzerland; along the Central Andes in countries like Argentina, Chile, Ecuador, Peru, Bolivia, and even Mexico; or in parts of Nepal and Pakistan. These glaciers, although perhaps in some cases less enormous or less well-known, are actually far more important to our daily existence and to the existence of millions if not billions of people.

This is not a scientific book full of scientific facts but rather a social, cultural, and political introspection into our *cryosphere* (the world of ice) that brings critical complex and rather obscure scientific information about ice and glaciers into perspective for our nonscientific lives. It reflects on the way we organize *or do not organize* to protect some of our planet's most important and extremely vulnerable natural resources.

In sum, this book is about glaciers, their beauty, their hidden mystery, and the evolving politics and social awareness that is emerging to ensure their protection and survival in an environment making them increasingly vulnerable and on route to extinction.

Glaciers

Dynamiting Glaciers

The following plan describes the method and management disposition of the glacier sectors that must be removed during the life of Pascua Lama, as the open pit area is extended towards the position of the glaciers in the Rio El Toro river basin. It is estimated that 10 hectares [25 acres] of glaciers must be removed and adequately managed to avoid the instability of slopes and environmental impacts. The thickness of the glacier sectors that must be removed is estimated at 3 to 5 meters [10–16 ft]. . . . mining equipment shall be employed as needed for each glacier sector to be managed (basically bulldozers and/or front loaders). . . . If necessary, controlled explosives shall be used, of small size, to remove the ice.

—FROM BARRICK GOLD'S "GLACIER MANAGEMENT PLAN"
—THE PASCUA LAMA MINING PROJECT
(ARGENTINE-CHILEAN BORDER; ENVIRONMENTAL
IMPACT STUDY, ANNEX B, 2001; UNOFFICIAL
TRANSLATION FROM THE ORIGINAL TEXT IN SPANISH)[1]

On September 6, 2006, Romina Picolotti, Argentina's Secretary of Environment and a career environmentalist, sat reviewing briefing documents to prepare for a meeting regarding the world's first binational gold mining project straddling the border between Argentina and Chile. Barrick Gold, the world's largest gold mining corporation, had discovered a massive gold and silver reserve in one of the highest, coldest, most desolate and remote areas of the Americas, the Central Andes mountain range, running from Venezuela, through Colombia, Ecuador, Peru, Bolivia, and down to the southernmost tip of the Americas shared by Argentina and Chile.

The Andes are among the highest mountains in the world, with the tallest peaks in both the Southern and Western Hemispheres. Peru's ranges surpass well above 6,000 meters above sea level (nearly 20,000 ft), whereas the highest mountain of the Americas (the Aconcagua) in Argentina towers at nearly 7,000 m (nearly 23,000 ft). At 6,960 meters (22,835 ft), the Aconcagua,[2] which means "stone sentinel" in the precolonial Quechan native tongue, is covered in snow in the winter and surrounded by massive glaciers year round, some of which are up to 8 km (5 mi) long.[3] The Central Andes are spotted with generous amounts of perennial ice surviving throughout the warm summer months to provide critical meltwater reserves for many millions of people and communities below on each side of the massive mountain range.

In Argentina and Chile, the Andes form a natural and political border that has been disputed since colonial times, even to the brink of war. So conflictive was this debate that the Chileans and Argentines requested that Queen Victoria of England resolve their dispute. Eventually it was King Edward VII, in 1902, who interceded to help ease tensions over border claims.[4] In sum, the highest peaks and the water runoff are used to mark the disputed border in a sort of connect-the-dots fashion. When a section of land is in doubt, which way the water flows determines who gets it: either to the Atlantic (making the territory Argentine) or to the Pacific (in which case it goes to the Chileans). As late as the 1980s, tension reigned along the border, with each country permanently contesting swaths of land no more than a few square meters in size up and down the Andes range.

But ill feelings over the border began to change in the 1990s, thanks to the steady inflow of small groups of foreigners with four-wheel-drive pickup trucks, shovels, and drilling equipment. They came in droves, literally looking for treasure. The Inca, the precolonial inhabitants of what is today most of the Andes region, tell tales of past Inca rulers escaping to these highlands during Pizarro's conquest, there stashing away fabulous amounts of gold far from human sight in a veritable El Dorado, hidden for a future resurgence of the Inca empire.[5] These stories are not simply myth. Modern gold prospectors knew from colonial transcriptions and other locally written history that not only had the Incas found gold in these terrains, but that the Spanish had also unearthed precious metals in many locations throughout modern-day Argentina, Chile, and Peru. In fact, using very old but reliable technology, many gold mines had already been explored and exploited. New, larger, and more heavy-duty equipment, along with more advanced prospecting techniques, could retrieve further treasure, and that was precisely what the mining industry had in mind. El Dorado was not a myth but a reality waiting to be tapped. A fair portion of this treasure was located in an area miners already knew well: the El Indio Gold Belt.

According to everyone's favorite encyclopedia, the El Indio Gold Belt[6] is a mineral-rich region formed during the late Miocene period. It spans the border between Argentina and Chile and contains large quantities of gold, silver, and copper. In the 1990s and 2000s, this area attracted a mad rush of miners to the Andes, much like that which had occurred in the Californian gold rush of the mid-1800s. But these miners weren't coming with just the clothes on their backs and shovels, but rather with the full financial backing of some of the world's most lucrative and speculative investment markets. It wasn't the 49ers in denim jeans who came to the Central Andes, but a new type of gold digger dressed in Armani suits and Italian neckties. It was the risk-taking stock market players who would drive this gold rush from start to finish.

By 2006, Romina Picolotti, an environmentalist and human rights advocate turned political figure, was in the middle of revamping Argentina's Environment Secretariat,[7] an office long-abandoned in the trenches of inefficient bureaucracy with little political will amongst Argentina's governing leaders to enforce the environmental law and policy already on the books. Not a member of any political party but a technical expert on environmental policy and founder of a globally recognized environmental nongovernmental organization (NGO),[8] Picolotti had gained recognition from Argentina's neo-Peronist President Nestor Kirchner in a case involving pulp mill contamination in another border dispute, this one to the East between Argentina and Uruguay.

Argentina had presented a legal complaint against Uruguay to the International Court of Justice in the Hague over the neighboring country's unilateral decision to install two paper pulp mills on the Uruguay River, the natural and political border between the two countries. Counter to what most believed, Romina Picolotti was against the use of the Hague tribunal for litigation in the case because the international tribunal was too conservative, and she feared that the case would be drawn out indefinitely. By the time a decision would be reached, the pulp mills would be already functioning. She preferred a much more incisive strategy, one aimed at freezing the financing on which the company depended to build the mills. Her analytical, level-headed strategic approach to engage the pulp mill case, as well as the fanfare that surrounded her recent winning of the 2006 global Sophie Prize, given each year in Norway to a single environmental advocate for innovative contributions to sustainable development, won her Kirchner's favor and an appointment to head Argentina's environmental policy agency. It was at this same time that former US Vice President Al Gore was going around showing his PowerPoint presentation about the perils of climate change. The world was going green, and Argentina had its environmental policy leader firmly in place. But then another border dispute distracted Environment Secretary Picolotti.

On September 6, 2006, she would meet a group of about fifteen environmentalists from San Juan Province, an area of the country resembling the American Far West in the northwest region of Argentina, at the upper limits of wine country, where Pascua Lama was located. But more had come to raise their concerns over the arrival of mining to their province. Some fifty environmentalists, some by trade, others simply by conviction, had trekked to the country's capital on a bus to meet with public authorities about problems they had with mining operations in their province, including the fear that glaciers were being impacted in order to extract gold. They didn't know it, but the Environment Secretary had a heavy hand in making their trip happen.

She needed help in her quest to exercise environmental oversight of mining operations and to draw attention to the impact that the sector was having on vulnerable resources and communities. Without public pressure and with a manifest and a very strong position of promoting mining, the central government would not formally engage in an effort to curb mining operations or even place environmental limits on mining policy; Picolotti needed the public to ask for her help in order to legitimately intervene. In Argentina, a federal country, mining is a local affair, and its environmental impacts are generally under the jurisdiction of provincial governments. Local communities concerned with the environmental impact of large mining operations began to create local assemblies aimed at bringing attention to the risks of promoting large-scale mining investments in lands with very little water and sensitive ecosystems. They were also concerned with the very scant or nonexistent public control over general industrial environmental impacts. These communities were fighting a battle against a Goliath multibillion dollar industry (with full support of the State) already expanding at a rapid pace.

With practically no financing, little in the way of formal organization, and mostly on volunteer time, these NGOs began to mark their territory and carve out a space in civil society from which they would take a stand against mining. But the small volunteer groups in San Juan couldn't conceivably afford the costs of travel en masse to the capital to voice their concerns to the national government.

Picolotti had come into contact a year earlier (before she became Environment Secretary) with some of these local community stakeholders advocating around mining impacts during a regional civil society workshop on the risks of the extractive sector. Public environmental protests in the capital city could help her tremendously by gaining attention from the national media and putting pressure on the federal government to take action—and particularly on provincial governments to let her engage as the Federal Environmental Authority at the provincial level. Picolotti was eager to step in but needed federal jurisdiction, which she didn't have.

She had her team scope the terrain in San Juan. They needed to identify an organization that could handle the logistical and administrative responsibilities of a coordinated advocacy campaign of the type Greenpeace might carry out. She was certain that she could muster support if they found the right actor to organize and lead a civil society movement. It had to be a formally established organization able to receive donations and/or grant financing, with legal skills and strong knowledge of the public administrative process.

The group that most stood out was the Fundación Ciudadanos Independientes (FuCI). Run by an experienced litigator and professor, Silvia Villalonga, FuCI was one of the few more organized civil society environmental groups in San Juan. Functioning after hours and on volunteer time, FuCI was a well-established NGO, with institutional identity and a legal persona, and it was already carrying out a strong campaign against the adverse impacts of mining operations.

In January 2006, someone slipped an anonymous envelope under the door of the office where FuCI operated containing evidence that Pascua Lama and Veladero were destroying glaciers. With legal capacity among its ranks, FuCI decided to file a complaint in the courts against Barrick Gold's Pascua Lama project. They were the ideal candidates to organize a collective initiative in Buenos Aires that would be focused on the environmental impacts of the mining industry.

The Environment Secretary first spoke to Greenpeace off record to see if it could mobilize more public advocacy around mining impacts; she specifically asked if it could finance groups in San Juan to come to Buenos Aires to meet with government officials and carry out peaceful *but very visible* protests. The Greenpeace route would not prove fruitful, but there were others, and eventually a donor appeared.

Through her networks in the environmental advocacy world, Picolotti contacted an organization (that will remain anonymous by request) working on forestry protection in Patagonia and convinced it to provide an anonymous grant to the Sanjuaninos if FuCI would coordinate a trip of local environmental actors to travel to Buenos Aires to call on the government for greater environmental controls over mining. The financing materialized, and things started moving forward quickly.

Soon thereafter, Villalonga, FuCI's director, received a call indicating that an anonymous source would make a donation to her institution if she could organize a trip for local environmental leaders to go to Buenos Aires to bring attention to the risks and impacts of mining on the environment and human health. Villalonga accepted the grant but had to open up a bank account to receive the funds because, at the time (and still today) FuCI worked strictly on a volunteer basis and did not even have an institutional budget or bank account to its name. That was the only bank account that the organization ever opened, and *that* grant the only contribution that would go into the account.

With added financial support from the Wine Growers Association of San Juan, which paid for a portion of the costs of hiring a bus, hotel, and food for a group of fifty to travel to the nation's capital, FuCI began organizing the collective initiative. It would eventually fill a bus with environmentalists to travel to Buenos Aires for what would be the onset of a national glacier protection movement that would resonate around the world.

It was quite a diverse mix, with a few very small environmental organizations, a group of mothers united against the environmental impacts of mining, a husband-and-wife duo who circulated information and videos about mining impacts at Barrick Gold's Veladero mine, a geologist who chose to specialize in glacier studies, small local wine producers, and several other odds and ends, all pushing to contain mega-mining investments and, particularly, to ensure that the mining sector would not destroy the province's water reserves. They carried the slogan now sounding across Latin America: "*Water Is Worth More Than Gold.*"

The group targeted meetings with the Environment Secretary (a logical but coincidental step), the National Parks Service, the Mining Secretary, Congress, and a number of political leaders whom they hoped would help disseminate their cause. The group wanted to express their concern that an international Canadian mining company was destroying precious glaciers and frozen grounds up in the remote Andes Mountains and that no one was doing anything about it. They wanted their government to act. Ironically, when Silvia Villalonga asked the Environment Secretary directly for her help in combatting mining, she was surprised to hear back that Picolotti wanted help from civil society. Villalonga walked away from the meeting feeling somewhat let down because she thought the Secretary was not mobilizing to support the cause.

These fifty environmental activists didn't realize it at the time, but they were about to change glacier history.

Barrick Gold's glacier problem had surfaced several years earlier in Chile, when indigenous community members in the Huasco Valley began to complain that new mining projects arriving in the area were threatening their ecosystems. Among these threats, they mentioned glaciers as some of the most vulnerable resources, but there were others, including the sensitive llama and vicuña populations (Latin American camelids that once thrived in these high altitudes until anthropogenic activity pushed them to increasingly remote locations); the high wetland systems called *vegas* that were also being destroyed; and access to traditional lands that indigenous groups used for subsistence livestock pasturing of llamas and cultural rites.

Local protests in Chile made their way to Chilean environmental organizations and then slowly began to permeate the border. The main problem for environmental groups on both sides of the border, however, was that, in Chile, the mining sector was simply too entrenched in the region to stop its advance over environmental concerns, whereas in Argentina practically complete environmental indifference existed for these projects. In the 1990s, San Juan Province made the decision to promote mining investments, offering lucrative tax write-offs to international mining companies to scour the Andes in search of precious metals and adopting national and provincial laws to this effect. Practically no concern or consideration was voiced or shown for the environmental impacts that could result from each project, and certainly almost no concerns were voiced about the *cumulative* environmental impacts that dozens and even hundreds of new mining prospecting projects might have on the natural environment.

The mineral deposits of the El Indio Gold Belt are generously distributed in the region along the border area. For the Veladero vein, located east of the border, the project would be administratively constituted in Argentina, just as the El Indio vein (a bit farther south) to the West had been developed in Chile. The problem was the Pascua Lama reserves. These were the most attractive reserves for the company, but administratively their exploitation would be problematic because they were located squarely on the border. "Pascua" is the name of the mineral area adjacent to the border but resting just inside Chile, and "Lama" is the name of the portion adjacent to the border but just across into Argentina. Mother Nature had not conveniently separated the gold for each country to more easily sort out ownership.

And although Mother Nature makes no political distinction as to where she develops her precious metals, man-made project equipment, staff, lodging, transportation, supplies, food, office space, files, accounting, and many other dimensions of the mining operations that would be required to extract the gold formally needed to take place "somewhere" and not merely in some intangible natural border area. Legally speaking, Pascua Lama had to be either Chilean or Argentine.

As one accountant I spoke to about Pascua Lama suggested, if one of Barrick Gold's trucks starts activities in an Argentine hangar in the morning, drives to the pit area in Chile for the first two hours of work, then dumps rock into a waste area on the Argentine side of the border, drives back to Chile for lunch, punctures a tire in Chile (a tire the company bought in Argentina) and has it replaced (with a tire purchased in Chile), makes two more back-and-forth trips between countries to pick up and dump rock, and then eventually ends up back at the Argentine hangar for the evening, to which country and accounting system do you report the depreciation of the tires of the truck?

Pascua Lama was a problem, not only because it was at an unthinkable elevation of above 5,000 meters (16,400 ft) and surrounded by ice, but because it was a legal, administrative, and accounting nightmare. To resolve this would require some careful and creative legal expertise and innovation.

Barrick Gold decided to place Veladero,[9] one part of the Argentine portion of the project, in motion first. They could set up the processing infrastructure in the Valle del Cura, in Argentina (the valley area adjacent to the border in Argentine territory) and begin to extract gold from one of the nearby veins found in one of the mountains adjacent to the Pascua Lama pits. They would have a full-fledged mining operation running by the end of the 2000s, when they hoped to inaugurate the border area portion of the project, Pascua Lama.[10] By then, they hoped to have sorted out the legal and administrative contingencies.

Pascua Lama would get an especially tailored administrative and legal regime, negotiated as a binational treaty between Argentina and Chile to collectively promote and facilitate a border mining operation. All of this got under way in the mid-2000s as Veladero was kicking off operations.

Veladero, an open pit mine located at 4,850 m (15,900 ft) above sea level, opened its doors in 2005 to produce some 5 million ounces of gold. Environmentalists complained early on about the project design, claiming that much of the infrastructure and the entire leach valley facility[11] where Barrick Gold gathered contaminated sludge and left colossal pools of contaminated water was placed on the area's most sensitive ecosystem, the *vegas*, effectively drowning this delicate highland wetland ecosystem. These *vegas* are virtual oases of humid wetlands found in high-altitude environments at the foot of glaciers and are fed directly by glacier melt. Staff at the Argentine National Park Service, which was against the project due to its direct impact on the San Guillermo Biosphere Reserve in general, and specifically on the *vegas* ecosystems, when off record, euphemistically referred to the project as "Argentina's sacrifice to Barrick" because of the large impact to the *vegas* areas below the Veladero mine.[12] But Veladero had not generated much protest about glacier impacts because no one really understood the glacier dimensions of either Veladero or Pascua Lama. People simply didn't know that there were glaciers up in these parts (Figure 1.1).

FIGURE 1.1 *Aerial view of Veladero's base camp with the majestic Los Amarillos Glacier in the background.*

Source: Barrick Gold. GIS: 29°24'38.59" S 69°53'17.12" W.

Workers of the mining company, some local folk who took their cattle to graze up in the mountains during the summer, and the few people who trekked through the Valle del Cura ("Valley of the Priest")—mostly mountain climbers and adventurers in the summer months—could see snow up on the mountain tops, and some mountaineers trekked to these perennial ice patches above 5,000 m (16,400 ft), ice that most locals refer to as "eternal snow."

But snow doesn't survive eternally. After a few days, it compacts to ice. If it survives the summer and then one more full-year cycle, it is definitely not snow, *it's glacier ice*. But, for the most part, the immediate areas around Veladero (save for the mountain peaks above the mine) didn't display any ice at all during the summer, even though freezing temperatures could persist for much of the year and almost certainly at night.

The group of environmental protesters arrived in Buenos Aires with a renowned geologist and glaciologist, Juan Pablo Milana, a local expert on glaciers working in San Juan and teaching glaciology at San Juan's National University. Picolotti received the group in her meeting room and invited them to sit at the long oval table, taking a seat among them as Milana unfurled faded and yellowed geological maps—like those found in an old library or in a surveyor's office, maps that no one has seen in many years and that have to be treated with utmost care because they fall apart with the slightest mistreatment. Ironically, environmentalists were not accustomed to meet with the highest environmental authority: the oval meeting room at the Environment Secretariat was rarely used to receive Argentina's leading environmental organizations. But, this time, *the Environment Secretary was one of them*, fighting as a member of civil society herself to get environmental agendas visibility on national and provincial political priority lists.

A large print of a multitudinous march of nearly 100,000 people crossing the international bridge to Uruguay at Gualeguaychú, a border crossing town between Argentina and Uruguay, hung on the central wall behind her seat. It was a community march against future industrial contamination on the river border by two pulp mills slated to be financed by the World Bank's International Finance Corporation (IFC) without having completed proper environmental due diligence studies. Picolotti often would purposefully sit underneath the photo when meeting with industry and other government representatives, purposefully making a statement about which side of the fence she stood on. If anyone in the central government was going to hear this community, it was the Environment Secretary.

"This is *Pascua Lama*," said Milana, authoritatively, "and these . . . [pointing to white blotches surrounding several polygons he had drawn carefully by hand to indicate the project pit area and the projected waste piles] *are glaciers*. This project is surrounded by glaciers," he claimed. "And this here is Veladero . . .

Barrick's other adjacent project, and it too has glaciers and periglacial areas . . . *that's permafrost or permanently frozen grounds*, which is also an ice-rich and very significant hydrological reserve."

The Secretary knew a lot about the environment: she herself was a career environmentalist and now ran Argentina's Environmental Agency. But she struggled for a moment to place what Milana was saying into context. *Glaciers in San Juan?* she thought. Argentina is famous for glaciers, but they're *in Patagonia.* There is a Glacier National Park there with lots of very large, spectacular glaciers seen by thousands of tourists each year from all over the world. She imagined the Perito Moreno Glacier, Argentina's most popular glacier, the one that all Argentines can name—most could name *only* that glacier, and even she had to struggle to recall the other less-well-known (but no less spectacular) glaciers nearby like the Upsala, the Viedma, or the Spegazzini Glaciers, to name a few. She had been to the Perito Moreno herself recently as Secretary of Environment, and she had visited this magnificent 30 km (19 mi) long glacier just as many other Argentines have, on an excursion as a child or on some discounted vacation package over Easter break.

The Environment Secretary listened to Milana go on about how Barrick Gold was planning to extract gold from *underneath* glaciers. The pit area in fact had three glaciers that would be completely destroyed, "dynamited" he said, in order to get at the gold. These were glaciers in Chilean territory, the Toro 1, Toro 2, and Esperanza Glaciers. They were literally just across the border, and Barrick Gold had sparked the reaction of local communities when they proposed destroying them. But that was a Chilean problem for the moment.

Glacier Ice and Mining Roads on the Conconta Pass

There were also many glaciers on the Argentine side of the border, at and near the project, said Milana. Many of these were along the access road to Veladero and Pascua Lama, a 180 km (112 mi) dirt road leading up from the town of Tudcum to the base camp at Veladero. Tudcum[13] is at 2,000 m (6,500 ft) above sea level, but the road rises dramatically only after 30 km (19 mi), reaching 5,090 m (16,700 ft) at the Conconta Pass.[14] Several small to medium-sized glaciers thrive along the highest peaks of this winding road whereas several rock glaciers (a type of glacier covered with rock) can be found at lower elevations at roadside.[15]

The community members sitting at the table described a road the company had introduced that ran right through the middle of three glaciers at the Conconta Pass.[16] Workers for Zlato, the contractor that carved out and maintained much of the road leading from Tudcum to the Veladero base camp, had distributed clandestine images of a bulldozer clearing ice and snow at the Conconta Pass. They had run the bulldozers through what seemed to be a healthy glacier (Figure 1.2).

At the time (and still today), it was hard to come by photographs of this ice. Now the road was closed to public traffic, and only Barrick Gold could give permission to visit the area. But some photos existed, and they had filtered into the public domain. One such photo circulated by an environmental advocate at the time (and taken from a glacier monitoring report produced by Argentina's National Glacier Institute [IANIGLA]) put the three glaciers in perspective and suggested that the road had in fact divided what was originally a single glacier system. Others spoke of Barrick Gold setting up furnaces to actively melt the ice away or using explosives to destroy the ice at the Conconta Pass with the excuse of attempting to avoid avalanches[17]—environmentalists argued that those avalanches were necessary to feed snow into the glacier system but that Barrick Gold was collecting the feed snow and removing it from the road so that their vehicles could pass, thus slowly strangling the glaciers. Stories abounded of glacier impacts at the Conconta Pass, whether grounded in scientific truth or not, and the photos of the bulldozer road clearing would be among the most controversial in the Barrick Gold glacier saga, appearing even today in numerous commercial mining magazines when referring

FIGURE 1.2 *Controversial moment when Barrick Gold workers carve out a mining road at the Almirante Brown Glacier at the Conconta Pass.*

Source: Anonymous. GIS: 29°58'43.87" S 69°38'03.42" W.

to Barrick's glacier problems at Pascua Lama and Veladero. The images clearly showed that Barrick Gold was removing snow and ice from the road (Figure 1.3)—but were they in fact destroying glaciers?

Because of the large controversy that emerged surrounding the Conconta Pass glaciers, Barrick Gold hired two scientists, Juan Carlos Leiva and Gabriel Cabrera of the IANIGLA, to study the three glaciers. The glaciologists repeatedly emphasized in this 2008 study (the first of several studies)[18] that their work was not an environmental impact study; that is, Barrick Gold hired them not to determine what had happened to the Conconta Pass glaciers, but instead to evaluate what state the glaciers were currently in. Immediately, this drew the attention of hardline environmentalists who claimed that the damage had already been done and that what was needed was an environmental impact study to determine why the glaciers were collapsing and not a current glacier bill of health. Environmentalists wanted to know what impacts the roads and bulldozers, as well as the systematic road maintenance that removed snow and ice, had caused to the glaciers, and what the continued use and maintenance of the road would mean for the glaciers' future health.

FIGURE 1.3 *Aerial perspective of the Conconta Pass glaciers and the controversial mining road. Environmentalists argue that this road cut the lifeline of the lower glacier, which disappeared by 2011. The Upper Almirante Brown Glacier (upper left), the Norte Glacier (upper right), and the Lower Almirante Brown Glacier (center). Photo is from 2004.*
Source: Golder Associates. GIS: 29°58′43.87″ S 69°38′ 03.42″ W.

In their 2008 report, Leiva and Cabrera said several things in Barrick Gold's favor and made other statements that were more ambiguous, but they laid out an unresolved puzzle, the debate of which would carry forth throughout the evolution of the conflict around mining and glaciers. Although they were not hired to produce an environmental impact assessment, they nonetheless drew attention to the potential impacts on the glaciers of dust from road use. The glaciologists considered that climate change was a leading factor for the witnessed glacier melt and deterioration in the Conconta Pass area. The three ice bodies recharged in wintertime with snowfall, while in summer they deteriorated significantly, said Leiva and Cabrera. In fact, the deterioration sometimes caused the bodies to dismember. The Upper Almirante Brown Glacier and the Norte Glacier suffered thinning and severe deterioration near their terminus ends adjacent to the road, whereas the Lower Almirante Brown Glacier (which the scientists claimed could no longer be called a glacier but which now should be called a *glacieret* or smaller perennial ice patch) would be dismembered into two distinct, thinning sections in the summer and recharge and rejoin its separate parts in the wintertime—that is, argued environmentalists, if the ice and snow from the glacier's feeding area were not removed.

But the question of the road's impact on the glaciers was not clear. The scientists suggested that clearing ice from the road didn't have a *measurable* impact on the glaciers, particularly since they were not continuous bodies of ice but rather three distinct glaciers (or two plus one glacieret). They did not write off, however, that the road might be having impacts related to dust particles soiling the glacier, thereby changing the albedo of the ice and resulting in increased heat absorption related to color alterations on the surface. Wear a dark shirt in the sun and you warm up. Deposit contamination on a glacier's surface and it too warms up, thereby spurring melt. While the glacier experts didn't say it, environmentalists interpreted between the lines of the report. The changing of albedo could accelerate glacier melt and the retreat of the three glaciers' ice, and in such a context, a reunification of the glacier system, or an improvement in the health of the individual glaciers after a good snowfall, was nearly impossible. Barrick was ensuring that any chance that these glaciers could survive in an adverse global climate was practically nil.

As the years passed, we began to see a progressive deterioration of the Lower Almirante Brown Glacier. The scientists' 2008 study showed a sequence of photographs that noted a substantial decrease in the thickness of the lower glacier. By 2011, the Lower Almirante Brown Glacier would completely vanish![19]

Was the visible deterioration of the Lower Almirante Brown Glacier due to Barrick's road? Or was it climate change? Barrick Gold and San Juan Province have repeatedly argued that climate change is melting glaciers, not mining. Who could argue, though, that the hundreds of trucks passing the site weekly were probably also affecting the glaciers due to fuel emissions and road dust? A random photograph viewed today on Google Earth (taken on April 25, 2011) shows at the base of the Conconta Pass (prior to climbing) at least a dozen large trucks winding up the road.[20] The glaciologists claimed they were not carrying out an environmental impact study, but they were clearly calling attention to the road dust and the risk it represented for glacier ice. Was the rapid retreat of the Lower Almirante Brown Glacier being spurred on by Barrick's road use and maintenance, by ongoing snow clearance and the effects of suspended dust particles changing albedo by darkening the glacier and thereby provoking accelerated glacier melt? Or was this the inevitable effects of climate change in action? Or was it a combination of these influences?

To Barrick Gold's credit, and to the credit of the provincial government of San Juan, the company had suggested, as a positive gesture, that it could regenerate snow accumulation on the Lower Almirante Brown Glacier, but Leiva and Cabrera found the proposal futile given the deteriorated state of the glacier, which they considered beyond repair. Environmentalists would say "too little too late": the damage had already been done, and the consequences were irreversible. The scientists also mention the provincial government's suggestion to reroute the road above the glaciers, which the glaciologists deemed would actually produce more impact than the current location because this would actually create more dust upwind of the glacier, thereby potentially increasing dust contamination on the glacier's surface.

Leiva and Cabrera mention in their study that a report they used in their own research by Golder Associates shows a photo from 1985 indicating that the three glaciers were linked in the past. But such photos are not publicly available, and neither the province nor the company has been willing to publish these images. In the end, whether you believe that the deterioration of the Lower Almirante Brown Glacier is because Barrick ran a road through the glacier and cut off its source snow or that it has vanished because climate change is melting all of the glaciers in the region, the important thing is that *all glaciers* in the area are water reservoirs. They regulate basin flow, and, as such, *all of them* should be protected, precisely because they are even more vulnerable today due to climate change.

It seems reasonable to argue that even if these glaciers will some day soon disappear due to global warming, while they exist, perhaps for several years or decades or longer, they are still critical and vulnerable water towers that must be preserved, particularly in arid climates such as that of San Juan Province, where glacier ice provides a considerable portion of water flow to local populations and agriculture during drier months. The conflict surrounding the Conconta Pass glaciers was in full swing when Marta Maffei, a Congresswoman from Argentina, crossed the border from Chile into

Argentina with Chile's newly emerged glacier protection bill in her briefcase, and it was at that time that Argentina's own glacier protection law would be crafted.

This all sounded bizarre to Environment Secretary Picolotti. Glaciers in San Juan, dynamiting glaciers to get at gold, a mining company bulldozing through glaciers, gold underneath ice. Was this possible? Was it a concoction from environmental extremists opposed to mining? How could anyone in his or her right mind plow a bulldozer through a glacier or imagine extracting ice from a massive moving ice sheet? This seemed impossible.

Then Silvia Villalonga, a lawyer from a small law firm in San Juan's capital 300 km from the alleged place of impact and president of the FuCI, pulled out a document in Spanish to share with the Secretary. The header showed the logo of a very recognizable company, Barrick Gold, owner of Veladero and proponent of Pascua Lama. The document was an annex to the company's Environmental Impact Study. It was titled: "Pascua Lama Project: Minera Nevada Company. Glacier Management Plan in the Rio el Toro Basin."[21]

Romina Picolotti, Argentina's top environmental authority began skimming through the short three- or four-page document. The voices in the room hummed and faded in the background as she read, incredulous:

> The following plan describes the method and management disposition of the glacier sectors that must be removed during the life of Pascua Lama, as the open pit is extended toward the position of the glaciers in the Rio El Toro river basin. It is estimated that 10 hectares [25 acres] of glaciers must be removed and adequately managed to avoid the instability of slopes and environmental impacts. [At that phrase, she thought to herself "how do you 'manage' a glacier, and how can you say you're going to remove ice from a glacier 'to avoid' environmental impacts? That's crazy!"] . . . The thickness of the glacier sectors that must be removed is estimated at 3 to 5 meters [10–16 ft].

Picolotti kept looking back at the cover of the document to ensure that it was truly a document produced by the company:

> the mining equipment shall be employed as needed for each glacier sector to be managed (basically bulldozers and/or front loaders).

She was completely stunned: bulldozing glaciers? Really? The words and images chosen by the company to describe the destruction of these glaciers were grotesque, even surreal:

> the chunks of glaciers shall be removed with the mentioned machinery until the surface is clear (principally rock).

Picolotti could not believe her eyes. *"Who authorized this?"* she thought. *"Aren't our glaciers protected?"* But it got worse:

> if necessary, controlled explosives shall be used, of small size, to remove the ice. . . . the chunks of ice that come apart that are removed, until the level of the terrain is reached, shall be "pushed" or transported by the same mining machinery to an adjacent area, nearby but outside of the boundaries of the development of the pit.

She clenched her fists, *"these bastards are moving glaciers,"* she thought, her anger mounting as the environmentalist inside her surfaced momentarily and shattered her normally calm and reflective Secretarial composure. She heard the voices around her describe other impacts, but nothing registered as she kept reading:

> the sites for disposal of the chunks of glaciers shall comply with the following basic conditions:

The list that followed described where and how the glacier ice would be disposed of. She thought, *"they speak of glaciers as if they were trash."*

Then another of the local representatives pulled out a neatly folded, colorful brochure with a cartoon depicting mining activity and handed it to Secretary Picolotti. "This is a brochure Barrick Gold gave out to communities in Chile showing how they are going to destroy the glaciers with bulldozers and dump trucks. They're going to haul off the ice and dump it somewhere else." She shook her head in disbelief.

The brochure indeed seemed to be official, with Barrick's logo prominently displayed along with the motto, "Barrick—Responsible Mining," shining brightly alongside cartoon depictions of a glacier being hauled off with bulldozers and dump trucks. In another image, the ice was transported to another location. She wasn't sure if she should laugh or scream. Instead, she took a deep breath and planned her next step.

She put Barrick's EIA Annex 1B and the brochure (Figure 1.4) down and looked back at the map that the glaciologist Milana was pointing to showing the multiple river patterns stemming from the Toro 1, Toro 2, and Esperanza Glaciers, all in the pit area. He was explaining how the glaciers all around Pascua Lama and Veladero were the lifeblood of the many rivers that stem from this part of the Central Andes.

All of a sudden, what Milana had said earlier began to sink in and gel. Barrick is saying it in black and white. She let out an uncharacteristically furious vent: "These sons of bitches are destroying our glaciers to get at the gold underneath. I can't believe this!" Quickly regaining her focus on the reason these fifteen people had traveled 1,000 km to come see her, she addressed Milana, "How big are these glaciers and how much water do they have in them? And where does the water run?" she asked.

"Well, the glaciers that are on the Chilean side run west, through the Huasco Valley home to the indigenous communities called the Huascoaltinos,

ENTREGADO INFORME DE VIABILIDAD DE MANEJO DE GLACIARES.

Con el objeto de entregar información de la viabilidad ambiental sobre el futuro Plan de Manejo de Glaciares, Barrick presentó a las autoridades un estudio sobre los efectos de trasladar parte de los hielos desde los glaciares Toro 1, Toro 2 y Esperanza hasta el glaciar Guanaco. Este documento es la base de lo que sería el proyecto definitivo de conservación de hielo de glaciares que entregará la empresa.

Hay que destacar que el traslado de los hielos durará entre cuatro y seis meses. Para apoyar la conservación del hielo se utilizarían cortinas para nieve, que son estructuras de madera que frenan el viento cargado de copos de nieve, los que caen formando un montículo de nieve y aumentando la acumulación de nieve en esa ubicación.

1 Se utilizarán excavadoras hidráulicas para cargar el hielo directamente en camiones.

2 Estos camiones lo transportarán a un depósito en el glaciar Guanaco, aproximadamente a cuatro kilómetros de distancia.

3 Al llegar al glaciar, los camiones descargarán el hielo directamente en el depósito, que se encuentra preparado para su conservación.

4 Una vez ahí, las cortinas para nieve lograrán que los bloques de hielo formen una masa continua.

BARRICK
Minería Responsable

FIGURE 1.4 *Barrick Gold's brochure surprised communities with the proposal to haul ice with bulldozers and dump trucks and transport it to a new location.*

Source: Barrick Gold.

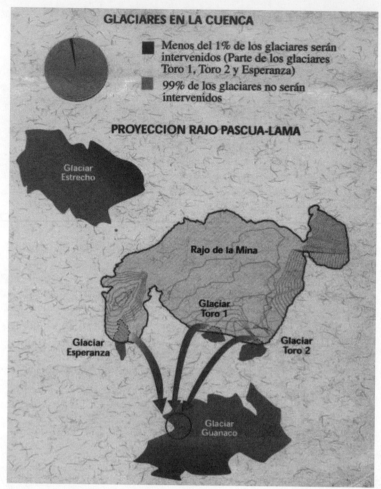

FIGURE 1.4 (*Continued*)

and then on to the Pacific. The ones you see on the Argentine side of the border run right into the San Guillermo Biosphere Reserve, a protected UNESCO site, and then on down to most of the rich agricultural valleys in San Juan and to the local population. And if you kept following the water that links up with a lot of other glacier runoff, eventually it gets to the Atlantic Ocean after traversing all of Argentina, West to East," replied Milana, as he described the terrain and the glaciers more specifically.

On hearing Milana's words, Picolotti grasped at a clear opportunity: international and transboundary impacts made glacier impacts federal jurisdiction. She could act.

"This glacier here . . . ," Milana pointed to a glacier labeled on his map as Toro 1, inside the projected pit area, "is about 300 m [984 ft] wide and 800 m [2,600 ft] long, or at least it was. It's shrinking."

"How does that compare with glaciers like those in the South, the Perito Moreno, for instance?" she asked.

"Well, the glaciers like Perito Moreno, or the Upsala, are colossal by comparison. You can't compare them in terms of dimension, but that doesn't mean that a small glacier like this one doesn't hold a lot of water. In fact, more people and agriculture depend on glaciers in San Juan than do people of the glaciers in Patagonia. Barrick is talking about removing 3–5 meters of thickness of ice. You can do the math: a tiny glacier the size of a football field, only 3 meters thick (tinier than any of these glaciers), holds more water than you and your entire family will ever consume in a lifetime.[22] But that's not the most significant fact. Glaciers regenerate and regulate water flow. Glaciers in areas like San Juan serve a critical and fundamental natural purpose. They capture water in solid form and release it slowly as we need it during warmer months and drier years. Each year, with new snowfall, they recharge and provide us with a continuous flow of water, for tens, hundreds, or even thousands of years. Their collective value, that is, if we take all of these little glaciers together, is incalculable. Without them, we wouldn't have water in the region in dry seasons."

Milana then took out an image from a mine in Kyrgyzstan, the Kumtor[23] mine run by Centerra, another Canadian mining company, where a sterile mining rock pile was invading a very active glacier. He asserted that this was the sort of problem Pascua Lama would run into (Figure 1.5). He also mentioned that, in Chile, similar impacts were occurring at Codelco's Andina

FIGURE 1.5 *The Kumtor mine in Kyrgyzstan, operated by Centerra, reveals critical impacts of sterile rock piles onto active glaciers. GIS: 41°51'21.74" N 78°11'40.22" E.*

project, right near the capital of Santiago[24] where the mining company had removed numerous rock glaciers (Figure 1.6).

Picolotti thought for a moment, mentally racing through the environmental code that, as a seasoned environmental lawyer, she knew extremely well. Argentina's law, she thought, doesn't protect its glaciers. They're completely vulnerable!

Abruptly, the environmentalist and public official took over, and she shot out a series of questions to the glacier expert who was suddenly the only person in the room she registered and focused on.

"How many glaciers are in this project area?" she asked, seeing at least six or seven glaciers highlighted on Milana's map in the immediate and central project area.

Milana answered, "Well, Barrick talks about a handful, but I've done some work in the area, mapping about fifty, just on the Argentine side and within the project's influence area. There are also many rock glaciers in the area that contain ice beneath the surface of the earth."

She creased her brow, perplexed, "Underground you mean? How do you know they're there? Do you have to excavate to find them?"

"No, they are identifiable if you know what you're looking for, a good satellite image reveals them perfectly. Sometimes you need to actually go to the site to verify ice content, but in other cases, you can guess it by the way the rocks move over the surface of the earth. Ice deforms due to its weight and the slope of the mountain, and rock glaciers, like normal uncovered glaciers, often creep downslope. You can see the movement by the positioning of the

FIGURE 1.6 *The Andina project by Codelco in Chile has destroyed several rock glaciers and operates in the proximity of large uncovered glaciers.*

Source: Alexander Brenning. GIS: 33°09′ 33.19″ S 70°15′18.98″ W.

rock on the surface. It mounds up in recurring crevasses and curves with similar arcs."

She was amazed: "You mean there are invisible glaciers?"

"Yes, well, they're not invisible if you know what you're looking for, and Barrick knows they're there. Their geologists do a thorough analysis before they even put shovels into the earth. They have to, otherwise a whole hillside can come crashing down if they don't do their homework and start moving around rock saturated in active ice."

"Incredible!" she murmured, "and so, how many glaciers are we talking about, you said fifty?"

He drew her attention to the map on the table, "Fifty in the immediate project area near the pit on the Argentine side, but in the entire project area, including in Chile and along Barrick's access roads on both sides of the border, we're probably talking about several hundred glaciers that could be impacted. No one has ever done an inventory of our glaciers. Neither have the Chileans. No one really knows how many glaciers are up there. But I've done

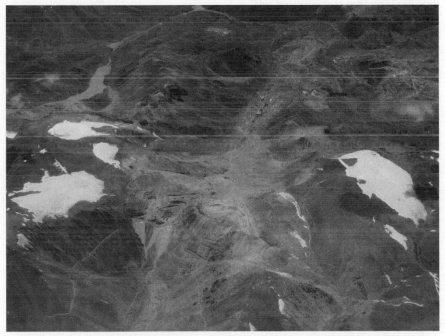

FIGURE 1.7 *Aerial view of Pascua Lama (foreground) and Veladero (background, right side) mining projects and their proximity to glaciers. Glaciers from top left and counterclockwise: Los Amarillos, Amarillos, Estrecho, Guanaco. The Esperanza, Toro 2, and Toro 1 Glaciers are visible in the center but are completely covered by debris and dust from preparatory stripping work at Pascua Lama.*

Source: Gerhard Huedepohl. GIS: 29°19′36.04″ S 70°01′ 11.74″ W.

rough calculations and my guess is that in San Juan Province alone, there are upwards of 10,000 glaciers."

Picolotti was astonished. In the mere thirty minutes since the meeting had started she had learned an incredible amount of startling information. There were glaciers, thousands of them, in more than just the Patagonia; there were glaciers in one of Argentina's driest and hottest regions, there were invisible glaciers, and there was a mining company proposing to dynamite glaciers to get at mineral deposits (Figure 1.7). Why hadn't the public heard of this before? Why hadn't the provincial authorities in San Juan Province, home to the project and with jurisdiction over mineral and environmental resources, intervened? Why hadn't the Federal Environment Secretariat in place before her term in office intervened when the project was first considered in Argentina? No one was talking about this in Argentina. Why?

Notes

1. See http://wp.cedha.net/wp-content/uploads/2011/11/Plan-de-Manejo-de-Glaciares-Barrick-english.pdf.

2. See Mt. Aconcagua: 32°39′12.38″ S 70°00′41.06″ W.

3. See, for example, the Horcones Glacier: 32°41′05.63″ S 69°58′12.47″ W.

4. See http://legal.un.org/riaa/cases/vol_IX/29-49.pdf and http://es.wikisource.org/wiki/Tratado_de_l%C3%ADmites_entre_Chile_y_Argentina_1881.

5. Scanu, 2012, p. 101.

6. See http://en.wikipedia.org/wiki/El_Indio_Gold_Belt.

7. Argentina's Environment Secretariat is formally called the Environment and Sustainable Development Secretariat. In Argentina, Secretariats are generally housed under Ministries, and the Secretary has a Minister as a direct supervisor. However, in 2006, President Nestor Kirchner elevated the Environment Secretariat to Ministerial Level by placing the Secretariat administratively in the President's Cabinet, thus answering directly to the president with no minister intervening. For this reason, the Environment Secretariat is sometimes referred to as a Ministry, and the Secretary of Environment is sometimes referred to as a Minister.

8. Romina Picolotti founded the Center for Human Rights and Environment (CEDHA) in 1999. She won the prestigious global Sophie Prize 2006, given to one environmental leader at the global level who has contributed innovatively to global sustainable development. CEDHA also won the Sierra Club's most prestigious international advocacy prize, the Earth Care Award in 2007, when it shared the stage and prize ceremony with Al Gore (John Muir Award) and Thomas Friedman (Journalism Award) for CEDHA's work at promoting environmental protection and human rights accountability for corporations.

9. The Veladero mine is at 29°23′26.53″ S 69°56′26.47″ W.

10. The Pascua Lama mine is at 29°19′04.66″ S 70°00′49.40″ W.

11. Veladero's leach pad is at 29°23′08.82″ S 69°56′09.89″ W.

12. This comment was made in an interview with a National Parks official who preferred to remain anonymous.

13. Tudcum is at 30°11′56.57″ S 69°16′11.98″ W.

14. The Conconta Pass is at 29°59′02.16″ S 69°36′41.30″ W; the controversial glaciers are just past the pass on route to Veladero at 29°58′34.04″ S 69°38′03.43″ W.

15. See roadside rock glaciers at the Conconta Pass at 29°59′25.79″ S 69°36′13.79″ W.

16. These are the Upper Almirante Brown Glacier, the Lower Almirante Brown Glacier, and the Norte Glacier.

17. Barrick publishes information on its website about the introduction of GAZEX detonators at the Conconta Pass used to avoid avalanches. See http://barricklatam.com/nuevo-sistema-de-control-de-avalanchas-en-el-camino-minero-de-veladero-y-lama/barrick/2012-05-29/135756.html.

18. For the Cabrera and Leiva study of the Conconta glaciers, see http://mineria.sanjuan.gov.ar/pascualama/Glaciarconconta0708.pdf.

19. At the time of submission, I spoke to Fredys Espejo, a former Barrick Gold worker who now is a staunch critic of the Veladero and Pascua Lama mines. He was allowed into the area in May 2014 on a routine civil society monitoring exercise to take water samples. Espejo passed through the Conconta Pass and noted that even the remaining two glaciers (the Upper Almirante Brown and the Norte Glaciers) are practically gone.

20. See 29°59′21.68″ S 69°36′30.69″ W; in Google Earth's *time feature*, load image from April 25, 2011.

21. See the unofficial translation of Barrick Gold's "Glacier Management Plan" at http://wp.cedha.net/wp-content/uploads/2011/11/Plan-de-Manejo-de-Glaciares-Barrick-english.pdf; for the Spanish original see http://wp.cedha.net/wp-content/uploads/2011/11/Plan-de-Manejo-de-Glaciares-Barrick.doc.

22. The math to sustain this is quite simple and also quite startling. A small glacier, 100 × 100 m—or the size of a football field—and only 3 m thick (most glaciers are much thicker) holds about 30,000,000 liters of water. That's 100 × 100 = 10,000 × 3 = 30,000 × 1,000 liters per m^3 = 30,000,000. (Note that one m^3 of ice is actually about 920 liters; we use 1,000 liters to keep the math simple and then use a conservative estimate in glacier thickness to compensate.) Divide 30 million by 200 liters per day, which is about what we should all have for our daily consumption, and that gives us about 150,000 days of daily water ration. Divide that by 365 days per year, and it makes for about 410 years. If a family of four all lived to be 100, then one small (tiny) glacier only 100 × 100 m and 3 m thick would provide them with a lifetime's worth of water!

23. Kumtor is at 41°51′21.74″ N 78°11′40.22″ E.

24. Andina is at 33°09′33.19″ S 70°15′18.98″ W.

2

What Is a Glacier?

It's mind-boggling (and a bit scary) to consider that while most of our planet's surface is covered with water, only about 2–3% of this water is actually freshwater—that is, *water that we can drink*. That means that most of the world's water (about 98%) is of no use for human consumption or for agriculture. But perhaps a more startling statistic that few actually realize is that of this minuscule percentage of water that is actually apt for consumption, three-fourths of it is packed away in dense millenary ice located in the polar ice caps; this is water that we will probably never see in fresh liquid form.

Except for documentaries we see occasionally on television about fearsome adventurers who traveled to Antarctica or to the ice sheets of the North Pole, most of us have never ventured (and probably never will) to the North or South Pole where this ice is located. These are rather inhospitable places of our planet that we could only tolerate on extremely nice days and only for a few days at best, if we were ever able to get there at all.

We hear about the polar caps melting due to climate change. We see images of penguins in the Southern Hemisphere or polar bears in the north suffering from a warming climate, and we even see entertaining animated movies about these obscure and rapidly changing environments and how odd creatures adapt or succumb to these changes. We hear from many media sources, from scientists and from environmentalists, that enormous ice masses at the poles are melting fast and breaking away into our oceans. James Balog, a photographer and cryoactivist, recently produced a documentary film called *Chasing Ice*,[1] which incredibly captured the *calving* (the collapse) of a chunk of glacier ice half the size of Manhattan Island, breaking off from the Ilulissat Glacier[2] and rolling into gelid waters off Greenland. Pieces of glacier ice more than 200 meters (650 ft) tall—as tall as skyscrapers—suddenly sank, vanished, resurfaced, and bounced around in the water as this colossal glacier crumbled into the sea. Since then, much larger calvings have been reported around the world.

FIGURE 2.1 *The Pircas Negras Glacier is a massive ice body in the Argentine province of San Juan providing water to communities in one of Argentina's driest regions.*
Source: Federico Barberis. GIS: 30°21'59.20" S 69°48'02.52" W.

We know that we are losing our glaciers to climate change, but most of us have never and probably will never see these melting glaciers up close. We think of them as extremely far away, much too far to visit or even see from afar before they disappear. For many of us, however, these amazing natural resources actually provide much more water than we might imagine, and many are actually much closer than we realize.

This book is an effort to bring us closer to *our* glaciers; the ones that make a daily difference to our lives and to the lives of many. It is about a resource dear to us all, but, for many, a resource we know little or nothing about (Figure 2.1).

Mountain Glaciers

Only about one-quarter of our planet's freshwater is actually in areas where we might be able to consume it as drinking water or use it for agricultural or industrial activity. Polar ice melts into the oceans, converting freshwater into salt water. It does not melt into rivers that feed our daily consumption; that is, we're not drinking water from rivers that originate in the North or South Pole. Instead, the water we consume in our homes and use for daily household, agricultural, and industrial purposes comes from other sources: rain, wetlands or marshlands, artificial reservoirs, and rivers that probably derive from nearby mountain environments. For most people, the source of the water they consume is close to home. For many of those people (a number probably in the billions), a fair portion of it, perhaps more than we will ever fully comprehend, at some point in the supply chain comes from glaciers and glaciated areas in mountain ecosystems not so far from where they live.

Much of our drinkable water is in mountain ice (mountain glaciers) located in more temperate climates such as in the Himalayas, in northern Canada, in the Rocky Mountains, the Alps, the Caucasus, or in the Central Andes, just to name a few ranges. This mountain ice, captured in dense bodies of perennial ice (ice that forms from winter snowfall and survives past the summer and into the next winter season), is also in glacier form, and, although we may not think of the ice up in California's Sierra Nevada, in the more obscure regions of the Aconquija Mountain Range in isolated Catamarca Province in Argentina, or in southeastern Armenia's Zangezour Mountains as "glaciers," that's in fact what they are, capturing snow, storing water, and critically feeding rivers and ecosystems downstream.

Glaciers Defined

Because this book is about glaciers, it would be appropriate to start with a basic definition of our topic. This is no small task, and, as we will see in the odd-numbered chapters covering the evolution of the Argentine glacier protection law, defining what a glacier is—as well as what it is not—became an extremely contentious glacier-melting debate!

Let's "Google" glacier and see what we get.

gla·cier 'glāSHər/ A slowly moving mass or river of ice formed by the accumulation and compaction of snow on mountains or near the poles.

Wikipedia suggests:

A glacier is a persistent body of dense ice that is constantly moving under its own weight; it forms where the accumulation of snow exceeds its ablation (melting and sublimation—or evaporation) over many years, often centuries.[3]

Another rather technical and more specific definition is offered by the US Geological Survey (USGS) one of the world's foremost authorities on glaciers:

A large, perennial accumulation of ice, snow, rock, sediment and liquid water originating on land and moving down slope under the influence of its own weight and gravity; a dynamic river of ice. Glaciers are classified by their size, location, and thermal regime.[4]

Argentina's national glacier protection law (which we will read more about in this book) sets forth the following legal definition, which is one of the world's first authoritative legal definitions of a glacier:

> All perennial stable or slowly-flowing ice mass, with or without interstitial water, formed by the re-crystallization of snow, located in different eco-systems, whatever its form, dimension and state of conservation. Detritic rock material and internal and superficial water streams are all considered constituent parts of each glacier.[5]

In all these definitions, and for most glacier definitions, a few elements recur:

> *Accumulation and compacting of snow* (compacted snow eventually turns to ice)
> *Movement* (most glaciers—*not all*—have some lateral or internal movement)
> *Size seems to matter* because definitions also vary according to what mass size is considered or is not considered a glacier; some definitions, such as the Argentine legal definition, accept any size or form, whereas others, like the USGS's definition, speak about "large" sizes, which is not very specific or practical if we want an exact measure.

As we can see, scientific definitions vary somewhat, which can lead to conflicting views of what actually constitutes a glacier, which in turn leads to differing interpretations as well as to a variety of legal implications about glacier protection and about what we can and cannot do at or near glaciers.

For our purpose, we stick to some basic concepts and definitions and work with these throughout the book. But the first thing the reader should do is to get rid of preconceptions of what a glacier is or is not, particularly if we have the large majestic glaciers of Greenland, Alaska, or Patagonia in mind. For the most part, we are not going to focus on such glaciers here.

So, for the sake of this book, we take into account the following seven points into our definition, always leaving some flexibility for anomalies and always with the assumption that we are defining glaciers with a view to protecting their hydrological value and function:

1. Glaciers are formed by snow that has accumulated, compacted, and survives for at least two consecutive summers (we refer to this as *perennial ice*, much as we refer to the perennial leaves of an evergreen tree); some glaciers form over many hundreds or thousands of years, but some may form for only a few years; these are still considered glaciers.

2. Glaciers may move laterally, as when they slowly slide down a mountain, but they also may be stationary—imagine the ice accumulated in the crater of a dormant volcano, which has no

lateral movement—and they may go through internal structural alterations due to their weight, their melting and freezing cycles, and their changing ecological balance or imbalances.

3. Glaciers melt in warm periods and provide water (thus reducing in size, through what scientists call *ablation*), and they grow during cold periods (adding or "accumulating" snow/ice) and thereby storing water. Glaciers actually have sections of accumulation and ablation that can be visibly and easily distinguished.

4. Although glaciers seek equilibrium (offsetting melt and accumulation), they are usually either decreasing or increasing in size depending on seasonal variations; while glaciers advance "down" a mountain, a glacier melting from climate change will actually appear over time to be moving "up" the mountain. This is because as the climate warms, the freezing area of the mountain moves steadily higher, melting away the lowest point of the glacier first.

5. Although glaciers can be of any size or shape, we generally refer to the existence of a glacier once the surface size has reached 10,000 m^2 (about the size of a football field—about 2.5 acres) or 100 ×100 meters; nonetheless, smaller glaciers, even tiny glaciers (sometimes referred to as *glacierets, ice patches,* or *ice fields*), taken collectively, can have very significant hydrological value. Some studies have even shown that smaller glaciers can provide more meltwater than larger glaciers.

6. Glaciers form where the natural surroundings are conducive to their formation and survival; scientists don't have a term to describe this area, so I've decided to call it the *glaciosystem*, drawing a parallel in relevance and borrowing from the more general concept of the *ecosystem*. See Box 2.1 for a detailed definition of the glaciosystem.

7. Glaciers may also have a significant amount of rock debris in their interior or on their surface, picked up from avalanches or simply by the glacier's movement over the terrain. These may be called *rock glaciers* or *debris-covered glaciers* (debris-covered glaciers and rock glaciers are not the same!). We'll talk more about these special glaciers later.

BOX 2.1 The Glaciosystem

The *glaciosystem* (or *glacier ecosystem*) is the glacier and its surrounding ecosystem that influences its constitution and composition with respect to its water and ice accumulation and ablation, thus determining its biological process, its natural evolution during its periods of charge and discharge, and which, if affected, could impact or cause the alteration of the glacier and/or impact the ecosystem in which it exists.

The glaciosystem (or glacier ecosystem) includes elements such as:

Solids: Geological/rock formations surrounding the glacier whose

(Continued)

BOX 2.1 Continued

characteristics and orientation influence the accumulation of snow, the valleys through which the glacier flows, walls, mountainsides, the slope on which the glacier advances, rock debris and other natural materials in its vicinity or in its ice, and the moraines formed and accumulated by its advancement

Biological: Flora, fauna, and other biological organisms in its immediate surroundings, underneath, beneath, and inside of its ice

Water, snow, and ice: Snow that accumulates in the glacier through precipitation; water that flows on the surface, inside, and underneath the glacier; ice of varying densities and in different stages of compacting; other glaciers that unite with the glacier from higher water and ice basins; other glaciers to which the glacier unites; frozen ground (permafrost) in the periglacial environment; natural or artificial lakes (dams) formed and nourished (even if only partially) by the glacier; and natural or artificial meltwater at the foot of the glacier

Air and atmosphere: The air surrounding the glacier, the atmosphere in the zone of impact that can be affected by artificial changes in the topography that alter the natural wind patterns that contribute to the natural accumulation of water and snow on the glacier by contamination of the air with particulates that are deposited on the glacier and that contribute to the natural evolution of the glacier.

The glaciosystem (or the glacier ecosystem) can extend to:

- all directions surrounding the glacier;
- snow and ice on the glacier and above or to the side of the glacier, as well as water immediately above, to the side, and below the glacier;
- the side of and on the valleys through which the glacier flows;
- the immediate proximity or at a significant distance to the glacier, depending on the specific case and on the relevance of an eventual impact in the ecosystem of the glacier.

Human populations (rural and urban), agricultural activity, and industries that are located in the vicinity of the glacier and that can be directly affected by changes in the mass of the glacier and the accumulation and ablation of snow might depend directly or indirectly on the glacier and its glaciosystem.

The health of the glacier and its glaciosystem are evaluated by measuring and monitoring the evolution of the following variables related to the glacier:

- Accumulation and ablation of snow/ice
- Line of equilibrium
- Mass balance
- Energy balance
- Temperature
- Caloric balance
- Water flow
- Albedo
- Impurities/Contamination
- Air/Atmosphere in the vicinity

The Function of a Glacier

Obviously, glaciers are important, but not only for the more obvious reasons. In this book, we'll get into the specifics of why they are important. Here are some obvious and some maybe less-obvious reasons:

- They store and provide water.
- They provide water during warmer months or during especially dry years.
- They retain and slow snowmelt, thus regulating water flow.
- They help contain global warming by keeping the local environment cooler.
- They reflect light and warmth back to the atmosphere because they are white.
- They provide natural beauty.
- They are critical to the survival of flora and fauna in high and dry mountain environments.
- They provide great scenery for tourism.
- They tell us about history.

Glaciers of varying sizes, some larger than small towns and others as small as a football field, provide many local environments with storage of critical water reserves, feeding rivers and underground water tables that in turn irrigate food crops and provide potable water for human and industrial consumption to millions of people.

The smallest of glaciers can store as much drinking water as most families will consume in a lifetime. A slightly larger glacier can provide entire small towns with water for many years. Larger glacier systems, made up of small to mid-sized glaciers, like those of the Himalayas, the Rocky Mountains, or the Central Andes, can provide water for millions or even billions of people for centuries on end.

The white color of glaciers, taken en masse, is also an important reflector of sunlight for our planet. The white of ice and snow reflects sunlight while the darker earth and oceans absorb it. If our glaciers melt away, the Earth's surface will absorb more heat, and global warming will accelerate. So, glaciers are helping keep our environment cooler. The more glaciers, the cooler we get; the less glaciers, the warmer. Think of glaciers as self-perpetuating and self-renovating Earth coolers!

Newly exposed and melted surfaces also expose organic material that was previously frozen. This releases captive gases contained in the organic material into the atmosphere (such as methane, which is up to twenty or more times more potent than carbon dioxide [CO_2] as a greenhouse gas), thus leading to rapidly accelerated global warming.

We are familiar with the function of humid tropical wetlands and their great importance to flora and fauna living at or near them. In high mountain environments where water is scarce, flora and fauna thrive near glaciated areas. In the Central Andes, for example, these are called *vegas systems*, and they exist thanks to the melting of glaciers and periglacial areas (more about periglacial environments later). The famed Nevada town of Las Vegas, derives its name precisely because this desert area, at the foot of snowcapped mountains, collects drainage water from higher elevations.[6]

What would mountain trekking be without glacier-capped mountains! Thankfully, humans are naturally nature lovers. We enjoy visiting mountain environments, and, for those of us who are lucky enough, we might even live to experience the adrenaline rush of visiting one of our Arctic and Antarctic regions. Glaciers provide natural beauty and awe-inspiring views that show us the wonders of nature and inspire our love for the Earth and the environment. Glacier pictures make for great coffee table books!

Thanks to glaciers, scientists have been able to learn about our past. Some glaciers are hundreds, even thousands of years old. They have been formed by the cyclical pattern of winter snow creating a thin layer of ice year after year. In between snowfalls, ice hardens; particles of dust from the Earth's atmosphere fall on the clear glacier ice and become trapped when subsequent snow falls the following year. After many hundreds or even thousands of years, the ice at the core of a glacier remains true to what the atmosphere was like at the time the core ice came into contact with the air. By taking deep samples of glacier ice from the polar ice caps, where thousands of years of our Earth's atmosphere are deeply buried, we can learn about atmospheric conditions of the past. Glaciers are one of Mother Nature's ways to tell us what our planet was like long ago.

Our Depleting Icy Resource

As in the polar regions of our planet, glaciers in more temperate regions and climates closer to home are also shrinking as rising temperatures creep up mountainsides, pushing our glaciers to ever higher elevations as they seek cooler environments. Eventually, the cold climate that glaciers need to survive in high mountain ranges will simply vanish and, with it, the ice stored in mountain glaciers will disappear. At that point, we will no longer have glaciers to perform their cyclical role of capturing, storing, and regulating the release of water into our environments. Then, only winter snow melt will feed our rivers for short periods, and ecosystems will simply run dry in the summer until the following year, when precipitation might return in the form of snow or rain. Without glaciers, many of our rivers risk becoming "seasonal" rivers that will only provide water during parts of the year.

The mountains of Venezuela, Colombia, Bolivia, Kenya, Tanzania, Uganda, Indonesia, California, Montana, Colorado, Spain, and other parts of Europe, which once had significant numbers of glaciers contributing significant amounts of meltwater to local ecosystems, will soon be glacier-free. Glaciers of all sizes and shapes are suffering from rising global temperatures, and many are vanishing or have already vanished. Earth will probably not see these glaciers again for tens of thousands of years, until the next Ice Age returns.

Should this concern us? Certainly!

How Much Water Is Actually Contained in a Glacier?

A very small glacier the size of a football field (this is the lower threshold scientists use to inventory glaciers) likely holds enough water to cover the consumption needs for drinking and other daily use (bathing, cleaning, etc.) of an entire family for over a generation. A quick and approximate mathematical calculation reveals just how much water a small glacier can hold.

1 cubic meter of ice   = approximately 1,000 liters of water[7] (1 liter is
                               equal to about .3 gallons)

1 small glacier (100 m²)   = approximately 10,000 m³ of ice (that's
                               100 × 100 m)
                             = 10,000,000 liters of water (10 million liters)

1 person utilizes about   = 200 liters of water per day per person (drinking
                               + household)

1 small glacier can provide 1 person with 50,000 days worth of water that's
    10,000,000/200
                             = 137 years worth of daily water allowance for
                               one person
                             = 35 years worth of daily water allowance for a
                               family of four

But the great thing about glaciers is that they are a *renewable* source of water. That is, they don't simply melt away as they release water. Glaciers recharge each year. They're like a bottle of water that never empties! Glaciers capture snow in winter, melt off a portion during the summer months when the environment needs it most, and then recharge during the next winter cycle.

Larger mid-sized glaciers (still considered relatively small) can hold the freshwater equivalent of what an entire country population consumes in a year or more. Larger glaciers hold colossal amounts of water that feed ecological systems in ways we may never fully understand.

But whatever the size of a glacier, the important thing is to understand the critical role perennial ice plays in our mountain environments: the *rationing* and *regulatory* function that Mother Nature has devised to ensure that we have water when we need it! And it is precisely this function, which is so critical to our environment, that we must work to protect.

We should also not underestimate the value of very small glaciers because all perennial ice more or less performs the same function. Taken collectively, even very small glaciers, just meters long and thick and created during especially copious snowfalls, can play an important and significant role in providing melt-water during seasons with less snow. In fact, occasionally Mother Nature sends us a heavy snowfall or several successive years of snowfall; this not only charges up our glaciers, but also creates small *temporary glaciers* that may survive only a few seasons but that contribute water to our ecosystems. Not too long ago, at some time starting around the fourteenth to sixteenth centuries and on until the nineteenth century, a "Little Ice Age"[8] occurred for about three or four centuries that helped create and recharge many of the glaciers in existence today.

Types of Glaciers

Nature has devised many types of glaciers: hanging glaciers that perch on mountain tops, ice sheets that cover large flat expanses as big as cities or even countries, glaciers that draw ice from multiple basins, glaciers that mix with rock, glaciers in very high mountain environments, glaciers at sea level that engage with the oceans, or temporary glaciers that come and go in heavy snow years and remain through drought years when they are most needed. There are moving as well as stationary glaciers. There are glaciers merely a few meters thick, others that can be taller than the Eiffel Tower, and yet others that can be taller than a very tall mountain. Some glaciers move down mountainsides, others creep upward (this because they are melting at lower altitudes; over time, their lowest point actually moves up the mountain). Glaciers contract and expand daily as they freeze and thaw, producing roaring sounds of running water and fearsome cracking noises day in and day out as their structures adapt and shift to accommodate a constantly changing ecological environment.

There are white glaciers, blue glaciers, black and brown glaciers, and even *invisible* glaciers. Yes! There are invisible glaciers taller than tall buildings, glaciers more than 100 meters (300 ft) thick on which you could be standing and not see any ice whatsoever anywhere! In some cases, entire mountains freeze all humidity in their interior and perform hydrological roles similar to that of glaciers—the so-called *periglacial* or *permafrost* areas above the freezing line! (More on invisible glaciers in Chapter 4.)

Glaciers can be wonders of magnificent beauty, drawing tourists and nature lovers from around the world. They can be sites for recreational activity such as hiking or skiing. Many cultures pay social tribute to glaciers or worship them for their natural and spiritual value. Their use is multifold, and their function critical to humans and to the environment.

How Does a Glacier Actually Work to Store and Regulate Water Flow?

Glaciers are natural water towers, not too different from those water towers we might find on top of buildings in big cities or on hills above a community. They are a magnificent adaption by Mother Nature to regulate water flow and availability to downstream ecosystems, capturing snowfall in the winter months and saving it for hotter and drier summer months or for years with limited snowfall.

So how does a glacier come to be and how does it work?

THE DIAGENETIC PROCESS

A few days after a snowfall, fluffy snow begins to compact, and, eventually (after a few weeks), it turns to ice. This is a process known as *diagenesis*. This is a word that is also used for rock transformation through compacting, but here we refer to snow to ice diagenesis.

In Mariana Gosnell's wonderful book called *ICE: The Nature, the History, and the Uses of an Astonishing Substance*, she describes the process as follows:

> Snow changes slowly into ice, the colder the setting, the slower the change. First, new-fallen snow crystals settle within a fluffy snowpack, moving closer to each other and becoming rounded as molecules evaporate off their sharp points and condense in their hollows (where there are more molecules for them to join up with). These small, rounded crystals, which, once they have lasted through a summer melt season, are called *névé* in French and *firn* in German, bond together, or sinter, under the weight of more snow above them as well as from the continued migration of vapor molecules, until eventually all air channels are blocked. When there remain in the pack only individual air bubbles, unconnected to each other, the permeable snow has become impermeable ice.[9]

Christopher White describes the process in his book *The Melting Point* as follows:

> Firn is the previous season's snow that has thawed and refrozen and compressed; it is more than halfway toward ice. On a glacier, the process of turning snow to ice takes two or three years and runs through several stages. Snow crystals change shape, consolidate, thaw, and refreeze. They succumb under temperature and pressure. On the journey, firn is the last incarnation before ice. (White, 2013, p. 89)

White later describes the snow on the ascent to Swiftcurrent Glacier in Montana:

> I pick up a dab of snow on my fingertip and inspect it. ... the snow
> crystals are hardly bigger than a grain of salt. It's tempting to calculate how
> many snowflakes create this crevasse, that headwall, or the entire glacier.
> Untold zillions, I believe. The zeros are lost to the calculator and to the
> brain. But I try. A cubic foot of snow may contain up to ten million snow-
> flakes, but once it (and more snow) compresses and mutates to a cubic foot
> of ice, the count is likely in the billions. A glacier may contain tens of mil-
> lions of cubic feet of ice or more. All glaciers on Earth—nearly 400,000 of
> them—descend from snowflakes. The mind reels. Too many zeros. I give
> up. (White, 2013, p. 140)

The cycle of snowflake to ice looks something like Figure 2.2.[10]

In Figure 2.2, the snowflake on the left looks much larger than the snowball on the right, but actually (save for possible evaporation), they have the
same amount of ice. The difference is in the density. Freshly fallen snow is
extremely airy when compared to pure water. If water is at 100% density, the
fluffiest snow is at 6% (White, 2013, p. 88). Ice will fall somewhere in between,
approaching nearly 92% or 93% density at its most dense. The snowflake on
the left has a lot of air in it separating the ice crystals. The ice ball on the right
has compacted so much that the air pockets in the original snowflake have
been expelled, thus making the ice much more dense. Pick up a handful of
fresh snow powder and hurl it at a buddy and you'll have a fun snow fight, but
pick up a dense piece of ice and throw it at someone and it's going to hurt!
After snow has transformed to ice through the diagenetic process, the resulting ice is much less permeable to air, thus making the same volume denser,
heavier, and more impenetrable. That's why ice is hard and snow is not.

ACCUMULATION

If you look around on a snowy day when snow has accumulated on the
ground, streets, housetops, plants and trees, or cars, you'll notice that it accumulates more in some places than in others. This has to do with the snowfall
patterns, wind patterns, with man-made *and* natural structures, and with
the effects of external forces on the snow. Overnight, snow may accumulate
evenly on a lawn or large field, but, during the day, kids playing in the snow or
other trespassers will intervene and rearrange the snow cover. Snow on a road

FIGURE 2.2 *Snow to ice diagenesis.*
Source: Geoestudios.

will likely get flattened out by traffic, whereas snow in places receiving little or no disturbances may survive for a considerable time.

You'll notice also that in a given area—say your backyard—even if the kids haven't touched the virgin snow, there are places where it accumulates more than in others. This might be because of the way the wind blows through your neighborhood, bringing more snow to a certain corner of your yard. Or it may be because as snow falls off your roof, it accumulates next to the grill. Or maybe as snow burdens the leaves and branches of trees in your backyard, it suddenly and periodically falls in clumps and settles in a large mound next to the tree. For all of these reasons, snow accumulates in different ways in different places. Glaciers form in different places precisely for the same reason.

As long as the average daily temperature remains below 0°C (32°F), the accumulated snow will slowly turn to ice, and it will survive unless acted upon (such as you shoveling the snow out of your driveway). If left alone, it will begin to compact. As the sun comes out each day, some of the snow will melt, particularly during midday hours. The snow and ice that accumulates in areas receiving sunlight will partially melt, whereas snow remaining in the shade may experience only minimal melting or no melting at all.

MELT (ABLATION)

If you live in a cold area, maybe in the mountains where it snows often and where the weather is very cold, much of the winter snow in your yard will survive all winter. There will be parts of your yard that may keep many centimeters (several inches) or even more than a meter (several feet) of that snow untouched, while other parts may experience melting or other impacts that move, deteriorate, or completely eliminate the snow.

By the end of winter, if you live in a temperate climate, your yard may have accumulated lots of snow but lost most of it before the end of winter as the temperature starts to rise. If you live in a colder environment, you may reach the end of the winter with a considerable amount of ice still in your yard. As spring arrives, most of us will lose all of the snow that has fallen in our neighborhood. If some ice survives, we may be in the presence of a nascent glacier (Figure 2.3)!

What's important for glaciers is which is greater: the accumulation or the melt?

If more snow falls than the amount that melts away, that is, *if accumulation is greater than melt*, then the mass of the ice body increases in density and size. If the melt is greater than accumulation, then the ice will completely vanish.

And if the ambient temperature remains very cold even as summer approaches, the ice that formed during winter may have a chance of surviving into the summer. Generally, unless it's extremely cold (such as in the polar areas or high up on a mountain), much of the ice will melt over time when

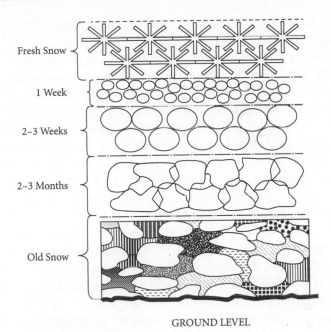

Fresh Snow

1 Week

2–3 Weeks

2–3 Months

Old Snow

GROUND LEVEL

FIGURE 2.3 *Snow to ice to glacier ice.*
Source: Geoestudios.

exposed to the sun, especially as the days grow longer and warmer and the ice is exposed to more sustained heat. But when the ice is hidden in a crevasse or spends much of the day in the shade, there is a chance it will survive. Sometimes it can survive through the spring and summer, into the fall, and, if conditions are right, even right up to the next winter. If it has survived through to the next winter, then new snow will be added to the ice, replacing the melted surface, and the cycle will begin all over again. If it survives for two consecutive summers, scientists consider it *glacier ice.*

The Glaciosystem

Glaciers form in areas where the environment is conducive to their formation; that is, where the ecosystem is *glacier-friendly.* Scientists haven't really come up with a term for this specialized glacier ecosystem, so I've decided to give it a name. I call this glacier-friendly ecosystem, the *glaciosystem.*[11]

Glaciosystems tend to be located in colder areas, areas with minimal sunlight, or where the prevalent winds deposit copious amounts of snow. In many mountain ranges, lowland areas may have very hot climates in summer where ice could not survive, but, as we begin to move up the mountainside, at very high altitudes (above 2,500 meters in the Central Andes [8,200 ft] or in a range like the Sierra Nevada in California), the temperature begins to fall rapidly. At nighttime, it is generally below freezing (below 0°C/32°F) even in the middle of summer.

If we take any given mountainside, the slope facing the sun during the day will invariably be warmer. Glaciers don't like the sun and warmth, and so you're more likely to find them on slopes facing away from the sun's prevalent direction. In the Northern Hemisphere, these would be north-facing slopes (which face the North Pole), whereas in the Southern Hemisphere these are south-facing slopes (which face the South Pole). If you're looking for glaciers on Google Earth, always keep this in mind! When it is cold enough, glaciers will form on either side of the mountain, but in many places, the north- versus south-facing slope trick is a useful rule of thumb to spot glacier ice and distinguish it from seasonal snow. You may also note that in a given area of a mountain, in the middle of summer, all of the glaciers appear at more or less the same elevation—say, for example, with their lowest points at about 4,500 meters (14,700 ft). In this example, it is very probable that the 4,500-meter line is the line at which glaciosystems will prevail in that area and that below that elevation ice will not survive the summer heat.

This can be useful information if you are trying to identify or inventory glaciers from satellite images. If we know that, in a given area, the glacio-system begins at 4,500 meters (14,700 ft), and if we look at an image from an intermediate season (say, springtime and find ice between 3,500–4,000 meters (11,500–13,100 ft), we can probably rule out that this is glacier ice. Mostly likely, it will have all melted by the end of summer. If we don't know what the glacio-system elevation limits are for a given area, we need to be especially careful when identifying glaciers in satellite images taken prior to the end of summer.

Because glaciers generally form on high mountainsides, unless they've been trapped in a crater or in a hollowed out cirque with no escape, they might be accumulating snow on an incline. As snow accumulates and as ice starts to form and gain mass, it gets heavy. If the incline of the mountain is too steep, the snow and ice will come tumbling down, as in an avalanche. But if the incline is just right, and the ice finds enough friction and resistance on the surface to detain its rapid slide (like rocks, boulders, or other rugged terrain), a permanent glacier may form.

The Composition of a Glacier

A cross-sectional view of a typical glacier will look something like Figure 2.4. In the image, the glacier is moving downslope from right to left. Snowfall feeds the glacier in the colder and higher altitude area near the right; this is called the *accumulation zone* because that is where the glacier accumulates snow/ice. As the glacier slides downhill, it will eventually reach a zone where the ice can no longer survive the warmer temperature (remember, temperature increases

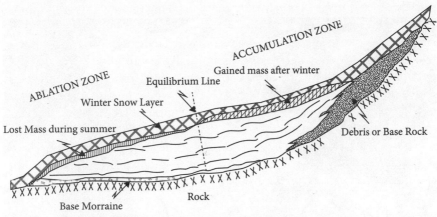

FIGURE 2.4 *The composition of a glacier.*
Source: Geoestudios.

as we move downhill). The lower portion of the glacier will thus melt as it reaches the warmer lower limits. This is called the *ablation zone* or the *melt zone*, and that is where the glacier loses ice.

Somewhere in the middle (not necessarily the geographical middle) is the *equilibrium line*, the point at which the glacier accumulates as much snow as is melted away (Figure 2.4). Remarkably, in Nature, the equilibrium line is actually quite visible to the naked eye.

In Figure 2.5, taken from Google Earth of the Bolivian Andes Condoriri Glacier near the city of La Paz, we can clearly distinguish the equilibrium line, which is the point where pale gray-colored ice becomes shiny white. That's the transition area from the ablation zone (gray) to the accumulation zone (white). You can see the equilibrium line for yourself by entering the following coordinate address in Google Maps right on your smartphone:

16 10 37 S, 68 15 06 W

Areas that receive copious snowfall in winter experience intense snow-melt in the springtime that brings significant water runoff for many weeks or months. Yet when much of the seasonal snow melts away, despite warmer ambient temperatures, the ice pack below remains cold. Some of this ice (the surface ice) melts slowly, providing a steady and continuous flow of water to rivers and streams during the rest of the yearly cycle and thus ensuring water provision for the many months after the seasonal snow has melted. This water provision is precisely what the glacier itself is providing to the ecosystem, above and beyond the seasonal melt.

When winter finally arrives anew, the cycle starts all over again and the glaciers recharge the ice they've released during warm months with new snow, compacted and transformed to ice over the colder months. This cycli-cal function is one of the most fundamental and significant roles glaciers play

FIGURE 2.5 *Equilibrium line in the Condoriri Glacier in Peru, between snow accumulation and melt, is clearly visible between shiny white and gray areas. GIS: 16°10'36.93" S 68°15'06.18" W.*

in our environment. It's what scientists call the *recharging* or *regulating* role of glaciers into ecosystems.

Climate change is placing this cycle at risk because it is changing the ambient temperature of the glaciosystem as well as snowfall quantity, and this results in more melting than accumulation. This is making glaciers retreat around the world. As average temperatures rise at lower elevations, they are also rising in higher mountain environments. Thus, the line at which the yearly average temperature is freezing is slowly creeping up the mountain. Above the freezing line, we can find perennial ice where glaciosystems are glacier-friendly. But because that line is moving upward, so are glaciers, slowly melting and creeping upslope until the environment one day will no longer sustain glaciers from year to year.

Where Are the Vast Majority of Our Planet's Glaciers?

There are glaciers on nearly every continent, even in Africa, in Indonesia, and on Australian soil. Although the largest and most extensive glaciers exist in areas such as Antarctica, Greenland, Northern Canada, and Central Asia, very large glaciers also occupy parts of Russia, the United States, China, Peru, Iceland, Scandinavia, Chile, Argentina, France, Switzerland, Kyrgyzstan, and New Zealand (this is just a sample list, there are others!). Aside from the more spectacular and colossal glaciers we are accustomed to seeing in scientific documentaries (like those of Antarctica and Greenland), there are also many smaller glaciers distributed in other parts of the world, more in number, smaller in size, but more relevant in terms of hydrological contribution to human consumption and to sustaining the flora and fauna of mountain ecosystems.[12]

These smaller glaciers are actually closer to home, closer to human populations and to areas where their melt is useful and relevant to us. Anyone who has visited a city like Geneva, Switzerland, or any other city at the base of glaciated mountains has crossed bridges spanning rivers that carry a constantly flowing and massive amount of water melting from glaciated mountaintops.

Awareness of relevance is an important first step in addressing any fundamental social problem. We protect what we know, understand, and value. But if we don't even know where glaciers are and much less why they are important to us, then it becomes clear why we are not protecting them.

Knowing *where* our glaciers are, *which* are most at risk, and *why* brings our changing environment home to our daily lives. The year this book was published followed the second straight year of very meager rain and snowfall in California, perhaps one of the lowest ever recorded. This year, Californians will depend more than ever on melting ice to cover their water needs. For populations living downstream from cold mountain environments that receive abundant snowfall, if all of a sudden that snowfall ceases and you're not getting water from rain and seasonal snowmelt, you will depend to a much greater extent on the Earth's cryosphere (the frozen surfaces of the Earth) to provide water.

The glaciers most important to the daily existence of millions (and even billions) of people living downstream from their melt are the temperate glaciers and debris-covered glaciers (as well as another variety called *rock glaciers*) found in lower latitudinal mountain environments of the Himalaya, the Andes, and of the mountain ranges of North America, Europe, Central Asia, and elsewhere.

Himalayan mountain glaciers provide meltwater to highly populated countries, such as India, Pakistan, Bangladesh, Bhutan, Nepal, China, Burma, Thailand, Laos, Vietnam, and Cambodia. The glaciated Hindu Kush, Pamir, and Tian Shan Mountains span from Pakistan through Afghanistan, Tajikistan, Kyrgyzstan, and parts of Uzbekistan, feeding water to millions.

The Caucasus Mountains, also with generous glaciated areas, provide drinking water to people of Georgia, Armenia, and Azerbaijan. Even some parts of Turkey, including for example Mt. Ararat[13] or the Pontic Mountains along the Black Sea[14] or the Mercandagi Range,[15] are spotted with glacial ice.

Japan is rich in glacier resources, including not only on Mt. Fuji,[16] but also in the Japanese Alps as at Mt. Tateyama.[17] Iran has glaciers not too far from the capital city of Teheran in the Elburz Mountains.[18] Even Lebanon (on Mt. Lebanon) is known to house some perennial ice that would qualify as glaciers by some definitions.[19]

Africa, hardly known for snow and even less for ice, has glaciers, and not just in the Kilimanjaro area. Kenya, Tanzania, and Uganda (bordering Congo) can boast impressive glaciers on the Mt. Kenya peaks and in the Ruwenzori Mountains of Uganda.[20]

Even Indonesia has glaciers! In fact, the Firn Glacier[21] and the Carstensz Glacier[22] of Puncak Jaya Mountain towering at 4,884 meters (16,000 ft) above sea level, along with the African glaciers, are a few of the world's last remaining Equatorial glaciers.[23] Yes! There are glaciers "on the Equator"!

Earlier in this section, I mentioned glaciers on Australian soil. Well . . . the Australian continent hasn't had glaciers for some time now, but a small group of islands called the Heard and MacDonald Islands (which is Australian territory) in the South Indian Ocean have some magnificent Australian glaciers.[24] Type 53 05 30 S, 73 31 03 E and see them on your phone (you may need to zoom out to see them properly).

The Americas are one of the world's heavily glaciated regions, with glacier areas stretching from the northern regions of Canada to the most southern confines of South America in Patagonia (Chile and Argentina). Canadian glaciers are colossal and are located across the country from east to west.

FIGURE 2.6 *The Marmolejo Glacier on the Argentine-Chilean border feeds Mendoza's wine country downstream.*

Source: JDTaillant. GIS: 33°44'18.39" S 69°50'31.66" W.

But glaciers also provide water to many other North Americans who are less known for drinking glacial water, including Mexicans, right in Mexico City![25] Glaciers provide water to millions at the foot of the Rocky Mountains. Californians benefit from glacier melt from the Sierra Nevada range,[26] as do inhabitants of Washington[27] and Oregon,[28] which are both rich in glacier resources. Even Nevada, more known more for its scorching deserts, has glaciers![29]

Latin America is another glacier-rich area. Glacial ice provides water to people in Mexico City, in Ecuador, Peru, and Bolivia, and even to communities in Venezuela and Colombia. Chile and Argentina are South America's most glaciated countries (along with Peru), with countless glaciers all along the central and southernmost Andean regions, as well as in other mountain ranges separate from the Andes, but also with very high elevation (Figure 2.6).

Glacier Vulnerability

Our glaciers are disappearing due to global ecological imbalances and that fact alone should alarm us because these mountain glaciers are the principal water source for millions of people and for downstream ecosystems. Part of this deterioration is a natural phenomenon that occurs over cycles spanning tens of thousands of years. Eventually, most of our glaciers will disappear; then, hopefully, many thousands of years later, they will return. They will probably outlive us (as a human race) and, in time, reemerge when we are long gone. In the meantime, however, we should be able to enjoy our glaciers and the water they provide us for many hundreds and even thousands of years.

However, global anthropogenic climate change is melting our glaciers away at alarming rates. (See Chapter 6 for more information on climate change and glacier melt.) This is due to the sort of global atmospheric contamination we have now heard about time and time again. Addressing global climate change and glacier melt at a global level seems monumental. But there are also many other types of impacts, many of which are anthropogenic and very local, that directly affect glaciers and that can be addressed and reverted.

First, let's consider how glaciers can be impacted. Heat is the obvious culprit— that is, *climate change*. Global warming is destroying our glaciers because it generates severe imbalances favoring more ice melt than accumulation. But there are several other reasons our glaciers are vulnerable, and it's not only due to generalized global temperature increases. Here are the most common reasons for glacier vulnerability:

- Generalized and sustained *temperature increases* can accelerate melt rates that exceed snow/ice accumulation; this may occur naturally in long-evolving geological eras measured in tens of thousands

of years, or it may occur abruptly due to anthropogenic global
warming, for example.

- Naturally caused *atmospheric contamination* can impact glaciers
  and accelerate glacier melt; a natural occurrence such as a volcanic
  eruption can deposit volcanic ash on a glacier, change its albedo
  (its reflective capacity), and, much like the results of wearing a dark
  shirt on a hot day would increase our surface temperature and make
  us hotter, glaciers also can suffer the same impact from changing
  surface color. These impacts are generally short-lived and, as long as
  the ash does not recur yearly, subsequent snowfalls can reestablish a
  glaciosystem's equilibrium.
- Anthropogenic atmospheric contamination, such as the black
  carbon emitted from vehicles burning diesel fuel near glaciers,
  can also soil and darken glacier surfaces causing the same albedo
  change and subsequent temperature impacts. It used to be that cars
  and trucks did not circulate at such high altitudes where we find
  glaciers and thus this direct impact was not as severe; however,
  times have changed and today humanity reaches confines it never
  once explored. Furthermore, wind currents can take fuel emissions
  from highly contaminated valleys up to remote glacier locations.
  These emissions-related impacts generally occur consistently over
  time, and, as such, the glaciosystem may not be able to recover from
  recurring glacier surface soiling.
- Dust from dirt roads or from nearby industry that emits particles
  into the air can cause glacier surface contamination and also
  affect albedo or surface conditions in such a way as to alter glacier
  equilibrium and induce glacier melt. These impacts are also
  generally recurring over time (Figure 2.7).
- Impacts to glaciosystems can alter those elements of the environment
  that are conducive to glacier formation. For example, if prevailing wind
  patterns cause snowdrift to settle in a given area, and we alter those
  wind patterns by removing a mountaintop in an industrial mining
  operation, we can adversely affect the glacier's snow accumulation.
  Geological alterations tend to be permanent, as are their impacts.
- Altered microclimates through the introduction of increased
  industrial activity can have influence on glaciosystems, thereby
  altering glacier equilibrium.
- Placing heavy weight on glaciers, such as rocks removed from
  another area (e.g., sterile rock removed in a mining pit excavation
  or from road work), can alter glacial structural balances, modify a
  glacier's advancement path, and thus impact that glacier.
- Running a road through a glacier (and *yes*, it has been done) can
  greatly impact a glacier's natural equilibrium. If you cut out a piece of

FIGURE 2.7 *The Toro 1 Glacier in Chile has been completely covered by debris from mining pre-stripping blasts at nearby operations at Barrick Gold's Pascua Lama project.* Source: Anonymous. GIS: 29°19′54.56″ S 70°01′09.36″ W.

glacier to make room for a road in the accumulation area of the glacier, you can affect the snow/ice intake and alter equilibrium; if you cut a road through a lower section, you can cause accelerated advancement of higher ice through the empty space and possibly cause the glacier's sudden and accelerated (maybe even catastrophic) collapse.
- Drilling into a glacier, removing a part of its mass, can obviously impact a glacier.

The Danger of Deteriorating Glaciers

When glaciers melt, they're dangerous. The melting ice alters the structure and integrity of the glacier. They shift, they bend, they break, and they collapse, crashing down on anything below.

Glaciers can be deadly, with chunks of collapsing ice the size of skyscrapers falling into glacier lakes below and forming enormous waves that rush down mountain valleys to take out anything in their path. Chapter 8 discusses these glacier lake outburst floods (GLOFs)—I call them "glacier tsunamis"—more in detail.

Glaciers can also dam very large lakes at high altitudes. If these dams melt, they can release colossal amounts water suddenly into nearby ecosystems.

Glacier Protection

So, now we've talked about the critical role glaciers play in natural ecosystems and the fact that they are both vulnerable and dangerous. Glaciers are our water reserves, and, for many communities, they act as naturally created water towers, rationally regulating water flow into local environments. We also realize that glaciers are at risk from natural evolution (long term) and also from anthropogenic activities that include not only climate change-related impacts but also many other local impacts that can also be caused by anthropogenic activity. We have also seen, long before reading this book, that one of the most visible signs of climate change is melting glaciers.

Our societies have developed laws, regulations, and policies to protect a vast number of natural resources including our air, forests, oceans, plants, flowers and trees, fishes, certain endangered species, flora, and fauna because they are important, because they are key to ecological systems and balance, and because they are important to our human existence.

We have also, over the years, developed carefully designed water policies and, in some cases, laws that control, protect, and monitor water resources. Over the past decade, global society has even developed the notion and legal concept of a universal "right to water." Along with air, water is our most important natural resource; without it, we cannot exist.

We would think that, because glaciers house the most important natural resource to our existence, and because they are vulnerable and because we are unduly impacting them, as a society, we would have taken steps by now to protect them. And yet, despite the obvious vulnerability of glaciers and the fact that they are critical to the sustainability and livelihoods of dense human populations, until very recently, *not a single country anywhere in the world had enacted a national law to protect glaciers*, the veritable water towers and reserves of our planet. Glaciers, for the most part, are not protected by law or by even by public policy.

How is that possible? In an age of deteriorating planetary ecology—and with glacier retreat as one of the most visible signs of climate change—wouldn't the Canadians, the Swiss, the French, the Americans, the Chileans, or Argentines, or the Chinese, Pakistanis, or Nepalese, or the Californians who have been environmental mavericks on so many fronts and with many well-known glaciers in their environments—wouldn't someone have identified this delicate and fundamental natural resource and concluded that these resources should be protected not only for their majestic beauty, but also because they are fundamental sources of freshwater for local populations?

What about the United Nations, so set on advancing climate negotiations through the UN Framework Convention on Climate Change (UNFCCC) and raising concern over glacier melt? Wouldn't they have enacted a global treaty by now to protect glaciers? How about ski countries? Switzerland? France? The United States? Wouldn't they want to protect their glaciers, if only to protect their winter tourism?

The fact is that *no country*, until the early to mid-2000s, had considered even basic public policy to protect glaciers. We're watching our glaciers retreat. We're seeing some of our most significant freshwater reserves vanish, we're causing this impact globally and locally, and yet we are doing very little or nothing at all to protect this resource. How can this be?

This startling, perhaps unbelievable fact is one of the main reasons I've decided to write this book. When I discovered that glaciers were, for the most part, completely unprotected, I sensed an urgent need to do something about it.

Conclusion

The essence of this book is about imminent glacier vulnerability and the emerging politics of glacier protection, its recent origins, and the steps societies are taking to protect our glacier resources.

Climate change is causing irreversible damage to our environment and to our glaciers, but the future disappearance of our glacier reserves is not a short- or medium-term unavoidable consequence. Critical mountain glaciers can and should be around for many years to come, and they can and should continue to provide drinking water to millions—even billions—of people on most of our planet's continents. Glaciers *can and should* be protected. Glaciers can and should be monitored and their needs attended to, and we should all know more about our glaciers, the role they play in our ecosystems, their vulnerability, and what we are doing to harm these monumental water resources. We can even "create" glaciers, and there are already promising experiments under way to do so (see Chapter 8). So, why wouldn't we take steps to ensure the very best protection for glaciosystems and for the glaciers they contain?

This book looks at glacier vulnerability and specific anthropogenic activities (such as mining operations in the Central Andes) that have placed enormous glaciated areas at risk. I examine the birth of the world's first national glacier protection bill, which surfaced in Chile but could not muster up congressional support, and how it was quickly taken up in Argentina, just across the border, where it met with an intense mining lobby that reversed a unanimous two-house congressional vote by securing a presidential veto only days after the law was enacted. In Chile, as this book went to press, there was a renewed discussion in Congress to introduce a glacier protection law at the national level. I look at how glacier awareness and education is changing the way communities understand their local ecosystems *and glaciosystems* and how the love and awe of glaciers is transforming societies and the way people are organizing to learn about and undertake cryoactivism to protect glaciers—glaciers that, in many cases, they never even knew existed. In Kyrgyzstan and Pakistan, for example, there is a move to introduce legal protection for glaciers into national law. As this book went to print, the Kyrgyzstan Parliament had approved a glacier protection law that was awaiting presidential approval, whereas in Pakistan there is already mention of glacier mitigation in the country's national climate change policy.[30]

In sum, this is a book that offers us an invaluable lesson and opportunity to learn about one of the most inspiring and yet vulnerable natural resources Mother Nature has placed on our planet in the hope that more people will become sensitized to glacier vulnerability and contribute in their own ways to healthier and more long-lived glacier resources.

Notes

1. See http://www.chasingice.com.

2. The Ilulissat Glacier is at: 69°09′27.50″ N 50°53′33.26″ W; for news on the filmed breakup, see http://www.huffingtonpost.com/2012/12/12/ilulissat-glacier-break-calve-video_n_2287987.html.

3. See http://en.wikipedia.org/wiki/Glacier.

4. See http://pubs.usgs.gov/of/2004/1216/text.html.

5. See unofficial translation at http://wp.cedha.net/wp-content/uploads/2012/10/Argentine-National-Glacier-Act-Traducción-de-CEDHA-no-oficial.pdf.

6. On origins of Las Vegas, Nevada, see http://en.wikipedia.org/wiki/Las_Vegas.

7. Ice and water volume differ by about 8%. Water is denser than ice and therefore a cubic meter (m^3) of water contains a bit more water than a cubic meter of ice (not much more). But, for the sake of keeping the math simple, the calculation presumes a one-to-one ratio of ice to water, which actually results in a slight overestimation of water content in ice (by about 8%). To offset, I've presumed that the glacier is actually *thinner* that it likely is; hence, in the end, these calculations are probably very conservative.

8. See http://en.wikipedia.org/wiki/Little_Ice_Age.

9. See Gosnell, Mariana, 2005, p. 88.

10. Diagrams of snow to ice diagenesis are courtesy of Geoestudios.

11. In 2012, with the need to address glacier protection from a policy perspective and not finding in academic works about glaciers the necessary terminology to address a glacier's specific ecosystem, I introduced the term *glaciosystem* with the view to offer society, policy makers, and the scientific community an understandable conceptual framework with which we could develop environmental policies and laws to protect glaciers. The term is purposefully geologically broad, and the definition aims to integrate natural

elements associated with glaciers and their surroundings, a characteristic that is useful from the point of view of public policy makers who seek to establish geographical limits to protect glacial and periglacial resources.

Part of the challenge in developing laws and policies to protect glacial and periglacial resources stems from the fact that sometimes the impacts to these resources do not come from impacts directly to the resources themselves but rather to the ecosystems that are needed in the first place for those resources to form and survive. For this reason, I believe that it is imperative to broaden the existing academic concepts defining glaciers and periglacial environments to incorporate a greater glacier ecosystem because this ecosystem is, in the end, absolutely necessary for the glacier to exist. Different terms have been employed in the past, such as "cryospheric environments" or the "glacier continuum," but these are each limited in their own scope to address the concerns raised from a public policy perspective. This definition of "the glaciosystem" has been shared with a significant number of glacier specialists in Latin America and beyond and has undergone several draft versions incorporating many of the comments received. The present version reflects the largest possible consensus among the actors who responded to the draft through February 2012. We invite all interested parties to contribute their comments to improve this definition.

12. For ample data on glaciers, see *The National Snow and Ice Data Center*: http://nsidc.org/cryosphere/glaciers.

13. Mt. Ararat: 39°42.528′ N 44°18.184′ E.

14. Pontic Mountains: 40°49.809′ N 41°8.907′ E.

15. Mercandagi Range: 39°32.404′ N 39°32.792′ E.

16. Mt. Fuji, Japan: 35°21′55.09″ N 138°43′40.87″ E.

17. Mt. Tateyama, Japan: 36°34′31.27″ N 137°36′56.93″ E.

18. See 35°57.788′ N 52°6.979′ E.

19. See 34°18′17.93″ N 36°06′59.36″ E.

20. See:

> Kenya: 0°9.247′ S 37°18.850′ E
> Tanzania: 3°4.401′ S 37°21.188′ E
> Uganda: 0°22.830′ N 29°52.430′ E.

21. Firn Glacier, Papua Indonesia; see 4°03′52.40″ S 137°10′51.60″ E.

22. Carstensz Glacier, Papua Indonesia; see 4°04′56.78″ S 137°10′42.67″ E.

23. Paleoclimatologist Lonnie Thompson, who carried out glacier studies in the Puncak Jaya in 2010, estimated that at the current glacier melt rate (7 meters of thickness per year), the Puncak Jaya glaciers would disappear by 2015, about the time of this book's publication.

24. See http://en.wikipedia.org/wiki/Heard_Island_and_McDonald_Islands; see also 53°05′30.37″ S 73°31′03.55″ E.

25. Mexico glaciers; see 19°10.717′ N 98°38.593′ W.

26. Sierra Nevada, California, glaciers: 37°5.957′ N 118°30.730′ W.

27. Washington glaciers: 48°6.865′ N 121°6.850′ W.

28. Oregon glaciers: 44°6.095′ N 121°46.314′ W.

29. Nevada glaciers: 38°59.308′ N 114°18.621′ W.

30. See http://wp.cedha.net/?attachment_id=14236, pp. 5, 14, 32.

3

The Birth of Cryoactivism

Juan Pablo Milana and the environmentalists of San Juan Province left a memorable impression in the mind of Argentina's Environment Secretary, Romina Picolotti. But she realized that fighting a glacier protection battle against the very well entrenched mining industry could be a defining confrontation for her tenure as Secretary. If she lost the battle, it would mean her inevitable resignation as head of the agency.

In her short time inside politics, she had already learned to pick her battles carefully because the political stakes were always at the highest level. The loss of *any* battle, however small, could be the end of her political favor with President Nestor Kirchner. And, in her case, because she was a public figure brought to the administration on technical expertise and not because of any political track, that would probably mean the end of her political career. Environment was not a priority issue for Argentina, although, as in many parts of the world, it was slowly gaining social recognition and consequently political force (these generally come in that order). As such, any politically unmanageable problem from the low-profile Environment Secretariat could mean unnecessary and unwanted political conflict for the executive branch. It would not be tolerated. Furthermore, both the president and his eventual successor Cristina Fernandez, for whom Picolotti would continue as Environment Secretary, were from provinces heavily entrenched in industry—the extractive oil and gas industry—and their outlook on development was mostly aligned with and tied to the oil sector. They believed that large tracts of land without industry and development (of which Argentina has many) is land gone to waste. One preposterous plan to emerge in their home province of Santa Cruz, site of Argentina's Glacier National Park, was to build a massive dam (Condor Cliff Dam, renamed the Nestor Kirchner Dam) by flooding a large glacier lake (Lago Argentino) well above its natural water line to harness hydrological power for downstream communities. This would flood and disturb numerous glaciosystems.

But Nestor Kirchner had confided to Picolotti when he first brought her to his administration as Environment Secretary that his generation did not understand environmental issues. He asked her to help him bring environment policy into his administration, and he earnestly wanted to address mining. One of the first things he asked her to do was visit the Cerro de la Vanguardia Mine in Patagonia, where gold was being extracted, to see what the mines were like. He wanted to be more environmentally responsible, but he didn't know how. It would be her job to lead the way for Kirchner's administration.

Picolotti came from an environmental advocacy platform, and mining had been on her radar screen for some time. From the nongovernmental organization (NGO) she founded, the Center for Human Rights and Environment (CEDHA), she had begun a few initiatives to address mining, including holding several group meetings with stakeholder communities, particularly in the northern Catamarca, Tucuman, and La Rioja Provinces, as well as in Patagonia. CEDHA had even opened up an office in Rio Negro Province, in the heart of Patagonia, and had established a close relationship with local indigenous communities engaged with several issues related to the extractive industry, particularly with oil and gas. Just before assuming her role as Environment Secretary, CEDHA had published a book in the form of a National Environmental Agenda Proposal; in it, one of the principal chapters focused on human rights impacts from the mining sector.[1] In that chapter, written in 2004–2005, CEDHA reviewed several mining projects, including Barrick Gold's Pascua Lama project, and glaciers were briefly mentioned as one of the natural resource risks of the project. This was the first time that CEDHA mentioned glaciers as a resource vulnerable because of industrial activity in Argentina. But CEDHA's staff found it difficult to build advocacy around mining issues. Mining as a sector is vast and distributed geographically throughout the country, posing not only numerous logistical advocacy challenges that were difficult to address from Cordoba and Rio Negro, where CEDHA had offices, but also profound social issues that needed a very diverse set of local, skilled actors to address the multiple social, economic, and environmental dimensions of the extractive industry.

Now, as Environment Secretary and heading national environmental policy at the federal level, Picolotti's leverage point on mining could not be better. This was the platform she had been waiting for to delve into mining issues, and she did so. Her main hurdle now was sorting out jurisdictional authority. In Argentina, as in many countries, mining and the custodianship of natural resources is in the hands of local government (in this case, provincial governments). The federal government could only engage if there was an international transboundary issue at stake, a conflict over natural resources between provinces (such as a river that flows between two provinces), or a national park at risk. The oversight of mining operations, including environmental impact studies, generally rests in the hands of provincial agencies and authorities. The national government merely attends to broad investment regimes and their administration, national fiscal flows related to taxation, the export of minerals, and the import of supplies for the industry.

Glacier vulnerability at that point was not attributed to the mining sector; in fact, despite global attention focusing on melting glaciers, Argentina had not registered glacier vulnerability at all at the national level. The only exception to this came out of local opposition to Barrick Gold's Pascua Lama project, and this was by civil society actors, not by government authorities. (It is important to note that no country was engaging glacier vulnerability at the time as a policy matter. As mentioned earlier, CEDHA had only tangentially picked up the issue in its 2005 publication, but only as "one more" issue in a series of concerns over mining impacts. The social perception of glacier impacts from mining had really only taken on force in Chile, and only in select environmental advocacy circles in San Juan Province that did not yet have a national voice. None of the most prominent environmental groups in Argentina said anything at all about glaciers.)

What was now before her desk provided some opportunities to introduce glacier protection policy. Few people, including congressional representatives, realized that there were any glaciers at all in provinces such as San Juan and even less that mining operations were placing them at risk. Picolotti presumed that getting a glacier law passed in Congress would be rather straightforward. If she could prop up a good foundation and argue for the need to protect glaciers because they were a national strategic priority, she could conceivably use the issue as a Trojan Horse to exercise her environmental oversight of mining and assume federal jurisdiction over at least some portion of mining operations. Once inside the mining sector, she could maneuver more effectively into other areas. Glacier protection provided an inroad to broader federal environmental oversight of mining.

Picolotti immediately convened her legal team and other agency specialists to discuss ways that they could protect Argentina's vulnerable glaciers from unnecessary anthropogenic harm, including, but not only, from mining operations. In the back of Picolotti's mind was her work on global climate change policy. She knew that global climate change was causing glacier melt and recession. A human rights violations case brought in 2005 by the Inuit indigenous peoples of Canada against the United States for transboundary contamination of glaciers and its subsequent impacts on indigenous peoples posed the question to the Inter-American Commission on Human Rights of the responsibility of states for transboundary atmospheric contamination causing severe climate change impacts to the way of life of local indigenous tribes.[2]

The question still in everyone's mind was how to address a broad global climate problem with multiple sources of contamination (originating in a variety of countries) and the responsibility for that contamination in terms of the very local implications this contamination has for local communities.

Who is to blame for the Inuit's environmental harm from climate change? The legal difficulty of grounding local and very specific responsibility over an act that involved hundreds, thousands, and even millions of source points of contamination to a specific state or groups of actors seemed to be an insurmountable legal challenge. Some of these questions would later arise when Barrick Gold began to defend itself against accusations that the mining company was causing glacier impacts at the Pascua Lama and Veladero projects. Were these impacts due to climate change, or were they due to local anthropogenic activity? And does the difference and distinction matter in terms of environmental responsibility?

The big difference with the glacier impact case in Argentina (from the less tangible causal relationship of the Inuit case) was that there were very specific local activities occurring that could be directly attributed to mining and that had direct impact on glaciers. Barrick Gold's specific mining activity was having a direct and identifiable impact on specific glaciers. Yes, climate change was also impacting glaciers of the Central Andes, including the glaciers impacted by Barrick Gold at Pascua Lama and Veladero, but the mining activity was also impacting them in very noticeable and attributable ways, namely due to emissions and dust from local transit as well as impacts due to physical intervention in the ice. Global climate change and related glacier vulnerability was a given, and no one would argue against that. The local environmental policy dimension in the case, however, would focus on the incremental impacts of already vulnerable resources. Glaciers were already receding due to climate change, making them vulnerable resources that warranted special *and extra* protection. Looked at in another light, if glaciers are already vulnerable to climate change, and considering that glaciers are such an important hydrological resource, then they should have the highest possible protection. And, in that respect, Barrick Gold's specific and direct impacts were of even greater concern.

The problem was that glaciers were not on the radar screen as vulnerable resources. Their obscure nature, their faraway location, and our societal ignorance about the role glaciers play in our ecosystems would make the argument difficult to sustain in the courts and for public policy. Due to this social alienation from glaciers, impacts to them went unnoticed and were not prioritized in terms of policy or protection. Picolotti needed something more tangible to work with.

The Glacier Law

The first presumption that Picolotti's legal team made was that existing law could suffice to protect glaciers because glaciers, in the end, are ice and ice is water—at least in one of its forms—and Argentina has many laws that protect water. That presumption started to fall apart as the Environment Secretariat's legal team began studying glacier dynamics and began to realize that the laws developed to protect water were insufficient to protect glacier ice. Ice, in fact, is not water—not from a molecular standpoint and certainly not from the perspective of the existing water laws. Even if we approached glacier protection through the protection of snow, which is sometimes considered in water laws, *snow is not ice* because ice is formed only after snow has transformed and compacted, thus changing its molecular properties. This is a process that science refers to as *diagenesis*[3] (see Chapter 2).

We protect what we know and understand. No one really understood glacier relevance at the time of the drafting of Argentina's first water laws. Water was and is a key economic natural resource dispensed by public authority, and the "use" of water is, in the end, the most significant aspect of the water laws in place. In other words, water laws generally lean toward protecting the use that individuals and industries will make of water.

The *source* of water may also appear in legislation as a key resource, but, again, this generally focuses on the protection of natural and artificial lakes where water is stored for a regulated and rationed use, and from which the water is then channeled through extensive river systems. The storage and protection of *water quality* in those sources of water are also often key elements of water laws. So why no one ever considered where the water "originally" came from is somewhat of a mystery. Perhaps no one really ever thought the issue through very well.

The existing water laws didn't study the source of the sources, which in many cases are glaciers and periglacial areas. In a province like Mendoza, where many of Argentina's most important glaciers that provide water to communities are located, the water law[4] that has served as a model for other provinces—one dating back to 1884—doesn't mention glaciers. In fact, the water law doesn't even mention the words *snow* or *ice* or make reference to the heads of water basins in the high mountain environment. San Juan Province's water laws, where much of the mining impacts to glaciers are located, also fail to mention ice, snow, or glaciers—which was also inspired by the Mendoza law.[5]

The Environment Secretariat's team was headed by Mariana Valls, daughter of one of Argentina's first environmental lawyers, Mario Valls. Mariana Valls was Picolotti's top legal advisor. Picolotti had instructed Valls to focus on strengthening Argentina's environmental regulatory framework, including such issues as forestry, environmental insurance, and others. Glaciers were placed on the list as a topic to attempt to get into the legislative machinery. There were already other priority items in motion that the team would address first, such as a forestry law, as well as environmental insurance. But glaciers would get their turn soon enough.

Several months would go by until that opportunity arose. To Valls's surprise, in October 2007, Congresswoman Marta Maffei submitted a bill to the Lower House to protect glaciers. This seemed too opportune. Where did this initiative come from? How did Maffei come up with this draft law, tailor-written for the circumstantial situation facing Argentina? The answer lay just across the Andes Mountains, in Chile, where the regulatory debate over glacier protection was already well under way (it still is, by the way).

Local advocates in San Juan, Argentina, like Villalonga and the glacier expert Milana, had recently learned about the brewing conflicts in Chile's Huasco Valley, home to the Diaguita-Huascoaltino indigenous peoples, as well as to local agricultural associations. These communities, downstream from new and very large mining operations, were concerned about the arrival of large-scale mining in the area, including projects like Andina (Codelco), El Indio (Barrick Gold), El Morro (Goldcorp), and Pascua Lama (Barrick Gold). They began to see their water resources turn turbid as mining activity increased, and they realized that their hydrological lifeline was at stake. They sensed that glaciers up in the high mountain environments were in danger, but they didn't know the true implications of the operations to these obscure water towers.

While two other large mining projects in Chile had already destroyed significant glacier resources (Andina by Codelco and Pelambres by Luksik), focus turned instead to a project that was still not operational and only in the design phase: Pascua Lama, the binational border project where Barrick Gold had discovered precious metals squarely on the border between Argentina and Chile. Residents in the Huasco Valley, near the communities of Del Carmen, Vallenar, and Copiapó, sensed that their rivers would be adversely affected by the arrival of Barrick Gold to the region. Most did not associate this impact to glaciers, but there was a sense that water was the central problem with mining in these highlands.

But, by then, the design and administrative architecture of the world's only binational mining project was already well under way. With due diligence obligations on both sides of the border, Barrick Gold published its environmental impact studies in Chile for Pascua Lama in the early 2000s. These studies barely mentioned the presence of glaciers and then only to suggest that ice would have to be cleared from the surface to reach targeted minerals, first because the ice sat on gold reserves, but second because the instability of glacier ice was a risk to the working environment. That is, *glaciers were a nuisance to mining* and had to be removed to get at gold and silver deposits.

Barrick Gold had already explored the region extensively and had drilled widely in the glaciated area, carving out roads through snow and ice in order to move heavy drilling machinery across the difficult terrain above 4,000 meters (13,000 ft). As concerns over eventual and potential environmental impacts began to brew, stakeholders quickly focused on Barrick Gold's proposed destruction of three glaciers, the Esperanza, the Toro 1, and the Toro 2

Glaciers. These were three relatively small glaciers nestled atop the highest peaks of the Andes in the region and were practically on the border with Argentina. The gold was underneath the ice, and the glaciers had to be cleared to get at the minerals.

The fact that Barrick Gold was announcing that the gold was underneath the ice implied several things: first, that Barrick Gold *had already perforated glaciers* to take mineral samples. It also implied that Barrick had freely navigated on and around glaciers to take those samples. Images still available today on Google Earth from the mid-2000s offer testimony to this exploration and snow and ice clearance (see Figure 3.1). The multiple zigzag lines in the image are exploratory mining roads in the area. Three glaciers are outlined with dark lines. These are the Esperanza and the Toro 1 and Toro 2 Glaciers. The border runs just behind the outlined glaciers.[6] In this image, we can also see roads running through or adjacent to the Esperanza Glacier and to the two Toro Glaciers.[7]

Local communities in the Huasco Valley in Chile, Chilean environmental groups, and several politicians became furious with the idea that a mining company would destroy glaciers to get at mineral deposits. Destroying glaciers with dynamite to get at gold sounded as preposterous then as it sounds now. It was simply wrong by nearly any standard and would never survive public opinion, which meant that politicians would also fall in line to protect

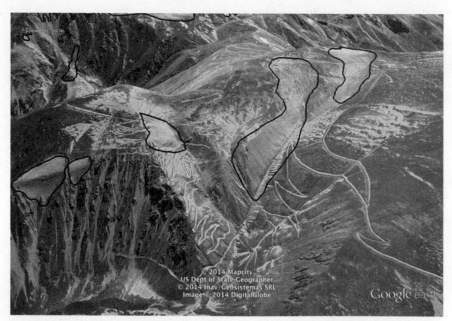

FIGURE 3.1 *Google Earth image from 2005 shows exploratory mining roads at and around Toro 1, Toro 2, and Esperanza Glaciers at Pascua Lama site. Glaciers are outlined. GIS: 29°19′22.99″ S 70°01′59.72″ W.*

glaciers, however lucrative the investment might otherwise be. This sparked interest by some environmental groups to begin working to protect glacier resources, and they began looking outside of Chile for guidance on possible laws and regulations to do so.

Surprisingly *no country* had actually focused laws or regulations on glacier protection, except in some cases indirectly through water laws or broad environmental laws recognizing state ownership of water and ice. A few indirect exceptions exist in legal jurisprudence that we might mention.

Some advancement occurred during the twentieth century to at least bring glaciers into the public domain. In 1921, during an era when water shortages had become an issue for many European nations and when glaciers were dangerously falling to pieces in the Swiss, French, and Italian Alps, a water tribunal court ruling in Italy established that "Alpine glaciers must be considered of public domain, and must be inscribed in the public water registry, along with the waterways they feed."[8] In 1986, the canton of Ticino, Italy, in Article I of a new public domain decree, established that glaciers and snow are property of the canton.[9] The 1991 European Alpine Convention makes only one mention of the word *glacier* in the "considerings" section of its Protocol for the Conservation of Nature and the Countryside, in which it states, "aware of the imminent importance held particularly by *glaciers*, Alpine meadows, mountain forests, and the waterways of the Alpine territory"[10] (emphasis added). France passed the Mountain Law (Loi de Montagne in 1985) and later established an urban code (L. 145 7 I,2 code de l'urbanisme) to "designate the most notable areas, sceneries and environments of the natural patrimony and of the mountain culture, notably gorges, grottos, *glaciers*, lakes, peats, marshes, places to practice mountain trekking and climbing, canoeing and kayaking, and define the modalities of their preservation"[11] (emphasis added).

In Latin America, Peru[12] and Colombia[13] had briefly mentioned glaciers in water laws, but merely to establish, as in some jurisdictions in Europe, that water in glaciers belonged to the State. Ecuador stated that water for irrigation could be taken from glaciers.[14] The United States and Canada mention glaciers in regulations only briefly in national parks codes, simply referring to the establishment of parks that may contain glaciers or rules that visitors must follow when visiting national parks, but presenting nothing specifically protecting glaciers from a legal standpoint.[15]

Austria is the only country where some more specific wording on glacier protection appeared in environmental law, but only at a subnational jurisdictional level. The provincial government of Carinthia mentions glaciers in their Nature Conservation Act of 2002, focusing on glacier landscape and watershed protection.[16] No jurisdiction however, took on glacier protection from an integral perspective or at a national level. That is—until a mining company one day decided it could dynamite a glacier.

In August 2005, Leopoldo Sánchez Grunert, a Chilean congressional representative concerned with the need to address global climate change, citing numerous commitments by Chile to global international treaties, and very concerned with the witnessed deterioration of Chile's glaciers up and down the Andes Mountain Range, submitted a motion to Congress to prohibit investments that would impact glaciers, for which he proposed to add a new subparagraph to Article 11 of the National Environmental Law (las Bases Generales del Medio Ambiente). The article he proposed read:

> All considering, not even by submitting to the system of environmental impact study, shall activities be permitted or carried out in glacier areas, unless these are exclusively dedicated to scientific research or to tourism benefits or to the benefit of capturing the natural melt of ice and to the drainage of its water. In these cases, projects shall be submitted to the system of environmental impact study.[17] [unofficial translation]

Although the motion was never approved by Congress and was finally archived in 2009, a process was set in motion by the proposal of this single subparagraph, to introduce glaciers into environmental public policy. Another congressional representative, Senator Antonio Horvath, and several environmentalist groups, such as the civil society organization Chile Sustentable (Sustainable Chile), followed up Sánchez Grunert's initiative and took further action on the matter. In 2006, these activists formed a working group to focus on glacier protection. Chile Sustentable published information about glacier risks from mining operations, as well as the first descriptive and properly researched arguments indicating that glaciers should be protected by law and by public policy as vulnerable and strategic hydrological natural resources.

Not long after Grunert, Horvath, and Chilean environmentalists began to introduce glacier legislation proposals to the Chilean Congress, the glacier buzz slowly made its way across the Andes to Argentina and beyond. Discussions taking place in global forums highlighting climate change and the risks posed by large-scale industrial investments like Pascua Lama inspired Alejandro Iza and Marta Brunilda Rovere of the International Union for the Conservation of Nature (IUCN) to conduct research and publish a compendium of applicable laws and regulations that could be used to protect glaciers. They were concerned not only with climate vulnerability but with the vulnerability of local anthropogenic activity, and they spoke of the importance not only of glaciers, but also of "glacier ecosystems"—areas near glaciers that were important to their sustainability, and drew attention to the significant services that glaciers provide to societies. They noted the lack of regulatory instruments to protect glaciers, and this, in the end, was the inspiration that drove them to propose some underlying regulatory and legal dimensions needed for glacier protection. They published their work,

Aspectos Jurídicos de la Conservación de los Glaciares (Juridical Aspects of Glacier Conservation),[18] that same year. It was a review of existing legislation in the hemisphere available to protect glaciers (or, perhaps better said, gaps that existed in laws that left glaciers unprotected). Iza and Brunilda focused on seven countries: Argentina, Bolivia, Colombia, Chile, Ecuador, Peru, and Venezuela. Their basic conclusion was that none of the study countries provided an adequate legal and regulatory framework or public policy to protect glaciers.

The research under way in Chile took this discussion to the next level, bringing the glacier debate out of the realm of science and into a domain focused on the law and public policy. *Chile's Glaciers: Fresh Water Strategic Reserves for Society, for Ecosystems and for the Economy*, by Chile Sustentable,[19] was eventually published in book form in November 2006, thus becoming one of the world's first publications (outside of scientific glacier academia) focusing specifically on public policy and glaciers, the importance of glaciers and their anthropogenic vulnerability, and their parallel importance as a strategic water reserve as well as their value for economic and industrial development. It was indeed the first publication to propose public policy and a specific legal framework (not merely an addition to an existing environmental norm) for the protection of glaciers.

Chapters 4 and 5 of this visionary publication, as well as several of its annexes, examined and proposed model text for legal and regulatory frameworks for protecting glaciers. Chapter 5 was a self-standing model bill that could be taken *as written* to Congress in Chile or any other country to promote glacier protection. The authors defined glaciers as strategic resources that needed public protection. The proposed model legislation laid out various dimensions of glacier protection that would inform central components of the ensuing debate. The authors proposed to regulate the protection of glaciers:

> as objects of strategic security to address the maintenance of ecosystems, the needs of human populations and productive activity, especially for agricultural production, the generation of hydroelectric power, mining processes, tourism, for the preservation of environmental values, scenery, and the environmental services they produce for the conservation of biodiversity, with the objective of maintaining these as hydrological reserves and water basin recharge, flows, and aquifers during summer and periods of drought.[20] [unofficial translation]

The initial paragraph of this model legislation provided a broad-based approach for the discussion that would follow, one attempting to frame glacier resources in three distinct dimensions: *environmental*, *social*, and *industrial*. Perhaps noteworthy in this approach is the "environmental services" tone of the text, in an attempt to anticipate potential resistance from key economic sectors that would surely oppose a glacier protection law that limited industrial activity such as mining in areas with glaciers.

The authors, Roxana Bórquez (a researcher), Sara Larrain (an NGO activist and independent presidential candidate in 1999), Rodrigo Polanco (a lawyer), and Juan Carlos Urquidi (president of the National Environmental Commission), laid out as no one ever had the reasons that relatively small, unknown, and obscure high mountain glaciers were a strategic resource and the hydrological lifeline for local communities and agriculture.

The authors told Chileans and the world that:

> Glaciers ... are the principal freshwater reserves of the planet and because of this, they are the principal insurance of our ecosystems and for the subsistence of human populations. Glaciers are strategic reserves not only because they provide water to our basins in summer, but also because they are the only source of recharge for our rivers, lakes and underground water in dry areas and during dry spells. (Bórquez et al., 2006, p. 5)

Like Sánchez Grunert, the authors made the critical leap of logical reasoning that no one else was making at the time: we were all focused on climate change as probably the most pressing issue of our time. We accepted that we needed to introduce national strategies to address climate change. A critical part of that strategy needed to be to protect the natural resources at risk due to climate change. Glaciers were perhaps first in line. We identify glacier melt as one of the most telling evidences of climate change; however, to this day, glacier retreat is seen simply as a symptomatic variable in our present environmental global context. Few are speaking of or taking actions to protect glaciers. This needed to change, suggested the authors of *Chile's Glaciers*.

Concerned with the uncontrolled advance of mining operations in the area (in terms of risks to glacier resources), Chile Sustentable began pushing for the adoption of a full-fledged glacier protection law in the Chilean Congress, one written along the lines of the recommendations they had made in their recent publication. They channeled their idea in the form of a bill tabled by Santiago City resident Senator Antonio Horvath, a civil engineer with a strong environmental background. Horvath proposed the first text for a glacier law to Congress in May 2006.[21] The introductory text to the law read emblematically:

> Chile, in Aymará [the original local indigenous language] means icy confines, that is that from the times of indigenous peoples, our country has been associated to the Andes, to snow, glaciers, and low temperatures. Glaciers are an element that identifies Chile in the Southern Hemisphere of the Planet. (Chile Sustentable, 2006, Annex 2, p. 111)

Chile is for the most part a very narrow country, flanked by the Pacific Ocean on one side and high mountain environments on the other, with merely a few hundred kilometers of land in between. The country has also been for a better part of its modern history since colonization by the Spaniards in the fifteenth century, a mining country. The Andes are part of the Chilean identity. People orient themselves in relation to the mountains. The mountains are simply a part of local identity, determining many facets of local life and economy, including water provision, energy, minerals, agriculture, livestock husbandry, tourism, and recreation. A Chilean friend once described the relationship of people in Santiago (the Chilean capital) with the Andes, suggesting that when people leave their home or wherever they may be outdoors, they look for the mountains to set their bearings; the mountains reassure them, giving them a sense of place, protection, and identity.

Prohibiting activities such as mining, agriculture, or tourism in glacier areas would constrain, proportionally and in percentage terms, large swaths of lands in many cases already under usage. A blanket prohibition of activities in glacier areas, such as would apply in the Argentine case, would place a fair portion of Chilean territory—at some points as much as one-quarter or more, off limits to human activity. Politically, this would be a difficult proposition, presumed the authors, and so they suggested a text that would allow room for exceptions to glacier protection, but under the necessary protective considerations.

Horvath's bill, called Valoración y Protección de los Glaciares[22] (Glacier Protection and Valuation), devoted a considerable section of text to definitional issues, providing a broad and encompassing coverage of the law to include glaciers of varying sizes, including debris-covered glaciers and rock glaciers. Glacier definitions in Chile would be as tricky and complex as they would be in neighboring Argentina.

Horvath also laid out various "intervention" typologies, including tourism and scientific activity, road intervention, mass removal, dust and liquid contamination, weight loading, and other industrial activity, and called for appropriate studies to be carried out in the event that these interventions occurred. He also called for a glacier inventory and monitoring to take place, as well as fines to be applied to violators of the law in the event that impacts to glaciers occurred. Some of these concepts would be carried over to the Argentine glacier protection law.

Unknown to the local population, large mining ventures to extract copper were already removing massive quantities of glacier ice. The Chilean national mining company, Codelco, was removing ice-rich earth in the Aconcagua River Basin at the Andina project (see Figure 1.6).[23] Pelambres, a copper mining project further south, undertaken by the Chilean magnate Luksik, and on the border between Argentina and Chile, had already removed significant portions of active talus rock glaciers—a type of glacier mixed with rock at the base of steep mountainsides, typical of the Central Andes region.[24]

The case of Pelambres was being documented by two glacier specialists, Alexander Brenning of Germany and Guillermo Azocar of Chile, both working at the time at the University of Waterloo in Canada. They were some of the first glacier specialists specifically looking at mining impacts to glacier resources. There had been some previous academic work in the 1980s on glacier impacts from mining (Robinson and Dea, 1981), but these focused more on the engineering challenges of mining on frozen grounds. Brenning and Azocar were among the first academics to home in on glacier impacts from mining from an environmental perspective. They did not know it at the time, but their research and publications would greatly inform the glacier law debates in Argentina and contribute directly to the birth of the eventual Argentine glacier protection law.

Brenning and Azocar published their findings in 2008 and 2010, showing extensive damage by the Luksik Group at two glaciers at the Pelambres project (Azócar and Brenning, 2008). They mentioned in the Pelambres paper that satellite imagery they had reviewed also showed mining impacts to periglacial areas just across the border in Argentina at the El Pachón project,[25] which was at the time under exploration by Xstrata Copper of Australia. This revelation would provide CEDHA with an important clue about where to target CEDHA's future cryoactivism work in 2010–2011.[26]

There were actually dozens of sites in both Argentina and Chile where one could find mining impacts from exploratory work to glacier and periglacial environments, but no one was monitoring any of this activity nor was there any sense that glaciers were at risk due to mining activity. All attention at the time was focused on Barrick Gold. Why Barrick? That's a good question. Perhaps because it is a large international mining company that had already drawn attention because of other controversial projects around the world. Perhaps it was because of the timing. New community-based social movements around mining conflicts were surfacing throughout the region at about this time, and this was the one that focused media attention in Chile. Also, Barrick's shortsighted handling of community opposition and government concern to potential glacier impacts caused a big stir in environmental activist and policy circles, which also placed great scrutiny on Pascua Lama.

As environmental and community groups began to pore through Barrick Gold's environmental impact studies—and thanks to Chile's decision to place all documentation related to environmental impact assessments (EIAs) online and open to anyone interested in finding such information—a controversial document published in February 2001 surfaced. It was the annex to Barrick's environmental impact study on Pascua Lama entitled "Glacier Management Plan in the Rio Toro Basin." The first paragraph of the annex read:

> The following plan describes the method and management disposition of the glacier sectors that must be removed during the life of Pascua Lama,

as the open pit area is extended towards the position of the glaciers in the Rio El Toro river basin. It is estimated that 10 hectares [25 acres] of glaciers must be removed and adequately managed to avoid the instability of slopes and environmental impacts. The thickness of the glacier sectors that must be removed is estimated at 3 to 5 meters [10–16 ft]. (Barrick Gold, 2001, unofficial translation by CEDHA)[27]

Later in the annex, Barrick Gold indicates that it will "remove chunks of glacier ice with bulldozers and front loaders until the area is clear of ice," and that, if necessary, "explosives will be used" to remove the ice.

Several things jump out from this crude description of how Barrick Gold planned to remove glacier ice and clear the surface in order to advance its mining activities. First, and perhaps most surprising (aside from the fact that the company is proposing to dynamite glaciers) is that the company suggests ironically that *it needs to destroy glaciers* in order to *protect the environment*.

However odd this statement may seem in 2015 (or even in the early 2000s, when it was originally published), it is very revealing about the mindset and motives of the company, not only due to the insensitivity shown by Barrick Gold to nascent public opinion and concern over glacier vulnerability, but because it also underlines the way in which industry understood, until that point, the risks of working in the cryosphere (i.e., the frozen environment). That viewpoint would change significantly over the following years as the discussion on glacier protection legislation and cryoactivism evolved.

What Barrick Gold is really saying in a rather blunt but honestly transparent manner is that industrial works in frozen grounds are dangerous. Humidity in frozen grounds (in permafrost) typically can freeze and melt in daily or seasonal cycles. Water volume expands when frozen, and ice volume contracts when it melts. The removal of massive quantities of frozen earth, as proposed by Barrick Gold to get at gold deposits under glaciers and in periglacial areas, could result in drastic changes to the equilibrium of the frozen ground, causing rapid thawing and instabilities of the earth and the ice accumulated in the form of glaciers around those grounds. Ice and earth in an area pertaining to a periglacial or glacial environment could conceivably become dangerously unstable if altered and collapse in a landslide-like fashion, sliding down mountainsides onto workers and/or company infrastructure. In fact, something similar was about to occur at another of Barrick's projects in Argentina (Veladero) precisely due to the company's placement of a colossal sterile rock pile on frozen grounds (more about that later).

So, the reasoning for destroying glaciers that Barrick Gold puts forward in its "Glacier Management Plan," strictly from a worker and infrastructure safety standpoint, is to avoid risks to the "working environment" where the company proposed to carry out operations. Barrick Gold hence needed to destroy glaciers in order to protect the working environment, as ironic and

counterintuitive as that may seem to environmentalists and local communities. This reasoning, however logical from a worker safety perspective, might
have taken place in the 1970s and 1980s in mining operations in the Rocky
Mountains, but it simply would not hold up in a modern and rational discussion about sustainable development and corporate accountability in the
2000s. Times had changed. Thankfully, for local glaciers!

Mining companies routinely hire geologists to carry out full geological surveys that, when in permafrost zones, include glacier and periglacial inventories; until 2004, no one had considered that when these glaciated resources
were disturbed or destroyed, they implied the destruction of valuable natural
resources. Incredibly, for example, in Barrick Gold's EIAs (see Chapter 4),[28]
when the company identifies what environmental impacts would be caused
by the Pascua Lama project, glaciers were not even mentioned as an affected
resource. Barrick does mention the potential impact to waterways (the runoff
of the glaciers) but not to the source of the water, which it was set to destroy.
However incongruent it may seem, the glaciers simply were not considered a
vulnerable natural resource or one that should even be protected for its hydrological value.

In any event, Barrick Gold moved forward with its "Glacier Management
Plan," presenting it to local authorities in Chile and clearly not anticipating
the reaction that would come of its proposition to dynamite glaciers to extract
precious metals. And that reaction would not be confined only to Pascua
Lama. Barrick's "Glacier Management Plan" was, in retrospect, the inspiration for modern-day glacier protection policy and legislation and much of
current-day cryoactivism in places as far away as Kyrgyzstan.

The straw that broke the camel's back was what Barrick Gold did once
communities and the Chilean government reacted adversely to its "Glacier
Management Plan." Consulting glacier and periglacial experts who were carrying out experiments in glacier areas along the Central Andes, Barrick Gold
came up with the incredible idea of *transplanting* the glaciers. If the ice were
so important as to warrant its protection, the company thought it could sal-
vage the ice, dig it up with bulldozers and shovels, and haul it off in dump
trucks to another site where it could be preserved.

The idea seemed preposterous to communities and to environmentalists,
or at the very least too incredible to believe. Can you really *move* a glacier? Is
that physically possible? And if so, can the dynamics of a glacier and its glaciosystem be replicated at a different location, in this case, to another glacier?
Glaciers move, they creep downhill, and they seek ecological balance with
their surroundings. They are in unique "glaciosystems" that can hardly be
recreated with natural precision.

Barrick's intentions were published in a colorful cartoon brochure distributed to local communities depicting bulldozers and dump trucks plowing into glaciers and hauling off chunks of ice (see Figure 1.4). The figure's caption reads: "(1) Hydraulic excavators will be used to load the ice directly onto trucks; (2) These trucks will transport [the ice] to a deposit on the Guanaco Glacier, approximately four kilometers away; (3) Once at the glacier, the trucks will offload the ice directly on the deposit, that is prepared for its conservation; (4) Once there, snow curtains will ensure that the ice forms a continuous mass."

Needless to say, the brochure added insult to injury for environmentalists. The government of Chile reacted strongly against Barrick's "Glacier Management Plan," prohibiting the company from touching the Esperanza, Toro 1, and Toro 2 Glaciers. Barrick Gold was forced to redesign the Pascua Lama project and forego, for the moment, the precious metals that it had discovered underneath these three glaciers. The pit area was reduced to exclude the Esperanza, Toro 1, and Toro 2 Glaciers, and the project began to move forward with its preparatory stages. The conflicts surrounding Barrick's glacier impacts, however, would not go away.

Glaciers Go Presidential

The Chilean glacier debate would go presidential in 2010, when then-president Michele Bachelet proposed a constitutional reform that included articles that would better protect water resources. In this reform, there would be explicit mention to glacier protection establishing that glaciers were indeed a public good that should be protected. In her address to Congress, Bachelet expressly referred to Chile's glaciers as the country's "white gold." Under this proposal, Chile would have become the first country to bring glacier protection not only into law but also into the constitutional hierarchy.[29] But the reforms were stalled and never moved forward.

As expected, the initial debates in Chile from 2006 over whether to introduce glacier protection into Chilean law centered on the motion presented in Congress by Senator Horvath, Girardi Lavin, Navarro Brain, Bianchi Chelech, Rodolfo Stange, and Kruschel Silva also stalled. The motion tabled by Horvath et al. to actually adopt a glacier protection law quickly ran into political opposition from the mining sector, primarily from companies such as Barrick Gold, Codelco, and the Luksik group, which each had projects with sensitive glacier exposure and much potential risk to glaciers or already existing impacts.

The mining sector, organized under its representative commercial bodies (such as the Chamber of Mining Companies), pressed Bachelet to drop the move toward glacier protection legislation. But perhaps the greatest pressure came from the government itself. The Chilean-owned copper mining giant Codelco had perpetrated some of the most prominent glacier impacts as far back as the 1980s in the Aconcagua River Basin with its Andina mining project to extract copper from the Central Andes in an area surrounded by glaciers and periglacial areas[30] (see Figure 1.6).

Despite approval in the Senate and in Congress's Environmental Commission, the bill ended up archived into oblivion. Instead, Bachelet succeeded in a more modest effort channeled through the National Environmental Commission (the CONAMA) to require specific glacier-focused environmental impact studies for projects in glaciated areas. This legislation would trip up Barrick Gold some years later when the Chilean environmental authority ordered a full suspension and closure of Pascua Lama's activities due to glacier impacts. The government also created within the General Water Department a specialized Unit on Glaciology and Snow (Unidad de Glaciología y Nieves),[31] mandating the agency to carry out Chile's first official glacier inventory.

Another secondary effort that did bear fruit was the country's adoption of a National Glacier Protection Policy in 2008.[32] Although not a law, Chile's National Glacier Protection Policy laid out in detail the importance and significance of glacier reserves for the country as a strategic and vulnerable resource. The policy defined glaciers broadly, drawing attention not only to the magnificent and colossal glaciers of the Chilean Patagonia, but also to smaller glaciers, glacierets, and even perennial snow patches, as well as periglacial forms (such as rock glaciers) that were all actually more important to communities and to agriculture than were the Patagonian glaciers. Glaciers of any size and form, even perennial snow patches, were considered significant hydrological reserves that should be protected due to their significant value to water basins.

The Chilean 2008 National Protection Glacier Policy reads:

> Glaciers are understood to be all perennial ice masses, formed by the accumulation of snow, whatever their dimension or form. (Chile NGPP, 2008, p. 3)

> Glaciers are renewable natural resources, and they are the support of nature. For this reason, it is important that we recognize that glaciers offer valuable environmental services (for example, climate regulation and water provision), contributing to the support of natural ecosystems and to activities such as tourism, scientific research and sports, etc. all of which implies the need to preserve and conserve glaciers, given their particular characteristics of regeneration and their fragile predicament to new climate scenarios of our planet. (Chile NGPP, 2008, p. 2)

> As such, the environmental value of glaciers, as renewable natural resources, is intimately related to their physical-natural characteristics, as we have said, for their cumulative capacity or as a fresh water reservoir in

a solid state, and for the gradual liberation of freshwater to basins, with the beneficial consequences for natural ecosystems and for human life, such as to local communities and indigenous populations which depend on them. (Chile NGPP, 2008, p. 2) [unofficial translation]

A clear schism was opening up between staff in Chile's water policy agencies and mining interests. Although a glacier protection law would not be politically viable at that moment, the beginnings of a strong, broad, and functional glacier protection regulatory framework, one that would eventually begin to create headaches for mining operations like Pascua Lama, was being set in place.

Glaciers Go Gaucho

Although all legislative attempts to get glacier protection on the books in Chile in the mid-2000s failed, a fortuitous event would bring the initiative across the Andes into Argentina and would eventually lead to the world's first federal glacier protection law, as well as to a series of subnational laws that to this day comprise the world's only glacier-specific legislation.[33]

At the end of the 1990s, an Argentine labor union leader turned Congresswoman, Marta Maffei, visited Chile and was sensitized to the plight of farmers and indigenous communities of the Huasco Valley, witnessing the problems they faced with deteriorating water reserves. Maffei returned to Argentina but kept in contact with her Chilean counterparts, and, in the mid-2000s, she traveled again to Chile. On this visit, she met Sara Larrain, an environmental leader at the helm of Chile Sustentable. Larrain had been a presidential candidate in 1999, and she served in a variety of national and international advisory panels and as a co-founder of Chile's Greenpeace office.

In her meeting with Larrain, Maffei learned about the glacier protection bill proposed to Congress by Chile Sustentable through Senator Horvath and others, and their concern with the impacts glaciers would suffer from projects such as Luksik's Pelambres, Codelco's Andina project near Santiago, Barrick's El Indio, and now also Barrick's Pascua Lama, which was the region's first binational project and which would be of interest to Maffei because one-half of it was in Argentina, her own country.

The proposed model glacier legislation attempted to find common ground with a Congress that was very mining-influenced—after all, Chile is a mining country, and much of the economy is grounded on copper, lithium, and other mineral extraction. The working group born from Chile Sustentable negotiated an environmental services view of glacier protection, one that would take into account the multistakeholder usage of the environmental resources that glaciers provide to the ecosystem; namely, water to be consumed by the ecosystem, by people, for agriculture, and by industry.

This "services" angle is how the Chilean debate around glacier law, policy, and regulations evolved, incorporating glacier protection as a function of environmental, social, and economic value.[34] The working group's recommended model legislation for the protection of glaciers states in its first paragraph:

> Article 1. *The present Law regulates the protection of glaciers as objects of strategic security to respond to the maintenance of ecosystems, to the necessities of human populations and to productive activities, especially for agricultural production, for the generation of electricity and to mining processes, and tourist activity.*[35]

Horvath's eventual glacier protection bill would not use this text but instead proposed a series of "glacier interventions" with corresponding limitations, which included the type of impact caused by mining operations. The bill proposed that, in the event of such interventions, appropriate impact studies should be carried out.[36]

But Maffei had a very different context to work with in Argentina. There was no mining lobby against glacier protection (not yet). The Pascua Lama conflict with glaciers had not yet attracted media attention. No one was talking about the need to protect glaciers. So, she returned to Argentina with a copy of Horvath's proposed legislation to protect glaciers and set immediately to work on her own glacier protection law from a much more conservationist angle.

She would eventually adopt several parts of the Chilean version according to the specific needs she saw in Argentina, but she would expand on others and be more specific on the issues that mattered most to her, namely, focusing on avoiding industrial impacts to glaciers.

Like most Argentines at the time, Maffei didn't know the first thing about glaciers, where they were (except for the obvious ones, such as in Argentina's deep south—the Patagonia region), or what their specific characteristics were in the Central Andes. It made sense that, as in Chile, Argentina's glacier resources were vulnerable and probably receding, although the media had not covered this phenomenon there. In fact, the one glacier that Argentines do know is the Perito Moreno, and, ironically, it's one of the few glaciers in region (and in the world) that is actually not shrinking but expanding! Some have wrongfully referred to the advancement of the Perito Moreno Glacier as an indicator that there is no rule of thumb and that climate change may not actually be causing glaciers to recede, but such arguments have been categorically debunked by academia. Scientists largely agree that the Perito Moreno's advance is an anomaly of nature that is far from the norm. Glaciers are clearly and unequivocally retreating worldwide.

Maffei was certain that she didn't want to take the Chilean environmental services approach to glacier protection. She wanted a much stronger law, one that would protect glaciers from the sorts of conflicts she saw were occurring in Chile. She wanted to keep mining away from glaciers, and, for that, she wanted an outright ban on mining in glacier areas.

Maffei turned to her closest advisor, Andrea Burucua, and together they began research. Their first step was to figure out whom they should consult in Argentina. Fortunately, Argentina had one of the region's only specialized glacier agencies, the Argentine Institute for Snow Research, Glaciology and Environmental Sciences (IANIGLA). They also approached the Argentine Council for Science and Technology (CONICET), which comprises the major-ity of specialized academic researchers working on glaciers for the IANIGLA.

Ricardo Villalba, Darío Trombotto, and Jorge Rabassa, lent their assis-tance to drafting the first version of an Argentine glacier protection law (which would come to be known as the *Maffei Law*),[37] focusing specifically on protecting Argentina's glacier and periglacial resources. Maffei and her team also met with geologists and glacier experts at the National University of San Juan, including the renowned Juan Pablo Milana who is one of the most knowledgeable persons on glaciers in Latin America and specifically on glaciers in the province of San Juan. Milana would become an invaluable source of information about glaciers and mining because most of Argentina's mining projects in glacier areas are precisely in San Juan Province. They also consulted with Alejandro Iza of the International Union for the Conservation of Nature (IUCN), who would later publish one of the first legal studies on glacier protection.[38]

Dario Trombotto's contributions to the draft bill were particularly impor-tant because he is a geocryologist (a person studying the combination of geology—the study of rocks, and cryology—the study of ice). Trombotto's participation in the Maffei glacier bill would lend an entirely new dimension to the text—a focus on the periglacial environment. The Chilean glacier pro-tection bill failed to consider the broader periglacial environment, which is an extensive frozen region of high mountain environments containing not only critical rock glaciers (which were contemplated in the Chilean bill), but entire swaths of frozen ground (some glacier free) that can be rich in ice and play a significant role in the provision of cyclical meltwater during warmer months and prolonged periods of drought. (See Chapter 4 for more on rock glaciers and the periglacial environment.)

Trombotto, who served at the time (and is still today) as Argentina's rep-resentative to the International Permafrost Association,[39] added the "perigla-cial environment" to the Chilean version of the law, and, with this, a much greater territorial region of very significant hydrological value would now be protected in Argentina.

The team advising Maffei approached the drafting of the law to protect glaciers in a way more similar to legislation being developed in Colombia for their wetland systems, which places categorical conservationist protection on *páramos* (tropical mountain wetlands). These sensitive wetlands, with similar hydrological value to the periglacial areas they were considering for the glacier law, provide critical water supplies to millions of people and ecosystems downstream.

The IANIGLA and CONICET experts participating in the draft text of the Maffei Law were scientists concerned naturally with the environmental dimensions and not political dimensions of the implication of the law. IANIGLA, which added the periglacial environment dimensions to the Chilean version, did not envision the industrial usage of the cryogenic environment. They saw and understood glaciers and periglacial environments as the top of the water chain, valuing them for their hydrological function as part of a greater ecosystem. They were not discussing their proposed definition with industry nor with any other actors outside of scientists working on glacier research. And Maffei had no intention of making that bridge either. Had she consulted with the mining sector, the law would have looked quite different and would have followed the environmental services angle of the Chilean glacier protection bill.

In Chile, the glacier protection bill faced heated debates with the mining sector about the proposed legislation and what it would mean for local mining projects like Pelambres, the Andina project, Pascua Lama, and perhaps others under way or envisioned in high-mountain frozen environments. Maffei and the IANIGLA drafters had no pressure to focus on anything other than glacier and periglacial environment protection, and that's precisely what they did.

Comparing its introductory articles with the Chilean bill, Maffei's law, when it went to vote, read:

> Article 1. *The present law establishes the minimum standards for the protection of glaciers and periglacial environments with the objective of preserving them as strategic hydrological resource reserves, and as providers of recharge water for hydrological basins.*[40]

In this definitional statement, there is no mention made of balanced "usage," service orientation, or industry. Glaciers and periglacial environments are simply important because they have an important hydrological value, and, as such, they should not be touched. Their view was strictly conservationist. Like the *páramo* legislation in evolution in Colombia, the glacier protection bill effectively set certain areas off limits to human activity. The geographical implications of the text were not very clear to anyone at the time, but what it achieved was basically to draw a line at a certain altitude, above which practically nothing was permitted.

Instead of allowing for certain economic actors (agriculturalists and industry) to make use of glaciers, Maffei's law delineated activities that could not take place on glaciers or in periglacial areas. Mining and petroleum exploration and extraction activities, for example, were expressly prohibited in glacier and periglacial environments. Like the Chilean bill, Maffei called for retroactive application of the law by establishing an obligation for projects already under way in glaciated regions to present specialized environmental impact studies to determine if they could continue or if they should be modified, shut down, or relocated. The implications for mining operations in places like San Juan were potentially disastrous. But the miners weren't watching Maffei and her team of experts.

The Environment Secretariat began receiving complaints from environmental groups in San Juan about what was happening at Veladero and Pascua Lama. The Fundación Ciudadanos Independientes (FuCI), which had organized the environmental bus tour to Buenos Aires, assisted by glacier expert Juan Pablo Milana, was the first to bring glacier concerns to the courts. FuCI confronted Barrick's Veladero and Pascua Lama projects in 2005, calling the attention of public officials and the Public Defender to what dust and acid drainage at Pascua Lama would do to glaciers and other environmental resources.

In 2005, FuCI presented a legal complaint against the government of San Juan for impacts to hydrological resources and to the "cryosphere"—*the world of ice*. In 2005, 2006, and 2007, the courts would receive a number of complaints having to do with glacier impacts. Glaciers had achieved a presence in the Argentine justice system. For the first time ever, FuCI and Milana had diverted public awareness from Patagonian glaciers to the other glaciers of the Central Andes. They spoke of some forty or fifty glaciers around Barrick's activities at Veladero and now at Pascua Lama that were at risk. And, among these, they began speaking of a special type of glacier, an enigmatic glacier no one had ever heard of—*rock glaciers*. These glaciers existed beneath the surface of the earth and were protected from the rising temperature brought about by climate change. And that wasn't all: Milana and his compatriots were telling us that the Andes were full of glaciers. Not in the tens, or even hundreds, but in the thousands.

Another individual also surfaced at the time. Ricardo Vargas, a tour guide who would take foreigners to visit the faraway ice-covered lands of the Valle del Cura and Sepultura now under exploration by Barrick and a dozen other companies. However, with Barrick's arrival to launch Veladero, the company had closed the roads to the locals and no longer let Vargas or other mountaineers up into the mountains. This decision by Barrick Gold not to allow public access to the project area or even to the 180 km access road, made Barrick many enemies, some of whom, like Vargas, would go out of their way to carry out activities to derail Barrick Gold's mining projects. Vargas, for example, sought legal help from a lawyer in San Juan. He joined forces with Diego Seguí, and together they filed a complaint to the National Supreme Court charging Barrick with a number of code violations, including the failure to conduct proper public consultations, failure to contract environmental insurance, and a statement of risks posed to the San Guillermo Biosphere Reserve, a UNESCO-protected site.

Actors from different parts of the province began getting involved in the growing pro-glacier and anti-mining sentiment, worried that Pascua Lama was merely the first of a dozen or more projects arriving in the area that would place glacier and other hydrological resources at risk. (That number, in fact, was closer to 150 or more projects, according to later declarations by San Juan's Mining Minister to local press.)[41] The social movement was "*sanjuanino*" with little or no involvement from national environmental groups. The Madres Jachalleras, the Frente Cívico por la Vida, and the Asociación de Viñateros Independientes (the Independent Wine Makers Association), as well as the civil society organization Inti Chutez all joined forces to protest Barrick's impacts on glaciers.

The San Juan-based Inti Chutez had obtained clandestine fresh video footage from Veladero showing massive blasting that lifted tons of debris and dust into the air. This detritus was picked up by strong winds, sending debris and dust onto the surface of nearby glaciers, soiling the pristine white ice, changing albedo, and ultimately leading to accelerated glacier melt. The Inti Chutez video was one of the first to call attention to the impacts of mining activity blasting to glacier resources.[42]

Juan Pablo Milana, the San Juan glacier expert, also spoke of an incorrectly drawn glacier in Barrick's EIAs for Pascua Lama. It happened to be a glacier right on the political border between Argentina and Chile. Barrick Gold had drawn the glacier fully within Chilean territory, with the glacier ice body coming up tight to the border. The crest of the mountain, which determines the water runoff to either side, serves as the natural and political border. The drawing by Barrick Gold implied that the glacier was on the slope facing Chile, and, as such, argued Milana, the glacier should drain only toward Chile. But Milana suggested that at least a third of the glacier was squarely in Argentina. He showed Google Earth images where one can see a significantly sized drainage line running from the glacier toward Argentina.[43] This, said Milana, was evidence that some portion of the glacier was in fact on a slope *below* the crest on the Argentine side of the border. The argument and evidence make sense; although apparently a minor issue, in fact, it is a major point because underneath this glacier was a significant portion of Pascua Lama's gold reserves.[44] The disputed glacier was the Toro 1, one of the glaciers that was the main focus of complaints in Chile.

Milana also alerted the Environment Secretary to the fact that Barrick had built one of its sterile rock waste piles on unstable frozen grounds of the periglacial environment.[45] This was dangerous because weight placed on unstable and moving periglacial ground could result in instability and possible collapse.

Diverse actors from local society in San Juan were each working in their own way to bring attention to glacier vulnerability caused by mining operations, and a process was under way in favor of creating glacier protection that would not ease until it went national. There was no turning back.

The cards were being dealt. Local environmental activists, the mining sector, and public authorities were at the table fighting over glacier reserves. This was the birth of cryoactivism, born directly from the conflicts surrounding the Pascua Lama project. Environmental groups on both sides of the Andes, without knowing it, were converging environmental advocacy strategies on glacier protection. Legal action was under way on both sides of the border to stop mining activities that placed glacier resources at risk. It was still largely unclear exactly where and how this impact was occurring or what the real extent of the risks and impacts were. But what was certain was that a rolling, growing snowball was forming that would take out whatever was in its path and that a cryoactivist movement would become firmly rooted in place.

For this reason, when the legal team at Argentina's Environment Secretariat came across the Maffei Law slowly making its way through the congressional bodies, they were ecstatic. This was precisely what they were looking for to create an environmental control entry point into the mining sector, *a federal law making glaciers a national/federal issue* and offering jurisdiction to the federal Environment Secretariat to intervene into the mining sector with environmental controls.

Picolotti had already amassed significant experience working with the various congressional bodies because she had personally created and negotiated in both houses the passage of the Water Basin Authority for the Riachuelo River as well as the Argentine National Forestry Law. She quickly sent out feelers to gauge the acceptance of the glacier bill that was still in the Environmental Commission.

Picolotti was on very good terms with Senator Daniel Filmus, the Chair of the Senate's Environmental Commission, which now had the glacier bill on its desk. Having received no opposition and nearly no debate whatsoever from the Lower House, the Environmental Commission could simply move the bill forward to vote, also without debate.

Picolotti had ensured that Filmus was aware of the law and would support it. As a Peronist party member, he was in line with the executive branch. Picolotti's engagement with Filmus on the glacier law, her attention to the bill, and her interest in the text sent a political and unspoken message to Filmus.

It sent a signal that the president was behind the bill (even if she had, in fact, no idea that it was moving forward). But in retrospect, from Filmus's perspective, there was no reason to doubt presidential support for the bill. The president lived in Patagonia, just kilometers from Argentina's National Glacier Park and would have a natural affinity to glaciers—or at least she should. Picolotti's presence, or so he thought, was merely a confirmation of the obvious. The president must be behind the glacier bill.

On one of the occasions when Filmus met with Picolotti to discuss the glacier protection bill at the Environment Commission, as well as several other bills, he asked her outright, "What do you want in the law?"

"I want the Environment Secretariat to be the Implementing Authority for the law. Not the IANIGLA, not the provinces, not the mining agencies. I want jurisdiction and power over glacier protection to be in the Environment Secretariat," responded Picolotti. She knew that, in the end, whoever has the implementing authority and jurisdiction decides which glaciers get protected and which get overlooked.

"How do we get the provinces to go with this? They'll want discretion to decide," responded Filmus, a career politician who realized that the most interested stakeholders that could be affected by the law would not want to turn over power or discretion to an agency they could not control.

"We'll pay for the glacier inventory out of the Environment Secretariat's budget," said Picolotti.

The glacier protection bill called for a nationwide glacier and periglacial environment inventory that would take time and money to carry out. The IANIGLA could not front the bill. The provinces had no installed capacity, no institutional base or budget to implement the law. Someone would have to step in and put up the money to cover expenses.

Picolotti could front the costs from the Environment Secretariat's budget. If handled properly, she could leverage jurisdiction and implementing power through assuming budgetary responsibility. A no-brainer for every-day politics: *you pay, you decide*. Filmus thought it was a reasonable offer and agreed to the strategy. The political pieces were in place for the law to move forward and for the Environment Secretariat to be the implementing agency.

But Picolotti realized that there were some fundamental issues in the text of the bill that would be of very contentious concern to the mining sector. If they understood the issues, they would react, and the bill would never survive the congressional vote or it would be watered down. She figured that if the law survived, the struggle would be between passing a conservationist law like the bill presently read or moving to a law that was more flexible, one crafted more along the lines of the "environmental services" form the bill took in Chile. But, even in Chile, the mining lobby was able to kill the bill, so there was no guarantee that anything at all would survive Congress. Although she preferred the conservationist tone of the present bill, she figured that it would be better to have a services-oriented law as opposed to no law. But, for the moment, there was no debate or reaction. *Where were the miners?* She thought. *Hadn't they read the bill?* Surely there was already a mining lobby firmly directed toward killing the bill. It could never pass in its present form.

There had been some minor discussions at the Senate's Environment Commission about the glacier protection bill, but it really didn't get at the heart of the matter and was focused mostly on jurisdictional discretion over the implementation of the law.

The Senators from the mining provinces had likely heard that there might be some minor controversy over mining operations and glaciers. But they really could not know how deeply the issue really went. They had surely heard of Barrick Gold's glacier problems in Chile with the Chilean portion of Pascua Lama (Pascua is the Chilean side whereas Lama is the Argentine side), and really only in as much as there was opposition to the destruction and/or moving of three glaciers. But few if any were actually looking at the glacier impact issue more broadly or more in detail.

San Juan's mining officials understood that the situation in Argentina was different from that in Chile. There were no glaciers on the Argentine side of Pascua Lama—*so they thought*. This opinion was grounded on the misinterpretation of a very brief descriptive glacier inventory produced by staff at Argentina's glacier institute, the IANIGLA.

Lydia Espizua, a geologist and glacier expert working with the IANIGLA, carried out a brief descriptive inventory in 2006 of a small portion of Barrick Gold's influence area at Veladero and Lama (the Argentine portion of Pascua Lama).[46] Her study area was limited to scoping out glacier presence in an area immediately surrounding the principal infrastructure of the projects and only on three of the many river basins running through Barrick's influence area, *but it was not an environmental impact study*. Although many believed that Espizua had carried out a full glacier inventory of Pascua Lama's and Veladero's influence area, as well as a glacier impact environmental study, what she actually did was simply look at the coincidence of the infrastructure map with the location of seven of the glaciers in the area, as well as with frozen grounds of the periglacial environment.

Her conclusions are thus quite misleading to a light reader who is looking for clarity on glacier impacts of Barrick Gold's projects:

> the works to be developed by the Pascua Lama project, in the Argentine sector, will not directly affect glaciers, snow patches, rock glaciers, while the impact on discontinuous permafrost is not significant. (Espizua, 2006, p. 44) [unofficial translation]

When the stories of Barrick's proposal to dynamite glaciers crossed the Andes, community concerns turned to the Veladero and Lama projects.

Public officials in the Ministry of Mining in San Juan found Espizua's report to be just the answer they needed to quell concerns, and, to this day, they hold up Espizua's report to suggest that glaciers will not be affected at Pascua Lama.

Other experts were not saying the same thing, however. Milana, who had met with Picolotti at the federal Environment Secretariat in September 2006, was saying just the opposite in fact.

Furthermore, Espizua's report *did* actually draw attention to several key issues that would come out when discussions over the content of Argentina's glacier law took place. Both Veladero and Pascua Lama were on permafrost of the periglacial environment. Espizua had inventoried considerable permafrost areas in the project area:

> The discontinuous permafrost area that will be affected [by project infrastructure] for these works is 300 hectares (740 acres), which represents 17% of the discontinuous permafrost area of the Turbio basin. ... [she then adds] The Pascua Lama pits, Penelope West and East, the conveyor belt (both superficial and underground), and the roads would affect 130 hectares (320 acres) of discontinuous permafrost. (Espizua, 2006, p. 44)

While public officials were saying that the Espizua Report was proof that there was no impact, in fact, these revelations about permafrost in the project areas, under the new glacier law, would make Pascua Lama illegal.

This was the first time public authorities began to consider the impacts of public works on permafrost grounds. This had certainly never been an issue for environmental agencies looking at environmental impact studies, and no one at these agencies had really considered the risks of performing activity or introducing infrastructure to permanently or cyclically frozen grounds.

Juan Pablo Milana, an expert on frozen grounds, had been contracted to review Barrick's proposed works, as well as environmental impact studies for the project areas. One of the points he noted in a confidential memo to the provincial governor was that Barrick Gold had placed a sterile rock pile dump on the periglacial environment at Veladero. Milana warned the governor that this was a great risk because the placement of large weight on structurally unstable frozen grounds could alter the stability of the slope, and the whole rock pile could collapse.[47]

In any event, the Espizua report, albeit mistakenly, calmed the concerns of public officials over glacier impacts by mining in Argentina with regards to Pascua Lama and Veladero.

Particularly, officials in San Juan Province and at the national Mining Secretariat were confident that, even with a strong glacier protection law in place, the province would pass the glaciers and mining test with flying colors, and there would be no conflict between glacier protection and mining operations.

The introduction of large-scale industrial activity at high elevations was very new to the region. No one had ever thought of building a mega mine at 5,000 meters (16,400 ft). Even the very high Grasberg mine (Freeport) in Indonesia rests nearly 1,000 meters lower at 4,100 meters (13,450 ft). But it was happening now, and this presented many questions and challenges to local scientists who were being called on to resolve some new dilemmas.

Glacier experts and particularly geocryology experts (experts of the periglacial environment) were suddenly in demand. Just as climate change-induced glacier lake outburst floods (GLOFs) pushed glaciological studies in places like Peru and Nepal,[48] in Argentina, mining investments were calling on glacier experts to provide technical advisory assistance to private companies. All of a sudden, miners were hiring glaciologists.

Juan Carlos Leiva, one of Argentina's most experienced glaciologists, recounted what it was like to carry out research activities at glaciers *before* mining companies started contracting services:

> To get up to an area like Pascua Lama, or even to the glaciers on the access road to Veladero, we have a good window for work in the summer time, when the weather is good. That's a three or four month period. You plan your trip several months earlier, and then pray that you'll have good weather when the day arrives. But the weather doesn't always collaborate with your agenda! You gather your equipment and spend a day driving to the base, from where you'll start your trek on foot. You hire mules, and then embark on a two or three day hike, on foot, up the mountain, to get to the glacier base. Sometimes good weather at the base, doesn't mean good weather higher up on the mountain—or that your good weather will last, and this could mean an abrupt and frustrating end to the expedition. If you make it to your study site, you set up camp, and then, weather permitting you spend two or three days carrying out your experiments and studies. Then you pack everything up and hike back down the mountain. This can take well over a week to accomplish, and if the weather doesn't collaborate, you may have to cancel and reschedule, and then you have to hope that you can find another convenient window of opportunity to repeat the expedition later in the season, before the cold weather sets in. And there's no guarantee that when you decide to embark again, nature will collaborate, often she does not. If you miss your window, you'll have to wait it out another year. With these limitations, there are only so many glaciers you can study each year. Sometimes just one. (Conversation with J. C. Leiva, October 2010)

With the arrival of mining investments, glaciologists are busy at work studying terrain, examining permafrost, and measuring glacier melt. They can work from fully functional basecamps within sight of the glaciers, with all of the modern amenities in place. Sometimes they can be flown directly to a glacier site by helicopter and carry out in a single day or in a few hours what would otherwise take days or weeks. They can also more easily coordinate their fieldwork with good weather days, leaving on short notice when the mining basecamp (very near the glaciers) announces good weather conditions.

High-altitude mining has increased research activity by glaciologists and periglacial experts at many project sites, including Pascua Lama and Veladero (Barrick Gold), Pelambres (Luksik), Andina (Codelco), El Pachón (Xstrata-Glencore), Los Azules (McEwen), and Altar (Stillwater), to name a few.

The Espizua report paved the way for the provincial mining authorities to accept the idea of passing a glacier protection law. Their principal concern was not that protecting glaciers would hinder mining, but rather that they did not want to give away environmental control to a federal authority—and particularly not to one like Picolotti, who came from a strong environmental advocacy background and who had already, as Environment Secretary, shown a certain disdain and distrust of the mining sector.[49]

Argentina is a federal country, much like the United States, one in which provinces have much autonomy over local matters, including environmental resources. Provinces review environmental impact studies and issue licenses for industrial projects. Except for a few exceptions, such as the Federal Minimum Environmental Standards Law, which sets out basic environmental thresholds that all provinces should meet and that are under federal jurisdiction, most environmental matters are addressed and resolved locally.

If the glacier law placed authority in the provinces, the issue of potential mining conflict with glaciers would be resolved *by the provinces*, and that was a good comfort zone for both public officials at the provincial level and miners. They knew that the provinces were hungry for mining investments and would ensure that glacier impacts would not be a barrier to mining investments, but they could not control a federal environmental authority, particularly one run by an environmental activist.

Picolotti had been struggling for nearly three years as Environmental Secretary to get jurisdiction to intervene on environmental impacts brought about by the mining sector, but had been unable to do so because of the federal nature of political relationships. The provinces were in the environmental driver's seat, not the national environmental authority.

In a place like San Juan, it was even more difficult to rein-in mining on environmental matters. The mining lobby was so powerful in San Juan, even inside government, that the province's environmental authority had been sidelined on mining issues. A specialized mining environmental authority, *within* the Mining Ministry, was created to review and authorize all environmental licensing. This institutional arrangement removed independence over environmental review and control of mining operations because the mining minister has undue influence and power over the public official who must evaluate the ministry's due-diligence performance on environmental matters. It is a formula for abuse.

Altering this arrangement was one of the principal objectives of the federal Environment Secretary. A federal-level glacier protection law was a Trojan Horse that could wedge in federal environmental oversight over mining operations. That was Picolotti's thinking as the vote on the Glacier Protection Bill went to the floor on October 22, 2008.

That morning, Argentina's Senators voted on a law to protect glaciers that they did not truly understand. No one imagined that most of Argentina's glaciers were in provinces that most Argentines did not even know had glaciers.

To speak of glaciers in Argentina was to speak of the Patagonian glaciers of the far south, but not of the little-known glaciers of San Juan, where much of the new mining was taking place. The Toro 1 Glacier[50] that was causing such a headache for Barrick Gold is exactly 2,400 km (1,500 mi) from the Perito Moreno Glacier,[51] the globally recognized glacier that is the only one that most Argentines would have recognized at the time of the vote.

Few Argentines realized that most of Argentina's glaciers were of the smaller type found in the Central Andes, in mountains of provinces most people had never associated with glaciers, like San Juan, Catamarca, Jujuy, Salta, and La Rioja. To some extent, Mendoza was better known for glaciers because of its copious snowfall, although not to the extent of the Patagonia region. Furthermore, even Mendoza's "glaciers" were not thought of as glaciers per se, but rather as "eternal snow." A national glacier protection law thus was not initially associated with these "other" glaciers in areas where water provision was actually infinitely more important than in the Patagonia, where glacier melt for the most part ends up in the sea. That was the scene when the legislators went to vote on what would end up being the world's first national glacier protection law.

A statement by Governor Beder Herrera of La Rioja, another glacier-rich province in northwestern Argentina, is a barometer for just how uninformed much of the population was about glaciers, even after the glacier debate had fully exploded in the national media. Questioned about potential mining impacts to glaciers in his province, Beder Herrera confronted the reporter who was questioning him on the proposed glacier law and said rather naively "stop screwing around with our mining, here in La Rioja, we have no glaciers!"[52]

A few months later, La Rioja's Environment Secretary, Nito Brizuela (Figure 3.2), said to a reporter that he did not know if the El Potro Glacier (the province's largest glacier) was actually a glacier or not.[53] He also expressed his doubts on national television as to whether there were glaciers in the vicinity of Famatina, a controversial mining project site where local residents

FIGURE 3.2 *La Rioja Province's Environment Secretary, Nito Brizuela, doubted whether the El Potro Glacier, the province's largest glacier, was a glacier or not. GIS: 28°23'04.72" S 69°36'16.24" W.*

quickly became avid cryoactivists with the passage of the glacier law. Brizuela said publicly that the IANIGLA would have to come study if the El Potro Glacier was actually a glacier or whether it was perennial snow. In fact, the Environment Secretary would learn, they are the same thing![54]

San Juan Province, surprisingly for many, along with Mendoza Province farther to the south, probably has the largest number of glaciers, easily exceeding 10,000 bodies of perennial ice. In provinces like San Juan, Mendoza, Catamarca, and La Rioja, the Andes are an everyday feature forming part of local culture and lore. They tower into the clouds thousands of meters over the valley floor. In the wintertime, they are completely snowed-over at higher elevations, forming an endless white terrain running from the north to the south.

In the summertime, however, most of the snow melts away, leaving only the highest and farthest peaks in white. Residents have lived in these areas for generations and yet no one has ever referred to these bodies of surviving winter snow as "glaciers." Just as in La Rioja Province, *sanjuaninos* generally call these white peaks *nieves eternas* ("eternal snow"). They are written into the literature and the mythology of the land, and they form a part of the cultural heritage of local communities, but they are ephemerally thought to be snow . . . delicate, soft and prone to melt, far from the coarse, thick, powerful, and dynamic characteristics of glacier ice, which was really the crux of the debate that was about to ensue.

On October 22, 2008, Argentina was about to awaken to a frozen world of ice. Representatives filed into the congressional arena thinking they were about to pass a law to protect environments thousands of kilometers away toward the Antarctic, in a world full of penguins and icebergs. No one in their right mind could envision not voting for this law, and so Maffei's Law on glacier protection would pass quickly through both houses of Congress. And that is precisely what occurred.

That morning, Romina Picolotti and I exchanged thoughts on the pending glacier law vote. The Senate would vote later that day on the law.

"I can't believe this bill has made it so far. The miners should have killed it by now, even the mining provinces have remained silent. It looks like it may pass unanimously without any amendments. If it does, I'm going to end up resigning," said Picolotti over breakfast. It was a premonition.

"Well, it would be as good an issue as ever to leave on principle," I responded, also surprised that none of the mining provinces were lobbying against the bill. I also thought about her comment on resigning if the bill went through, and I remember thinking that, indeed, if it did go through, it would mean trouble for her continuity.

Surely Congressman Daniel Tomas of San Juan, who presided over the Mining Commission, or San Juan's governor's brother, Cesar Gioja, who presided over the Senate's Mining Commission, would have recognized the implications for the bill for San Juan's mining and would have steered the wording to something much more service oriented, like the Chilean glacier bill.

In fact, unknown to nearly everyone (except for the miners, of course), the majority of San Juan's mining explorations under way, which included nearly 200 mining concessions handed out to mostly foreign firms, were in glaciated or frozen areas of the periglacial environment, above 3,000 meters. El Pachón (Xstrata Copper), Del Carmen (Malbex), Las Flechas and Vicuña (NGX Resources), Los Azules (McEwen Mining), and Altar (Peregrine/Stillwater), just to name some of the more prominent investments, were in glacier areas, and each of these projects would have potentially irreconcilable differences with the new glacier law if it passed. Ironically, one mining concession was called "El Potro Glacier," and much of it corresponded to the immediate area under the El Potro Glacier!

Surely, the law would not pass, at least not in its current form.

The World's First Glacier Protection Law

The mining sector representatives in Congress missed the ball on the glacier law debate. No one took note. No one read the fine print. No one really thought or knew what the law was really about, and they wouldn't for some time still.

The bill flew through the House of Deputies and then through the Senate, without so much as a debate. Maffei was vice president of the Environmental Commission in the Lower House and saw the glacier bill through with utter ease. In practice, with no opposition and unanimous support in the first house, the Senate would present little debate because the bill would not be opened up for review unless some Senator had an issue with the law.

The bill passed quietly through the various administrative steps in 2007 and 2008. There was no public debate about the content, there was no media attention, and it went through the mechanics of Congress seamlessly. No one knew of the bill, and, more generally, no one knew anything at all about glaciers and much less about periglacial environments and the critical role that these hydrological resources play in our sensitive ecosystems.

One person who did know very well by then was Picolotti. She had her team follow the bill through its various stages and was anxiously awaiting the final vote. The day had finally come, and the cards were all dealt. Nothing seemed to be in the way of the glacier law, and it looked like it would pass even with the conservationist tone submitted by Maffei.

In the afternoon of October 22, 2008, I received a call from Argentina's Environment Secretary, stunned by the Senate vote. Argentina had just passed the world's first national glacier protection law,[55] which also included protection of the periglacial environment. Incredibly, the law had passed unscathed! Some of the more salient articles read:

Article 1. *The present law establishes the minimum standards for the protection of glaciers and the periglacial environment with the objective of preserving them as strategic reserves of hydrological resources and as providers of water recharge for hydrographic basins.*

Article 2. *Definition. To the effects of the present law, glaciers are all perennial stable or slowly flowing ice mass, with or without interstitial water, formed by the recrystallization of snow, located in different ecosystems, no matter what their size, dimension or state of conservation. The rock debris material of each glacier is considered a constituent part of the glacier, as are the internal and superficial water courses. Likewise, the periglacial environment is the area of the high mountain with frozen grounds that acts as a regulator of hydrological resources.*

Article 6. *Prohibited Activities. The following activities are hereby prohibited on glaciers as they could affect their natural condition or the functions cited in Article 1, or as they would imply their destruction, moving, or interference with their movement, in particular:*

a) *The liberation, dispersion or deposit of contaminating substances or elements, chemical products or residue of any nature or volume.*

b) *The construction of architectural works or infrastructure with the exception of those necessary for scientific research.*

c) *Mining or hydrocarbon exploration or exploitation. This restriction includes activities in periglacial areas saturated in ice.*

d) *Emplacement of industries.*

Picolotti was ecstatic. "I just got a call from Mariana Valls [the head of Picolotti's legal team who had been sitting in on the congressional vote], the vote is official, . . . we have a glacier law! This is incredible!" she said, astonished that the law survived so effortlessly and with no amendments.

"No votes against?" I asked, surprised, thinking that it seemed almost too easy.

"No, there was no debate," she replied, echoing Valls's comments of a few moments earlier.

"And the miners?" I replied, just as surprised as she was, "Didn't they say anything?"

"Yes, . . . well, actually, Gioja [the Chair of the Senate's Mining Commission and brother of San Juan's Governor José Luis Gioja] stood up and left the room before they finished reading the text of the law. I think he freaked out," said Picolotti.

It was likely that, as the law was read out loud, Cesar Gioja must have realized it was a problem for San Juan. It was also likely that he hadn't read the law until then. No one brought it to his attention, nor had he personally gotten involved in the debates taking place in the Environment Commission of the Senate. As the Senate's clerk read Article 6, prohibiting mining and hydrocarbon exploration and exploitation where glaciers are present, including in the periglacial environment, he suddenly stood, gathered his things, and left the room. His walkout did not go unnoticed, and, as the fallout from the glacier law grew larger, this moment became a symbol of the growing crevasse opening between environmentalists and the mining sector.[56]

Picolotti paused in our conversation, reflecting on what was to happen next: "we'll have to wait ten days now, and hope it stays firm and that there is no veto."

Picolotti told Valls to gather the Under Secretaries, the legal team, and the communications team at the Environment Secretariat. She wanted to talk to them immediately about the vote and the next steps they would take. No one was to say anything to the media. The entire Secretariat would go silent until the law was officially promulgated in the following weeks. Any bad press could rattle the mining sector, and they could lose the law to a presidential veto.

"When the miners react, the president is going to flip. They should have seen it," Picolotti thought. She was sure that the glacier law was going to be strongly debated and expected the opponents to the law to either work to kill the bill or water it down to a service-oriented tone that would allow flexibility for miners to continue working in the high-altitude peri-glacial areas. In part, she had remained silent, expecting a strong fight at some point. She was actually pessimistic that the law would survive at all, which is why this vote and outcome were completely unexpected. When the miners didn't react, she didn't intervene because she was not going to be the one defending their interests. But the fight never came. As the vote results came in, she realized that this oversight on their part, and her silence, would bring problems for her. "When the problems for mining are revealed—which won't take long, this is not going to go down very well," she said to her team. "Brace yourself, we're in for some trouble!" she said in a worried tone.

Picolotti knew that if pressured by the mining sector, the president would probably bend and act in their favor, which meant a veto. It was the only thing that could stop the law at this point. The situation would reveal Picolotti's silence, and a presidential veto would mean that Picolotti, Argentina's environmental policy steward, would have to resign. And although that would be the ethical thing to do, it would also mean that the Environment Secretariat would lose critical momentum in picking up on environmental policy, on legal reform, and on compliance controls. Plus, with her resignation, the chances that the law could be salvaged for a revival before Congress were slim. She was slowly beginning to work her way into the mining sector, with grudging reluctance but with the overall acceptance of some of Argentina's most staunchly pro-mining provinces, provinces that had very little in the way of real environmental controls. She knew from the beginning that she needed to pick her battles carefully and that she could not afford to lose a single one. This might be the one.

The aftermath of the vote on the glacier protection law produced a chilling effect. There was no massive celebration, no festive reactions from environmental circles. No one really knew what had just taken place.

We expected that the governors of the mining provinces would react against the law. Although they might be able to use reports like Espizua's to suggest that mining projects were not affecting glaciers, they had no arguments regarding the periglacial environment. The new glacier law effectively meant that mining could not take place in periglacial areas. It was an unequivocal and unarguable point that would render projects like

El Pachón, Altar, Los Azules, and many others along the highest Andean peaks illegal.

But that didn't happen, at least not immediately. The governors were quiet. Something seemed wrong.

Notes

1. See CEDHA, 2005, p. 243.

2. For a press summary of the Inuit case, see http://inuitcircumpolar.com/index. php?ID=316&Lang=En; for the petition presented by Inuits to the Inter-American Commission on Human Rights, see http://inuitcircumpolar.com/files/uploads/icc-files/FINALPetitionICC.pdf.

3. See Anderson and Benson, 1963.

4. For Mendoza's water law, see http://www.hydriaweb.com.ar/pdf/leyes/provincia-les/codigo_aguas-mendoza.pdf.

5. For San Juan's water law, see http://www.hidraulica.sanjuan.gov.ar/LeyProv5824. PDF.

6. See 29°19′28.71″ S 70°01′45.22″ W.

7. Toro 1 Glacier can be seen at 29°19′56.05″ S, 70°01′08.59″ W. Toro 2 Glacier is at 29°19′48.25″ S, 70°01′27.83″ W. Esperanza Glacier is at 29°19′50.76″ S, 70°02′11.81″ W.

8. See http://query.nytimes.com/mem/archive-free/pdf?res=F40B14FF3B5810738DD DAE0994DB405B818EF1D3.

9. See http://www3.ti.ch/CAN/RLeggi/public/raccolta-leggi/legge/numero/9.4.1.1.

10. See http://www.alpconv.org/en/convention/protocols/Documents/protokoll_naturschutzGB.pdf.

11. See http://droitnature.free.fr/Shtml/LoiMontagne.shtml; http://droitnature.free. fr/pdf/Lois/1985_0109_Loi_Montagne_JO.pdf.

12. See http://wp.cedha.net/?attachment_id=14201.

13. See http://wp.cedha.net/?attachment_id=14205.

14. See http://wp.cedha.net/?attachment_id=14198.

15. See http://home.nps.gov/applications/npspolicy/getregs.cfm.

16. See http://wp.cedha.net/?attachment_id=14194.

17. See http://wp.cedha.net/wp-content/uploads/2014/10/grunert-mocion-glaciares-chile. doc; the bill is Number 3947-12 and can also be searched on the Chilean Senate website under "*proyectos*" at http://www.senado.cl.

18. See http://wp.cedha.net/wp-content/uploads/2011/04/glaciares-docs-paper-uicn. pdf.

19. See http://wp.cedha.net/wp-content/uploads/2012/10/glaciares-chilenos-borquez-larrain-et.al_.pdf.

20. Bórquez et al., p. 95.

21. See http://sil.senado.cl/cgi-bin/sil_mocionesaut.pl?58@Senador@Horvath%20 Kiss,%20Antonio#.

22. See http://wp.cedha.net/?attachment_id=14192.

23. See Google Earth at 33°09′39.57″ S 70°15′09.72″ W.

24. See Google Earth at 31°43′26.42″ S 70°27′11.56″ W.

25. See Google Earth at 31°45′07.89″ S 70°25′48.49″ W.

26. *Cryoactivism* is activism to protect the iced surfaces of the Earth, a term coined in 2011 by French glacier expert Bernard Francou to describe CEDHA's work in Argentina in the promotion of the glacier law and glacier protection.

27. For Barrick's "Glacier Management Plan" (in English), see http://wp.cedha.net/wp-content/uploads/2011/11/Plan-de-Manejo-de-Glaciares-Barrick-english.pdf; for the original Spanish version, see http://wp.cedha.net/wp-content/uploads/2011/11/Plan-de-Manejo-de-Glaciares-Barrick.doc.

28. See http://seia.sea.gob.cl/externos/admin_seia_web/archivos/3053_2000_8_3_PE.zip.

29. For Bachelet's original presentation to Congress proposing constitutional reform (2010), see http://centralenergia.cl/uploads/2010/01/Proyecto-de-ley-reforma-constitucional-de-las-aguas.pdf.; for Bachelet's current constitutional reform proposal (2013), see http://michellebachelet.cl/wp-content/uploads/2013/10/Nueva-Constitución-28-35.pdf.

30. See Google Earth: 33°09′39.57″ S 70°15′09.72″ W.

31. See http://www.dirplan.gov.cl/noticias/2009/noticia-07/Unidad%20de%20Glaciolog%C3%ADa%20y%20Nieves.pdf.

32. http://wp.cedha.net/wp-content/uploads/2012/10/CONAMA-2008-Pol%C3%ADtica-Glaciares-Versión-Final-Agosto-2008.pdf.

33. At the time this manuscript went to print, both Chile and Kyrgyzstan had bills and/or debates in Congress to pass national glacier protection legislation. Discussions were also under way (by the author) to move Peru to adopt a national glacier law.

34. See http://wp.cedha.net/wp-content/uploads/2012/10/CONAMA-2008-Pol%C3%ADtica-Glaciares-Versión-Final-Agosto-2008.pdf.

35. This is an unofficial translation; for the original text, see Bórquez et al., 2006, *Glaciares Chilenos*, p. 95.

36. See http://wp.cedha.net/wp-content/uploads/2014/04/mocion-de-horvath-mayo-2006.doc.

37. See http://wp.cedha.net/wp-content/uploads/2013/05/Proyecto-Maffei-Ley-de-Glaciares.pdf.

38. See http://wp.cedha.net/wp-content/uploads/2011/04/glaciares-docs-paper-uicn.pdf.

39. See http://ipa.arcticportal.org/about-the-ipa/country-members.html.

40. This is an unofficial translation; for original text, see http://wp.cedha.net/wp-content/uploads/2012/06/4777-D-2007-Glaciares.pdf.

41. See Inversor Energético y Minero, Año 5 Nro. 55, 2011.

42. See http://www.youtube.com/watch?v=y6FU4m_UQHM.

43. See 29°19′53.07″ S 70°01′05.52″ W.

44. See http://wp.cedha.net/wp-content/uploads/2011/11/informe-TORO-1-frontera-incorrecta-Milana.pdf.

45. See http://wp.cedha.net/wp-content/uploads/2011/10/Special-Report-waste-pile-collapse-ENGLISH.pdf.

46. See http://wp.cedha.net/wp-content/uploads/2011/11/Informe-Glaciares-Lama-Veladero-Espizua-2006.pdf.

47. Juan Pablo Milana kept this memo private, but has confided personally that he did indeed send it to Governor Gioja.

48. See Mark Carey, *In the Shadow of Melting Glaciers.*

49. See http://www.diariodecuyo.com.ar/home/new_noticia.php?noticia_id=173862.

50. The Toro 1 Glacier: 29°19′53.16″ S 70°01′09.23″ W.

51. The Perito Moreno Glacier: 50°29′14.48″ S 73°06′27.73″ W.

52. See http://www.lavoz.com.ar/noticias/politica/dejen-joder-con-criticas-mineria.

53. See Google Earth: 28°23′04.72″ S 69°36′16.24″ W.

54. See http://www.noticiasrioja.com/index.php/news/locales/5875-nito-brizuela-sostiene-que-famatina-no-es-un-glaciar.html.

55. For the original 2008 version of the Argentine National Glacier Law (this is not the law currently in effect, but rather the vetoed law), see http://wp.cedha.net/wp-content/uploads/2013/05/Proyecto-Maffei-Ley-de-Glaciares.pdf.

56. More might have been read into this walkout than it may have actually meant. Many presumed that the implications of the law were dawning on Gioja at that moment, and he was getting up to go start the anti-glacier law lobby. A much simpler interpretation—which in retrospect is probably more likely and that would not be at all out of step with modus operandi in Congress—is that he simply didn't have the patience to wait around for the formalities and that he just picked up his things and left as he might for any other session.

{ 4 }

Invisible Glaciers

Up in the highest reaches of the Central Andes, along the Sierra Nevada in California, along the European Alps, in some of the most unlikely places, including countries like Turkey, Bulgaria, Kosovo, Romania, Montenegro, Armenia, Azerbaijan, Afghanistan, Iran, and China, and in some more likely ones such as Mongolia, Russia, Nepal, Norway, Sweden, Argentina, Chile, and Canada, lie entire swaths of frozen lands containing enormous quantities of invisible water in a solid state, hidden from sight until the surrounding ecosystems call on these lands to provide summer meltwater. As much as 25% of the surface of the Earth's land experiences these frozen conditions, and more than 9 million people live in such environments. Even more live immediately below these lands, and yet most of us have never even heard of this frozen realm.[1] The Incas and the Aztecs are known to have used this frozen terrain to store and conserve food.[2]

I am not talking about the more obviously glaciated regions with visible white cover on high mountaintops (which also act as water towers and basin regulators), but rather of that strip of land that lies somewhere below the lowest limit of the visible glaciers and somewhere above the timber line.

No ice or snow may be immediately visible in this region, but, sure enough, the Earth is storing colossal amounts of ice, protected from the warm ambient temperature, for when the environment needs it most. We can think of this invisible frozen region as a buffer or *hydrological ice zone* that ecosystems call on for steady water all year round. It's what glaciologists call the *periglacial environment*.

The term itself is somewhat deceiving. *Peri* suggests "perimeter" or "surrounding," so we might guess that the *periglacial environment* is the area surrounding the glacier, a sort of buffer zone around the visible ice where logically some sort of cryogenic activity (freezing activity) is occurring. Although such activity may indeed be occurring around the fringes of any given glacier, this is not the area known as the *periglacial environment*.

Periglacial environments are much more complex than their name might suggest. They are self-defined areas with their own temperature-defined characteristics and attributes, and they exist independently of glaciers. They are grounds that can have enormous hydrological value, even greater than that of glaciers. They can exist in areas with visible glaciers or they can exist without any glaciers at all. Yes, we can have periglacial areas *without glaciers*. In fact, glaciers are by definition outside of the periglacial area; or, we might say that periglacial areas are defined as areas that are absolutely distinct from glaciated areas. In fact, in many areas affected by climate change, once glaciers are fully melted away, periglacial environments may continue to thrive for some considerable time longer. Periglacial environments will outlive glaciers!

Society knows very little about periglacial environments. Even among experts, it is an obscure realm of glaciology more intricately linked to geological studies than to the study of ice. Some geographers have taken an interest in periglacial environments, and, more recently, hydrologists are looking more closely at the water value of this swath of frozen land. Those who devote their research to this little-known region of the cryosphere generally call themselves *geocryologists*. In sum, they study the interaction of the cryosphere with the local geology or, more simply, the interaction between ice and rocks.

In Figures 4.1–4.3, ice abounds despite the fact that none is visible (except for some remnant snow or ice in the far background and highest altitudes of the image). The ice—lots of it—is all in the foreground!

FIGURE 4.1 *Periglacial environments in the Chilean Andes above 4,000 meters hide ice beneath the surface of the Earth. This image shows an ice-rich rock glacier flowing from right to left in the foreground.*

Source: Alexander Brenning. GIS: 29°09'05.08" S 69°57'16.05" W.

FIGURE 4.2 *Typical tongue-shaped front of an active rock glacier of the periglacial environment. Mendoza Argentina. Note the 30–40-degree inclination of the front slope and the surface crevasses and furrows.*
Source: Mariano Castro. GIS: 33°37'51.53" S 69°35'48.00" W.

FIGURE 4.3 *A debris-covered glacier near the transition point to a rock glacier reveals a massive ice interior. Mendoza Argentina.*
Source: Mariano Castro. GIS: 33°35'18.15" S 69°35'12.08" W.

How Does the Periglacial Environment Function?

What's truly fascinating about these periglacial areas is that, invisible to the untrained eye, they hold massive quantities of glacier ice hidden away beneath the surface of the earth; some subterranean glaciers can be the size of tall multistory buildings (or taller) and as wide as several city blocks (or wider). You could be standing on a million cubic meters of ice, but you wouldn't know it unless someone told you where to look. In fact, you wouldn't see any ice at all for kilometers around, if at all. And yet, it is there.

By the time you finish reading this chapter, you will be able to spot this ice and these *rock glaciers* will become, as if by magic, suddenly apparent where you could never imagine you'd find ice. These glaciers are so obscure that many communities living in their vicinity simply don't know of their existence. Animal herders and small farmers who have walked their surfaces for generations are often unaware of the ice beneath the ground. The revelation of the periglacial environment may have profound implications for people who live and depend on high mountain environments. As they discover these resources, they change the way they consider their habitat and the way they care for their ecosystems. But if we don't know about these water reserves, how can we take measures to ensure that they are not destroyed by anthropogenic activity?

In many high mountain environments, particularly those of the more temperate regions of the planet, despite warm or even very hot temperatures that may be characteristic of low-lying valley areas, temperature decreases rapidly as we climb higher on the mountain slopes.

It can be 38°C (100°F) at the valley floor and yet very cold, even below freezing, atop the highest mountain peaks. What is even more interesting, however, is what is occurring between these extremes. As we walk up a mountainside, temperature begins to fall as a function of vertical distance. In a place like the Central Andes, for every 100 meters we climb (about 330 ft), temperature drops about 1°C (that's about 2°F).

If we were to take temperature readings throughout the year, up and down the mountain, we could define temperature lines (called *isotherms*) and pin-point where and when the temperature is very close to the freezing point (0°C/32°F). We would find that during the winter months, the freezing isotherm is lower on the mountain and during the summer, it is higher up. Between the lowest point and the highest point of the freezing isotherms, the grounds are constantly in a freezing and thawing cycle.

This *freezing point isotherm range* is approximate and can vary substantially by season, sun exposure, or variations of temperature between years. As we will discover, the freezing point isotherm range is extremely important for water storage and meltwater regulation and critical for the periglacial environment.

Just as the freezing point line will be lower on the mountain in winter and higher on the mountain in summer, these oscillations also occur in smaller increments in daily cycles. Temperatures generally fall at nighttime and are higher during daytime. For this reason, the freezing isotherm is generally at a lower elevation at night than it is during the day. These cycles can even occur during daytime hours depending on sun or shade exposure. On any given day or time of day (since the temperature line may move according to the clock), anything located at an altitude above the freezing point will freeze and anything that might have been frozen but that is now below the altitude of the freezing point will thaw. These cycles can also occur in year-long periods, moving the freezing range lower on the mountain during an especially cold year or series of years and moving it up the mountain during an especially warm year or series of years.

Much as we take out our winter blankets when winter approaches, Mother Nature rolls out her ice blanket in the winter and freezes the periglacial environment. I will use the ice blanket analogy to explain what is happening in these extreme climatic environments. We can imagine that during the winter, Mother Nature sits at the lowest point of the glaciated environment (the visible white glaciers) and unfurls a thick frozen ice blanket over the mountain, covering the mountainside below with snow and ice. Anything under the ice blanket is instantly frozen solid. The lowest fringe of the ice blanket is the transition point between frozen and thawed states. The areas that remain under the ice blanket freeze and remain in this frozen state, and the area immediate below the lowest fringe of the ice blanket thaws. We can think of the lower fringe as the freezing point isotherm.

As winter recedes, this ice blanket is slowly drawn up, exposing new areas that were previously covered and frozen. Those newly exposed areas, rich in snow and ice, slowly thaw, releasing the ice and snow in the form of water. The ice blanket will be drawn further and further uphill as spring advances into summer. By the end of summer, the ice blanket reaches its most drawn up state. Then, as fall and winter approach, it is slowly unfurled again. This process refreezes grounds that are covered again by the ice blanket. This process recharges the ground with humidity in the form of new snow and newly frozen ice.

I use the ice blanket metaphor because it helps us better understand the function of the periglacial environment. The ice blanket as a whole (as well as the immediate area just above it) is the periglacial environment, stretching from the lowest points of visible glaciers to the timberline below. The ice blanket freezes the environment and captures humidity underneath. The lower fringe of the ice blanket (the moving freezing point) is cyclically exposing ice to warm temperatures, causing melt and releasing water to the lower ecosystem; it then recovers the ground to recharge it with humidity, and the earth stores that humidity as ice during colder months. We can also consider the thickness of the blanket because areas closer to the surface will be warmer and more apt to thaw than deeper areas, which are frozen solid and protected from the warmer surface temperatures. The portions that freeze and thaw cyclically are called *discontinuous permafrost areas* whereas the portions of the Earth that never thaw are considered permafrost—or permanently frozen.

Any part of this area at or below freezing temperature freezes. Any part of the terrain that is above freezing temperature thaws, releasing water into the environment below. The movement of the ice blanket's lower fringe happens continuously: seasonally, monthly, weekly, daily, and even hourly. This alters the frozen and thawed states throughout the periglacial environment, which is what generates the constant hydrological contribution of this frozen world. The frozen area moves progressively downward as winter sets in and moves progressively upward as summer arrives, even though the up and down cycles can and do happen at shorter intervals (daily or weekly), resulting in short-term freezing and thawing cycles near the fringes.

Frost Heave

One of the phenomena that occurs during these freeze-thaw cycles of the ground is the progressive vertical up and down movement of rock, water, and ice. Put a plastic bottle of water filled to the rim in the freezer and watch what happens. It bursts open (don't do this experiment with a glass bottle)! When water turns to ice, it expands its volume by 8% or 9%. When water in the ground freezes in the natural environment, it also bursts through the terrain, causing significant alterations in the placement of rock and earth around it. The natural expansive power of ice formation is so strong that few materials can resist it. Thus, the freezing of water between rocks and in rock pores pushes rocks apart and can also literally burst rocks open when humidity within the rock pores freezes.

When the ice begins to melt, it contracts by the same amount and water seeps downward through the cracks that have been opened up by the previous expansion. When the water below freezes again, its volume will expand anew, pushing rock and earth out and up. When another melt cycle sets in, the ice transforms to water again, creating new cracks and seeping downward. This is followed by refreezing, after which the expansion and upward push of material occurs again. After many cycles of this freezing and thawing, which can occur seasonally or even within very short daily cycles, rocks and soil will be pushed toward the surface. More rocks will be placed nearer to the surface, whereas more ice will remain below as water seeps down and then refreezes. This is a process that scientists call *frost heave, frost action,* or *cold-climate weathering.* When we examine rock glaciers of the periglacial environment, as well as debris-covered glaciers, we notice that the surface of these glaciers is fully covered by a mantle of rock even though the inner core far beneath may be solid ice. The process may take decades, centuries, or even millennia, but the periglacial area is busily at work—every season, every day, and every hour of the day—on this process and these structures.

You can reproduce the effects of frost heave in your refrigerator and create your own rock glacier. Simply place a generous number of small rocks in a plastic container and fill the container with enough water to barely cover the rocks. Place the container in the freezer and let it freeze completely. At first, only a few rocks will bulge out to the surface, if any. Take the container out and let it thaw for an hour or so, but never let it fully thaw. Return the container to the freezer and let it freeze again. Do this throughout the week, occasionally removing the container and letting it thaw to reproduce the same type of cyclical freezing and thawing that occurs in the periglacial environment. After several cycles of freezing, thawing, and refreezing, you will notice that the entire surface will be slowly covered with exposed rocks that have heaved up from below. At this point, in a real rock glacier, the rock cover begins to protect the deeper ice core even though the surface of the rock glacier keeps going through the freezing and thawing cycles. The rock cover impedes the warmer temperature above from melting the ice below, and as such, the ice in the core can survive even if surface temperatures would not permit it. Water from surface melt continues to flow downward and then refreezes in the core to become ice, pushing more and more rock upward. As the ice core concentrates and grows, it gets cooler and cooler, and the surface rock layer gets thicker and thicker to help to protect this inner ice core.

A typical refrigerator can help us understand the periglacial environment. Think of the freezer (above) as the glaciated environment (the ice cubes are the visible glaciers). And think of the warmer (but still cool) lower refrigerator as the periglacial environment. If we crank up the cooling thermometer to the coldest setting, some of the liquids in the refrigerator may freeze. That's similar to the fringe areas of the periglacial environment. But if we were to move the ice cubes from the freezer above to the refrigerator below, in a few hours they would probably melt. But if we place the ice cubes in a small plastic cooler, or wrap them with copious amounts of paper, or even place them in a plastic container with generous amounts of sawdust and rocks, the cubes will last much longer in the refrigerator. If we place the container of ice cubes and sawdust in the freezer overnight, it will refreeze the ice cubes and any melted water (mimicking the drop of temperature to below freezing in a typical periglacial environment). If we then bring the ice cubes back down to the warmer refrigerator during the day, we can actually sustain the life of the ice cubes in the refrigerator practically indefinitely without them ever fully melting. Over time, thawing will move the ice, sawdust, and rock around, causing frost heave and pushing the rocks upward. There will be enough cold in the refrigerator to ensure that the cubes will not fully melt during their daily warmer cycle. At nighttime, the melted water will be refrozen. You've just created a periglacial environment right in your kitchen!

The dynamics of frost action have drawn much attention from scientists in countries where periglacial areas are a way of life, such as Canada,

Norway, parts of the United States, China, and Russia, as well as from various government agencies (French, 2007, p. 50). This is because dealing with these contracting and expanding grounds can cause nightmares for public works such as buildings, roads, homes, and the like.

The hydrological implications of this cycle are fundamental to the provision of water to our streams, rivers, and water tables. This is the way that Mother Nature has devised to slow the rapid seasonal snowcap melt that occurs in the spring. With this cyclical feature built into the mountain environment, we can count on water capture and release in cyclical phases to take us through all of the year, with or without snowfall.

The higher we go in these frozen areas, the more likely it is that any accumulated ice remains all year round because the temperature is likely to be well below zero for most or all of the year. We must remember, however, that the glaciated environment (which is still farther above the periglacial environment) is also undergoing freezing and melting cycles. This cycle is visible above the surface of the Earth, whereas the dynamics of the periglacial area is concentrated invisibly below the surface. When water is released from areas above the periglacial environment, such as from the melt of glaciers or seasonal snow, through the displacement of ice sliding downhill, or from ice that breaks off glaciers, this water is transported by gravity through and into the periglacial area and may also act as a hydrological input to periglacial processes.

The hydrological balance of a periglacial area is extremely complex and very difficult to measure. However, the properties of the periglacial area are not that different from those of glaciated areas. A certain amount of water enters the periglacial environment throughout the year, part of it is stored as ice, part of it runs off immediately into lower elevations, and part of it is retained. The lower fringes release water into the environment through the various freezing and thawing cycles.

Physical Characteristics of the Periglacial Environment

What are some of the characteristics of the periglacial environment that can help us identify it? We've already mentioned one, *vegetation*—or better said, *lack of vegetation*. Vegetation in such frozen areas abruptly stops at a given height, as if someone had purposefully gone through and razed the land of all trees and plants. This barren area is usually located just below the visible glaciated areas and above the timberline. This swath of land where nothing seems to grow is several hundred meters wide—maybe more, maybe less. There is only rock, and the ambient temperature for the most part of the year is below freezing. Take a look at your favorite high mountain terrain and see if you can identify this strip of barren land. You're probably looking at a periglacial environment (Figure 4.1).

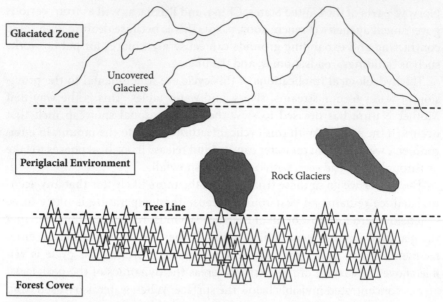

FIGURE 4.4 *The periglacial environment.*
Source: JD Taillant.

Many things happen when water freezes and thaws cyclically in these periglacial environments. First, plant life has a hard time adapting to the frozen periods of the year. Freezing temperature ruptures plant tissue by expanding water volume in the tissue of the plant. Just as plants in our yards might suffer or even die because the nocturnal temperature dropped below freezing for a few nights in a row, nature has a hard time sustaining plant life when the temperature drops below freezing for prolonged periods. This is why periglacial areas are generally and for the most part vegetation free (Figure 4.4).

What is particularly interesting in this barren fringe—and what provides us with clues to its character—is the visible results of the freezing and thawing cycles and the cold-weathering or frost heave that we described in the previous section. The resulting effect is not random. Patterns form.

Permafrost and Mass Wasting

If the surrounding characteristics of the geology are just right, rocks and ice will gather more in one place than in another, and the cyclical cold-weathering dynamics will accumulate in a similar manner. It could be that a steep ravine or a very slanted mountainside produces abundant rock avalanches. If a heavy snowfall occurs on the surface, the avalanches will mix snow, ice, and rock, and it will necessarily accumulate at the base of the slope, where the fall comes to a stop. There, a mixture of rock, snow, and ice builds up.

This mix becomes a dynamic *ice-saturated mass*, growing and expanding as new ice forms from snow and water that enters the mass and freezes. The ice and rock contract when parts of it melt off and then expand when refreezing occurs. If freezing temperatures are maintained throughout the year, these cycles reproduce themselves day after day, season after season, year after year, decade after decade, and even century after century, forming a thick ice-saturated rocky earthen mass. It can be in the form of continuous ground that is constantly stretching and contracting, or it can accumulate in a large body of rock and ice, in a crevasse, at the base of a mountainside, or in the valleys between mountainsides.

This rocky, ice-saturated, and growing surface layer is continuously shifting, and it can, if on a slant and if humidity in the ground is sufficient, begin to slip downhill when a critical weight is reached. The entire ground surface, mixed with rock and ice, may begin to slide downhill, just like a glacier creeps downward to lower elevations. The downslope movement of debris under the influence of gravity is what scientists call *mass wasting*. Mass wasting can occur in both frozen and nonfrozen terrain; however, periglacial areas are especially conducive to mass wasting because they offer broken fragments of rock (broken due to the freezing cycles), expansive and dynamic movement of rock and ice caused by the freezing and thawing cycles, and lubrication caused by meltwater and ice.[3]

Depending on the concentration and accumulation of this material, it may be present in thin layers, in concentrated piles, or in other forms, shapes, and sizes. The various forms are similar in that they contain ice and rock, are undergoing freezing and thawing processes, and are constantly modifying their form and shape. These areas store water and, when the temperature rises, they release water into the environment.

For academic reasons, scientists classify these periglacial environment features or elements by different names according to the various attributes that distinguish each. These distinctions are based mostly on location, ice content, form, and movement.

From an environmental policy perspective, the importance of these periglacial forms lies less in these distinctions of type and more in the hydrological value and properties of each, which are in fact very similar. Like typical uncovered white glaciers, they all capture water and snow and transform it into ice; thus, they all store water, and they all play a role in providing water to the ecosystem. A few of the more common elements found in the periglacial environment are:

- Active rock glaciers
- Inactive rock glaciers

- Talus lobe rock glaciers
- Protalus rock glaciers
- Permafrost[4]
- Solifluction (or ground creep)
- Relict rock glaciers

All of these (except for the last, *relict rock glaciers*) are of hydrological importance. The first four (with the suffix of *rock glacier*) are basically different variations of the same phenomenon: they are bodies of dynamic ice and rock located in or on the surface of the Earth; they are of diverse shape and size; and, to varying degrees, they move vertically and/or laterally.

Permafrost is a generic name given to permanently frozen grounds. The term indicates that we are speaking of grounds that are continuously frozen. Any humidity within the soil is in ice form. As science would have it, however, not all permafrost is *permanently* frozen, and this anomaly, from an environmental standpoint, is what most interests us because the unfreezing of frozen ground could release water into the environment.

Scientists classify various types of permafrost thus:

Permafrost Categorization[5]

Continuous:	frozen 90–100% of the time
Discontinuous:	frozen 50–90% of the time
Sporadic:	frozen 10–50% of the time
Isolated:	frozen 0–10% of the time

Curiously, even the continuous and permanently frozen grounds (the 90–100% of the time category) isn't always frozen and thus could be contributing meltwater to the surrounding environment.

The lower fringes of periglacial areas that thaw and freeze cyclically are generally what we refer to as having *discontinuous permafrost* or *sporadic permafrost*; that is, these grounds are frozen between 10% and 90% of the time, which is quite a large range. Another important characteristic of permafrost is that frozen soils are not a function of humidity. That is, completely dry soils can also be frozen simply because they are at a temperature below the freezing point. Another anomaly is that freezing may not occur at 0°C (32°F) because saline concentrations in the water may lower the freezing point to somewhere between 0° and −2°C.

The International Permafrost Association[6] comprises geocryologists who study the dynamics and characteristics of frozen soils. We will focus on the little-known hydrological value of permafrost areas when we discuss remote sensing of periglacial areas later.

Solifluction, gelifluction,[7] or *creeping ground* is a layer of permafrost that is actually moving over the surface. It moves because of a combination

FIGURE 4.5 *Creeping soils of the Central Andes visible as curved, ripple-like crevasses on the surface.*
Source: Juan Pablo Milana. GIS: 29°01′30.04″ S 67°50′07.03″ W.

of dynamics between the cryogenic processes that create contraction and expansion of the surface layer, the incline (gravity), and the characteristics of the environment. This is ice-rich ground that literally creeps over the surface of the Earth, creating visible folds on the landscape as the soils move forward that can be visibly recognized sometimes even in satellite imagery.

In some cases, non-periglacial mass wasting, particularly at the base of mountain slopes, may be confused with periglacial phenomena because the resulting rock pattern distribution on the surface may look similar to the inexperienced eye. Furthermore, relict frozen grounds, which were once actively sorting rocks into typical periglacial features, may, after the ice has gone, leave behind rock patterns that show a once active frozen ground and thus falsely indicate the presence of ice. Generally, relict permafrost will be distinguishable from active permafrost features because it has less marked and more smoothed-out features (Figure 4.5).

Rock Glaciers

Rock glaciers are cool (pun intended)! This is one of the reasons we're going to talk about them. They are also among the most significant and magnificent elements of the periglacial environment. They contain colossal amounts of ice and also help us define the limits of periglacial areas.

By definition, rock glaciers are formed by cyclical freezing processes involving the mixing, freezing, and thawing of snow, ice, earth, and rock emplaced in specific areas of the periglacial environment. They live and thrive in the periglacial environment and deteriorate when they slip too far downslope and reach warmer fringes, or when the warmer fringes creep up on them—that would be climate change!

Because rock glaciers are in this transitional area between the glaciated areas of a mountain and the vegetation line, and because they are right at the spot where everything cyclically freezes, they have fringes that are permanently melting away. Some parts of the rock glacier, however, such as those at higher elevations or in the deep core ice, are permanently frozen in the same way that our ice cubes in the refrigerator, stowed away in a plastic container among rocks and sawdust, remained partially frozen despite temperatures above freezing.

Another characteristic of rock glaciers is that they have different layers of rock on their surfaces as well as different layers of ice in their interiors. The ice nearer to the surface, because it is closer to the changing outside temperature (which generally surpasses freezing for considerable amounts of time), is experiencing cycles of freezing and thawing, while ice deeper inside the core of the rock glacier may be in a more permanently frozen state. The deeper you go into a rock glacier, the more solid the ice and the more protected it is from the outside ambient temperature and the less likely it is that temporary fluctuations of ambient temperature will melt it. It is precisely this cyclical freezing and thawing process combined with the stable ice core that gives rise to the rock glacier in the first place.

As climate change warms ambient temperature over the long run, however, and as the freezing isotherm range creeps up the slope of a mountain, if the temperatures don't fall again even the deep core ice of a rock glacier will slowly begin to melt. It may take a while, maybe years, maybe decades, and maybe even centuries, but eventually, if the rising temperature trend continues, they will melt away.

This *surface versus core ice* difference is key in the evolution and ecological balance of rock glaciers. The surface layer is the *active layer* because it is freezing, thawing, contracting, expanding, and moving according to variations in ambient temperature. It's also the layer releasing water. The cooler core, because it is more protected from ambient temperature variations, is more rigidly frozen and thus more stable. The ice core acts largely as a water reservoir, as well as the principle cooler for the rock glacier in general. In this regard, the rock glacier's ice mass, protected from the ambient temperature, acts as its own natural refrigerant.

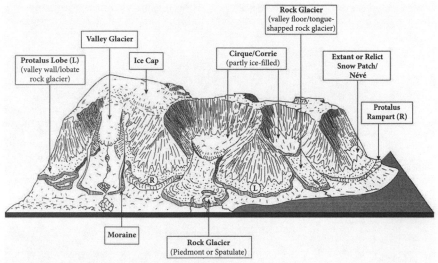

FIGURE 4.6 *Categorization of glacier types.*
Source: Whalley and Azizi (2003, p. 3).

In some cases, rock glaciers simply form at the base of a slope as rock and snow mix, freeze, and accumulate along the edges. These are the so-called *talus rock glaciers*. These don't necessarily flow down a valley base, and so they are generally not very long; however, they may protrude from the talus base a few hundred meters and encircle an entire mountainside by several kilometers. Sometimes the talus rock glaciers protrude further out, making a spatula-shaped or lobular form. These are referred to as *talus lobe rock glaciers*. provided by researchers W. Brian Whalley and Fethi Azizi, shows some of these types of rock glaciers (Figure 4.6). Whalley and Azizi by the way, are using rock glacier analysis to provide potential evidence that there was once (or still is) ice on Mars because some rock flow and accumulation on Mars look very similar to rock glacier features on Earth (Whalley and Azizi, 2003) Figure 4.7.

ACTIVE, INACTIVE, AND FOSSIL ROCK GLACIERS

The type of rock glaciers we've just described that are in constant activity and moving are referred to as *active rock glaciers*. They are located in mostly frozen terrain, which means that they go through freezing and thawing cycles, constantly receiving new water and snow, incorporate new rock, and move; therefore, they are considered "active."

As we move downslope, or as the freezing range moves upslope, active rock glaciers (Figure 4.8) can be overcome by warmer temperatures and

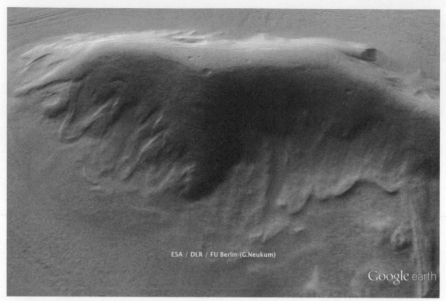

FIGURE 4.7 *Rock glaciers on Mars? The accumulation of rocks in rock-glacier-like formations suggest the possible presence of ice.*
Source: Google Earth (in Mars mode). GIS: 42°11′03.33″ N 49°59′11.61″ E.

FIGURE 4.8 *A relatively small rock glacier, which has partially collapsed, shows a massive ice core. A person stands on a rock for comparative size. Central Andes.*
Source: Dario Trombotto. GIS location unknown.

find that their new location is no longer in the permanently frozen zone; that is, they've moved out of the periglacial environment. When this occurs, they begin to melt, and they do not recuperate enough ice and water content to sustain their shape and size and so they begin to shrink. This is a degenerative state of the rock glacier; the rock glacier loses its dynamic active state to be replaced by a sedentary *inactive* state. Sustained warm temperature will eventually penetrate the surface and begin to reach the inner core of the rock glacier, slowly melting the core ice away. As frost heave action ceases and rocks stop moving, the crevasses, ridges, and curvatures of the surface will smooth out in time, and the vibrant rock glacier will become progressively calm, much like a deflating ball. Over time, which may be a few decades or sometimes centuries, the ice content of the rock glacier will melt until all of the core ice is gone. The rock remains, assorted in a similar fashion to the original rock glacier, but with much smoother and rolling contours. It may be covered with vegetation (not mandatory); it is now a *relict rock glacier* (or *fossil rock glacier*). At this stage, the relict glacier is completely devoid of ice and no longer displays any of the dynamics that once made the active rock glacier thrive. It's important to understand that these processes may not occur in a perfectly linear fashion because climate patterns may vary considerably year to year, decade to decade, and even century to century, regenerating active layers or progressively extinguishing them. To see various forms of rock glacier, see Figures 4.9–4.11.

FIGURE 4.9 *This spectacular three-tongued rock glacier in Salta, Argentina is 500 m wide (1,640 ft), 2 km long (1.2 mi), and more than 40 m thick (130 ft). Note the typical 30–40-degree inclined front slope. GIS: 25°00'37.58" S 66°21'17.00" W.*

A rock glacier on the north side of Engineer.

FIGURE 4.10 *Massive rock glacier at Engineer Peak in the San Juan Mountains of New Mexico, United States. Note the flow patterns, crevasses, and furrows, as well as the abrupt end of vegetation.*
Source: Michael Caton. GIS: 37°42′19.33″ N 107°48′34.09″ W.

FIGURE 4.11 *A rock glacier at the head of the Mercedario River (note the river's birth stream). The glacier towers well over 50 m (165 ft) into the air.*
Source: Juan Pablo Milana. GIS: 31°58′30.36″ S 70°10′32.81″ W.

GENESIS

Curiously, scientists don't all agree on the genesis of rock glaciers, particularly because sometimes they are connected to normal uncovered white glaciers or to debris-covered glaciers—which have some very different characteristics from typical rock glaciers. In simplistic terms and for the reader to have a basic understanding of the difference, a rock glacier is a mix of ice and rocks with a thick rock mantle on top, while a debris-covered glacier has mostly ice interior covered by a thinner mantle of rock. The presence of rock glaciers in the vicinity of, below, or adjacent to normal uncovered glaciers, as well as their physical connection to some debris-covered glaciers and the difficulty of fully grasping how these ice bodies actually link up, promotes debates about their origin. Are glaciers, debris-covered glaciers, and rock glaciers three different and distinct ice bodies intermingling simply because they are in contact—by coincidence? Or are they each simply different stages of evolution between types? When we see the three types in a transitional setting, moving from one to the other almost seamlessly, the arguments for evolution are compelling. But when we find them as stand-alone typologies in the natural environment, the sense that they each have their own unique origins and self-contained evolution can be just as convincing.

Some scientists suggest that the intermingling and physical coexistence of the three types are the result of the opportune natural circumstance and that one type simply provided feed for the other, including primarily the ice, rocks, and snow needed for the receiving body to thrive. In such a scenario, a covered white glacier might feed the ice needed by a debris-covered glacier, or a debris-covered glacier might be opportunely positioned so that at its terminus, it is feeding rocks and ice to a rock glacier further below.

What everyone can agree on with respect to rock glaciers is that they form where the surrounding environment is conducive to their formation (a stable, very cold or freezing climate) and that their basic ingredients are snow, rock debris, water, and freeze-thaw cycles, like that found on talus slopes. To produce a rock glacier, there has to be snow with a predominance of cold (freezing) temperatures all year round. There has to be an extraordinary accumulation of snow and rock at a given site. The site should ideally be at the foot of a very steep slope (without being too steep) that allows for a continuous and cyclical process of accumulation and mixing of rock, ice, and water, all melting, freezing, expanding, and contracting. If there is no continuous entry of new water, ice, and rock into the rock glacier, or if temperatures don't drop below freezing for most of the year, the rock glacier will eventually melt away completely during warmer periods. Likewise, if the temperature doesn't creep above freezing for at least some significant portion of the year, the surface snow would not melt, and we'd likely see a normal, white, uncovered glacier and not a rock glacier. Note that for many rock glaciers, in the summer there is no snow or ice in sight anywhere around the rock glacier (see Figures 4.12–16).

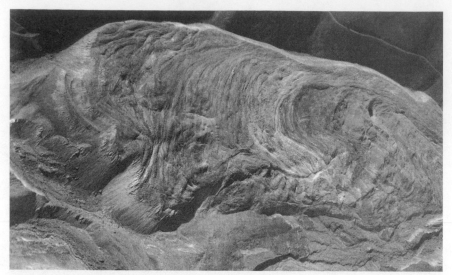

FIGURE 4.12 *Aerial view of a massive rock glacier in Argentina, more than 40 m (130 ft) thick and 2 km long, shows movement lines and furrows.*

Source: Juan Pablo Milana. GIS: 31°57′41.92″ S 70°03′12.02″ W.

FIGURE 4.13 *This rock glacier near Copper Mountain, Colorado (Rocky Mountains), is 700 m long (2,300 ft), 300 m wide (985 ft), and about 30 m (more than 100 ft) thick. GIS: 39°25′28.90″ N 106°06′58.52″ W.*

FIGURE 4.14 *Four rock glacier systems visible on these slopes of the Kaçkars Mountains of Turkey near the Black Sea. GIS: 40°49′59.61″ N 41°08′14.65″ E.*

FIGURE 4.15 *Talus slope rock glaciers in San Juan, Argentina (Central Andes), as well as a larger valley-flowing rock glacier in the upper right quadrant. GIS: 31°57′57.43″ S 70°10′42.58″ W.*

FIGURE 4.16 *Active talus rock glaciers in the Borohoro Mountains of Xinjiang, China; Heyuan Feng peak is seen in the background. GIS: 43°34'10.20" N 86°04'38.90" E.*

Once the accumulation of rock and ice reaches a certain size, weight factors start to play a determinate role in the dynamics of the overall mass. If placed on a slanted slope, lubrication at the base (produced by meltwater or by the contact of the ice with the warmer earth) will create the ideal environment for a massive (but very slow-moving) slide to occur. Thanks to the unevenness and coarseness of the surface, this mass will not quickly slide away, but instead it will creep slowly downhill. The downslope creep of a rock glacier is not as noticeable as that of a normal, uncovered, white glacier, which may advance many feet per day and many hundreds of feet per year. Instead, the rock glacier moves much more slowly, sometimes only a few meters per year, if even that much.

The rolling, flowing, churning movement of the ice and rock on the surface of a rock glacier forms typical patterns in the way the rock is sorted. These are similar to the patterns mentioned earlier for creeping permafrost soils. Crevasses, wrinkles, folds, and concentric curves appear in the way the rock is deposited and sorted on the surface. This reveals the forward and vertical movements of the rock flow. Seen from above, an active rock glacier appears to flow down a mountainside as lava would. In fact, many rock glaciers are confused with past lava flows from dormant or extinct volcanoes and vice versa.

Some argue that rock glaciers are not related to glaciers at all but that they form on their own and have independent dynamics and systems in comparison to normal, uncovered glacier processes. In this view, uncovered white glaciers are simply an accumulation of snow, transformed to ice, that may flow downslope, whereas rock glaciers only derive from a mixture of rock, snow, and the cyclical freezing and thawing processes (the cryogenic process) of the periglacial environment. Thus, these are called "cryogenic" rock glaciers.

Others point to long valley glaciers, which begin as uncovered white glaciers, pass through a debris-covered phase during which a thin rock mantle is added, and then become full rock glaciers with a much more significant presence of rock in the interior, nearer to their terminus—with the transition from debris-covered glaciers to a rock glacier oftentimes occurring rather seamlessly. In this example, we see an evolutionary explanation to rock glaciers, and these are referred to by some scientists as *glaciogenic rock glaciers*; that is, *deriving from glaciers*. Others discard this theory, suggesting that glaciogenic rock glaciers are, in fact, not rock glaciers but debris-covered glaciers and that the seeming transition of a glacier to a debris-covered glacier to a rock glacier is in fact evidence of two or three separate systems.[8] In this last example, the transition is accomplished by the fact that the rock glacier feeds off the rock being pushed by the upper glacier system, but ends up as a different and distinct system (Figures 4.17 and 4.18).

Regardless of their genesis, distinguishing a debris-covered glacier from a rock glacier may not always be that easy. Some signs are sometimes available.

FIGURE 4.17 *Debris-covered glacier in San Juan, Argentina, shows fascinating rock/ice marble-like swirl.*

Source: Juan Pablo Milana.

FIGURE 4.18 *Debris-covered glacier in the Chola Pass to Gokyo, in the Everest-Himalaya area. Note the hikers for size relevance.*

Source: Jorge Garcia-Dihinx. GIS: 28°00′22.92″ N 86°51′53.98″ E.

In debris-covered glaciers, the rock cover is oftentimes thinner than that seen on rock glaciers. The difference can be from just a few centimeters or meters on debris-covered glaciers to many meters on a rock glacier. Also in debris-covered glaciers, the thin mantle sometimes breaks open, revealing ice. This is less common in rock glaciers. Furthermore, debris-covered glaciers tend to form small lagoons on the glacier surface (called *thermokarsts*) that seem not to occur in rock glaciers.

Perhaps the real answer is somewhere in between, and certain rock glaciers form simply because of the accumulation of rock and snow in certain areas (such as at the foot of steep slopes) whereas others do in fact derive from glaciers as these massive ice bodies deteriorate in lower and warmer environments.

In Argentina, where the National Glacier Inventory is under way and scientists must register these ice bodies in an official inventory, much debate occurred at the Argentine Institute for Snow Research, Glaciology and Environmental Sciences (IANIGLA) over how to distinguish rock glaciers from debris-covered glaciers. In the end, they chose not to. The simplest solution was to create a new category when a transition is present: the "debris-covered glacier with rock glacier."

HOW BIG IS A ROCK GLACIER?

Rock glaciers, like regular, white, uncovered glaciers, come in all sizes, ranging from a few dozen meters wide and several hundred meters long, to several blocks wide and several kilometers long. In terms of thickness, they can be anywhere from 10 to over 100 meters (33–330 ft). That's taller than a thirty-story building! They can be of all sizes, but generally cover areas ranging anywhere from 1 ha (2.5 acres) or 0.01 km^2 (about the size of a football field) to much larger sizes of more than 100 football fields.[9]

Generally, rock glaciers that slither down contained slope-sided valleys have the tendency to form into glacier-like tongue-shaped forms or into spatulate (lobate) forms that flow down to the valley floor. They have a typical slanted front, cut sharply at a 30–40 degree angle.

HOW MUCH WATER DOES A ROCK GLACIER HOLD?

Very little research has gone into studying rock glaciers, and even less has been done to define the hydrological contribution of rock glaciers to ecosystems. We know that they contribute water because they are formed precisely because their genesis process includes an important melt stage and a continuous melt cycle. We just don't know how much water because few people have studied them.

As a global society, few rock glacier specialists exist. They are generally geologists who have taken a liking to the cryosphere. Some geographers have also jumped into the realm of rock and ice. More recently, with conflicts around mining impacts to glaciers and to periglacial environments, hydrologists have also come into the picture to study the ice and water discharge from glaciers and rock glaciers. Those who specialize in rock glaciers are sometimes called *geocryologists*. Perhaps with the increased interest in rock glaciers, we will begin to discover more information about the hydrological value of rock glaciers to the environments below them. Another point to consider however, is that rock glaciers are but one element of the periglacial environment, and although we tend to think that these massive ice bodies are the big hydrological contributors to the lands lying below them, it may be that the greater permafrost regions in which rock glaciers are located are actually the big contributors, making a full measurement of the entire frozen system extremely difficult. What is certain is that the periglacial area in general has enormous hydrological wealth, in storage and waiting to be used by Mother Nature when she needs it.

In Figure 4.8, of a section of a rock glacier that collapsed, we see significant ice content. At a glance, we might guess about 50% ice, 50% rock and earth. However, if we were to take samples of the earthy-rocky areas, we would find that once we've brought the mass of frozen rock above the melting point, much water will be released. Much of this seemingly earthy area is actually ice-saturated. Healthy, active rock glaciers can have as much as 80% ice content.

Let's make a quick and very conservative calculation of the water content of the fairly small rock glacier[10] in Figure 4.13 from the Rocky Mountains (see Box 4.1).[11]

If we consider larger rock glaciers, for instance the Calingasta Rock Glacier in San Juan Province of Argentina,[12] the hydrological relevance of these

BOX 4.1 Water in a Small Rocky Mountains Rock Glacier
(39°25′ 40.67″ N 106°06′59.67″ W)

The surface area as seen from above can be estimated as a rectangle of about 200 × 600 meters. (660 × 1970 ft)

The surface is thus = 200 × 600 m = 120,000 m² (144,000 yards²)

We consider that 1 m² of surface (1.2 yards²), viewed volumetrically is equal to 1 m³ (1.3 yards³)

There are approximately 1,000 liters (265 gallons) per 1 m³ (it is actually about 920 l (243 gallons), but we'll use 1,000 for easier math)

So, if we have 120,000 m² (surface), then we have 120,000 m³ in volume per meter of thickness

120,000 m³ × 1,000 liters per m³ = 120,000,000 liters (32 million gallons) per meter of glacier thickness (that's 120 million liters)

But, we will be conservative and assume only 50% water in each m³

(recall that many rock glaciers have up to 80% ice content)

So, 50% of 120,000,000 liters per meter of thickness is = 60,000,000 liters (16 million gallons) per meter of thickness

If we consider that this rock glacier is about 20 m thick (65 ft), but that maybe 10% of the top layer is pure rock cover, we can estimate about 18 meters (59 ft) of rock glacier thickness.

So, 60,000,000 liters × 18 meters = 1,080,000,000 liters—or 1.08 billion liters (285 million gallons) of water by conservative estimates in this small rock glacier!

That's an unfathomable amount. So, let's take it to a city-level analogy. If we could bottle that water for human consumption—say, what a person consumes daily (close to 2 liters per day *or 0.5 gallons*)—how much water would that be?

We'll take the city of Denver as an example. Denver has a population of 1.3 million people. At 2 liters of water consumed per day per person, the city population consumes 2.6 million liters (687 thousand gallons) of drinking water daily.

We have 1,080,000,000 liters available from the rock glacier. Divide the glacier's water by the number of liters consumed daily by the entire population and you get

1,080,000,000 / 2,600,000 = 415 days of daily water consumption for the entire city's population.

This small Rocky Mountain glacier could provide drinking water to the entire city of Denver, Colorado (1.3 million people), for well over a year!

FIGURE 4.19 *The Calingasta Rock Glacier in San Juan Argentina, 3.7 km long (2.3 mi) and 500 m wide (1,640 ft), is of enormous proportions. It holds enough water to provide all of Argentina's population with over a year's worth of daily water intake! GIS: 31°15'17.03" S 70°10'24.51" W.*

glaciers is mind-boggling (Figure 4.19). And we should keep in mind that we are not even considering the surrounding periglacial areas of frozen soils (not related to the rock glacier), which are also likely to hold enormous ice content. For our hydrological estimate, we'll only consider the rock glacier itself.

The Calingasta Rock Glacier is 3.7 km (2.3 mi) long and about 500 meters (1,600 ft) wide. It's about 40–80 meters (130–260 ft) thick, depending on where you measure it. If we assume 50% ice content, we get 1,850,000 m² (2,213,000 yards²) of ice. One cubic meter (m³) of ice is 1,000 liters of water (265 gallons), but we'll only take 50% of that, so that's 1,850,000 m³ × 500 liters, or 925,000,000 liters or 244 million gallons (that's nearly 1 billion liters; 1 liter is equal to about .3 gallons) for every meter-thick layer of ice. We will calculate, conservatively, that the Calingasta is only 40 m thick (although we know it is thicker), which makes for 37,000,000,000 liters, or 37 billion liters (9.8 billion gallons) of water contained in the Calingasta Rock Glacier (Figure 4.19).

At 2 liters per person per day of drinking water, if we consider Argentina's total population of 41,000,000 people, that's 82,000,000 liters of water consumed daily (27 million gallons). So 37,000,000,000 divided by 82,000,000 gives every Argentine about 451 days of water, or over a year's worth of potable water for the entire Argentine population. If we used the Calingasta Rock Glacier in the Denver example, we're talking about 50 years worth of daily drinking water!

ROCK GLACIER AND PERIGLACIAL WATER REGULATION

It is evident from these examples that rock glaciers and periglacial environments more generally are immense water towers, holding colossal amounts of water in relatively confined environments. In our examples, we've unrealistically bottled the glacier water for consumption, so that the reader could get a sense of the amount of water contained in the ice mass. But, in real life, glaciers don't get bottled up and distributed, nor do they melt away overnight. In fact, glaciers and periglacial areas work very efficiently at providing water over long periods of time, making them a renewable water source as long as the glaciosystems in which they are located remain healthy.

Although winter snow melts off quickly in springtime, raising our rivers and streams to very high levels, perennial ice releases water at a much slower but steadier pace once the snowmelt has gone away. It's as if we placed a large faucet in the middle of the mountain range, opening it very slightly so a slow but steady flow of water can enter the ecosystem.

This is what we call the *basin regulator function* of a glacier: the glacier provides the water basin below it with a reduced but steady flow of water throughout the year. This water provision and regulatory role of glaciers and periglacial areas is critical so that our ecosystems have water during warmer months and particularly during especially dry years.

If, for some climatic reason, we don't receive snowfall in a given year or for a series of years (as occurred for the past two years in California), the melt from glaciated and periglacial areas will be critical to providing desperately needed water to our rivers and streams.

The Argentine glacier protection law passed in 2010 protects glaciers and periglacial area, not only because they are significant water reserves, but because they play a critical role in regulating water flow.

Defining Periglacial Areas

Until now, we've defined the visible periglacial area as that area lying between the timberline and the glaciated zones. However, this is an extremely broad definition and not quite exact. There may be areas where low temperatures prohibit plant life but where no ice content is stored beneath the earth and thus no periglacial elements are present. In fact, there are more specific characteristics that will help us in defining more precisely just where this periglacial area is located.

In past times, before the advent of user-friendly geospatial technologies and programs like Google Earth, identifying periglacial areas was a difficult and cumbersome task. One needed good aerial photographs (generally taken by planes, often military aircraft) or, in the absence of such photos, lots of field work. Considering that in many cases these environments are quite hostile to human activity and very difficult to access, many periglacial areas simply went unknown.

There was one very simple and easily applied method of knowing just where periglacial areas lay. This is through the visible identification of active rock glaciers. Rock glaciers slither down mountainsides until they reach a point where they are no longer in harmony with the environment—that is, until they reach a point at which they melt away. This melt-away point for rock glaciers is commonly taken as the "lower fringe" of the periglacial environment.

In other words, find the active rock glaciers, identify their lowest points, and then connect the dots (Figure 4.20). This very simple method works rather well where there are active rock glaciers present. One has to be careful, however, not to confuse *active* with *inactive* rock glaciers, since inactive rock glaciers have already traversed the melting point and are located at lower elevations. Inactive rock glaciers are likely no longer in the periglacial environment, even though they still may have a massive ice core. For this reason, field testing, temperature readings, and visual verification were (and remain) necessary to gather precise information about the lower limits of these bodies of ice and of the periglacial environment more generally.

Another limitation to using the lower limit of active rock glaciers methodology is the fact that there may be periglacial areas without rock glaciers.

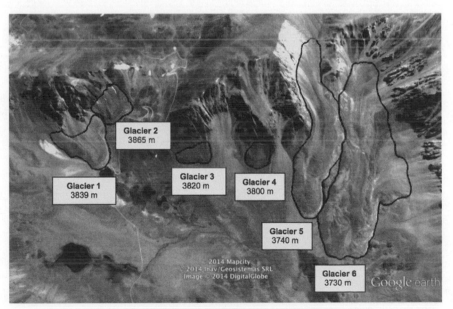

FIGURE 4.20 *The lower limits of the six rock glaciers in this image denote the approximate lower limit of the periglacial environment. GIS: 31°03'12.16" S 70°15'06.27" W.*

As we mentioned earlier, several cryogenic elements in the periglacial environment, including creeping ground and other areas of permafrost, may not show up visually in an aerial photo or even by on-the-ground observation.

Remote Sensing of Periglacial Areas

More recently, advances in technology, good georeferential data, and user-friendly programs like Google Earth have made the identification of frozen soils much easier. Although we cannot necessarily identify ice in the ground, we can very efficiently demarcate where periglacial areas are most likely to be.

The University of Zurich, Switzerland, has developed a remote sensing tool that processes existing reliable data on elevation and air temperature for the entire planet and determines the probability of the presence of frozen soils for any given area of the globe. This tool, in essence, is a global permafrost map. What's fantastic about this tool is that it can be downloaded in seconds to your computer and into Google Earth. With this program loaded onto your computer, any location visited on Google Earth automatically downloads the frozen terrain for the area and displays it (at your discretion) as a superimposed layer on your Google Earth screen. It's that simple![13]

In Figure 4.21, we put the remote sensing tool for frozen grounds to work, and we examined the region near Tehran, Iran, and in the Alborz Mountain

FIGURE 4.21 *Remote sensing tool by the University of Zurich maps permafrost instantly on Google Earth. Here, frozen grounds near Tehran, Iran. GIS: 35°58′00.07″ N 51°51′59.89″ E.*

Range. The highest peak in the range is Mt. Damavand at 5,670 meters (18,600 ft), definitely high enough for periglacial area presence. A very small amount of snow is visible in a small area around the summit of Mt. Damavand, as well as at another series of lower peaks about 15 km (9.3 mi) northwest of Mt. Damavand. We can guess that we might find in those vicinities periglacial environments. Without a remote sensing tool, we'd have to scour the mountain range in search of rock glaciers, if there were any to be found. If we couldn't find rock glaciers, only a site visit with good thermometer readings would reveal periglacial grounds.

If we had very good aerial photographs and were very good at finding rock glaciers in these images, and we had much time on our hands, we might discover that there are several active rock glaciers in the Alborz Range.[14]

With the remote sensing tool, the search is rendered substantially easier. By loading the frozen terrain layer, we can quickly find specific areas where temperature is at or below freezing. Purple and red areas are the coldest. Yellow areas are likely to be permafrost in favorable conditions (e.g., in the shade or on slopes facing north), whereas green designates areas of uncertainty.

When we load the permafrost layer, we see immediately that the rock glaciers we've mapped are all in these predictable areas, particularly in the yellow areas where, under favorable conditions, rock glaciers will thrive (because this image of the Alborz Mountains is printed in black and white, red and purple areas appear as concentrated dark spots at the mountain peak areas, green as a darker shaded gray, and yellow as very light gray).

Concluding Remarks on Rock Glaciers and Periglacial Environments

If before reading this book you had never heard of a rock glacier or a periglacial environment, you now know of a natural resource that for most of the world's population is a complete mystery. You can already identify rock glaciers in a satellite image, and you now have a newly acquired appreciation for this fantastic adaption that Mother Nature devised to capture humidity in the winter months for use in warmer and drier seasons.

The swath of land between glaciers above and the treeline below is an extremely active hydrological fringe of our mountain environments, playing a critical role in the storage and provision of water. Rock glaciers and other periglacial features like permafrost form because of the cyclical freezing and thawing cycles, and the resulting weathering process is precisely how this enigmatic area of our planet acts to store water.

As glaciers are cornered into ever-diminishing survival space, and as they end up melting completely away, periglacial environments will play a more and more important role in water provision to our ecosystems.

Rock glaciers are a fantastic feature of these swaths of land at high elevations, but they are not the only feature of these landscapes. Ice stored under the earth is a critical resource that we must protect and understand. Hopefully, this section has helped shed some light on this fundamental cryogenic phenomenon of our cryosphere.

Notes

1. See French, 2007, pp. xvii, 13.

2. See Corte, 1983, p. 265.

3. See French, 2006, p. 224.

4. In strictly definitional terms, all types of rock glaciers can be considered "permafrost," which can simply be referring to "frozen soils."

5. French, 2006, p. 94.

6. For the International Permafrost Association, see http://ipa.arcticportal.org.

7. *Gelifluction* is a type of *soliflucition*, more specifically implying the presence of ice.

8. See Barsch, 1996, p. 31.

9. See Barsch, 1996, p. 22.

10. The example rock glacier is at 39°25′40.67″ N 106°06′59.67″ W.

11. A proper calculation of the hydrological content of a rock glacier is a complicated process. The calculation offered here is simply a conservative approximation utilizing visual estimates for the sake of example. The numbers chosen are purposefully lower than the likely true measurements, so as to calculate a fair lower limit of the actual number.

12. See 31°15′57.33″ S 70°10′40.10″ W.

13. The map can be obtained easily and at no cost simply through an Internet connection. The reader can download this global permafrost map at http://www.geo.uzh.ch/microsite/cryodata/pf_global/GlobalPermafrostZonationIndexMap.kmz.

14. See, for example (1) 36°01′34.83″ N 51°33′05.00″ E, (2) 36°03′14.20″ N 51°32′40.04″ E, (3) 36°04′04.84″ N 51°33′25.60″ E, (4) 36°04′31.81″ N 51°32′57.25″ E, and (5) 36°02′13.92″ N 51°56′30.76″ E.

The Barrick Veto

Glaciers touch my heart, they are what I am, it's almost a personal
identity of the heart, of my roots, of where I belong.

—PRESIDENT CRISTINA FERNANDEZ
DE KIRCHNER OF ARGENTINA,
WHO VETOED THE 2008 GLACIER PROTECTION LAW
AFTER A UNANIMOUS CONGRESSIONAL VOTE[1]

Buenos Aires, Argentina—October 23, 2008. The team at the Environment
Secretariat could not believe the outcome of the congressional vote the previous day, October 22. Argentina had achieved the world's first national glacier
protection law, the Minimum Standards Law for the Protection of Glaciers
and the Periglacial Environment.[2] The law was strongly conservationist and
excluded all industrial activities on or near glaciers and in the periglacial
environment. It declared glaciers a strategic reserve, defined glaciers broadly
to protect even small perennial ice patches, and banned mining in glacier and
periglacial areas. Some of the more salient text read:

Article 1. *The present law establishes the minimum standards for the
protection of glaciers and the periglacial environment with the
objective of preserving them as strategic reserves of hydrological
resources and as providers of water recharge for hydrographic basins.*

Article 2. *Definition. To the effects of the present law, glaciers are all
perennial stable or slowly flowing ice mass, with or without interstitial
water, formed by the recrystallization of snow, located in different
ecosystems, no matter what their size, dimension or state of conservation.
The rock debris material of each glacier is considered a constituent part
of the glacier, as are the internal and superficial water courses. Likewise,
the periglacial environment is the area of the high mountain with frozen
grounds that acts as a regulator of hydrological resources.*

Article 6. *Prohibited Activities. The following activities are hereby
prohibited on glaciers as they could affect their natural condition
or the functions cited in Article 1, or as they would imply their
destruction, moving, or interference with their movement, in
particular:*

a) *The liberation, dispersion or deposit of contaminating substances
   or elements, chemical products or residue of any nature or
   volume.*
b) *The construction of architectural works or infrastructure with the
   exception of those necessary for scientific research.*
c) *Mining or hydrocarbon exploration or exploitation. This
   restriction includes activities in periglacial areas saturated in ice.*
d) *Emplacement of industries.*

It took a while for the implications of the law to sink in. In fact, at this point,
no one really understood the magnitude of its implications. If they had,
Congress would have never unanimously voted the bill into law. Not even the
mining provinces that stood much to lose with the passage of the new glacier law, which not only prohibited mining in glacier areas but that also for-
bade mining in the little-known but very extensive periglacial environment,
realized just how prohibitive this law would be for mining in high-altitude
regions of the Central Andes above 3,000 meters (nearly 10,000 ft). Nearly
200 projects were already in exploration in the provinces of San Juan and La
Rioja at or above this altitude. Many of these projects were suddenly at risk
and could potentially have to be abandoned.

But, in San Juan Province, where most of these projects were under way,
public officials were surprisingly calm. Felipe Saavedra, San Juan's Mining
Minister, had probably not read the law carefully; and, even if he had, with
Argentine Institute for Snow Research, Glaciology and Environmental
Sciences (IANIGLA) geologist Lydia Espizua's report in hand suggesting that
Pascua Lama's infrastructure did not affect glaciers or periglacial environments "significantly," he would have presumed that San Juan could work with
and around the glacier law. But no one in the mining sector had really looked
closely at the law, and they probably did not fully fathom its implications.

Although this perception would soon change as the general public discovered the conflicts between mining interests and glaciers, for the moment,
none of the provinces were reacting against the newly adopted glacier protection law. In fact, the major media did not pick up the news at all because
none of the big dailies had published any information whatsoever on the new
glacier law and much less so on its implications for mining.[3] If there was one
newspaper in the country that should have picked it up, it was San Juan's
Diario de Cuyo, a radically pro-mining daily very much in line with the provincial executive branch, but the only bit of legislative news *Diario de Cuyo*

published on the day following the glacier law vote was about an anti-tobacco law that was passed in the city of Buenos Aires![4] They too had missed it.

But federal Environment Secretary Romina Picolotti, who had worked quietly with the environmental commissions of both the House of Deputies and the Senate, knew exactly what was at stake. If the mining provinces realized the true implications of the law, they would have never let it survive or they would have modified the conservationist inclination of the law, replacing the prohibitions in the law with a more "environmental services" tone and approach that placed glaciers at the service of industry. Picolotti remained silent, waiting for the mining sector's attack. She expected the law to be watered down and that, in order for it to survive, it would have to make concessions to the mining provinces. Instead, it had gone through the various congressional steps without resistance, and it was approved without amendments.

But it seems that at least one mining company was already at work analyzing the law, and it wouldn't be long before they did react—and strongly.

It's hard to know for sure who exactly realized what was at stake and which actors took action to defend mining sector interests affected by the glacier law. In retrospect, many actors today claim that they didn't want this law the moment it went into effect, but the evidence doesn't show that any of these acted to leverage pressure for a presidential veto, which at this point was the only clear way to derail the law. The provinces saw no foreseeable conflict with the law simply because they did not understand its relevance to their industrial investment interests. They simply didn't understand glaciers or periglacial environments. Recall that the Environment Secretary of La Rioja wasn't sure that there were glaciers in his province.

Had provincial authorities understood the law at that moment, their concern would probably not have been the prohibition of mining in glacier areas but rather who had jurisdictional authority over glacier issues, which now, by law, was in the hands of the federal government and specifically the Environment Secretariat. The other dimension (in addition to the mining ban) that might have generated conflict was financing: that is, who pays for the implementation of the law? The Environment Secretariat had promised the necessary financing to carry out all implementation activities called for by the law, including a glacier and periglacial environment inventory that would be carried out by IANIGLA, Argentina's national glacier institute. The Secretariat would assume administrative and financial responsibility for implementation, thus providing technical and political relief for the provinces. Satisfied on this point, the provincial governments failed to comprehend the very conflictive problems the new law imposed on the emplacement of mining investments in glacier and periglacial areas.

The mining companies, on the other hand, had the greatest stake in the matter and stood to lose much if the law created barriers to their investments.

It was inevitable that they would be the first to look more closely at the fine print, particularly those companies like Barrick Gold and Xstrata Copper that had already invested considerable amounts of money in preparatory activity in glacier or periglacial areas. Their projects could be indefinitely stalled or even canceled if the law stood. They, more than the provinces, would know just where the problems lay with this new law.

All mining projects carry out geomorphological studies of the concession area that, necessarily in this case, includes a review of glacier and periglacial areas. The geomorphology of rock glaciers, for example, or frozen grounds is very visible in satellite imagery. Even mining companies with relatively little investment in the way of infrastructure would know the geomorphology of their concessions well, and they were quick to realize that if mining exploration was banned in the periglacial environment, many would simply have to pick up their bags and go home.

An example of the problem created by the glacier law is Xstrata Copper's El Pachón project[5] in San Juan Province. El Pachón is a US$5 billion copper project investment on the Argentine border with Chile. Xstrata had been exploring the concession for several years and was in the run up to produce its environmental impact studies and prepare the project for launch when the glacier law was first voted. In Xstrata's preliminary environmental impact assessment for the project, the company published a geomorphological map showing more than 200 rock glaciers present within the limits of El Pachón's concession, one of them squarely inside the eventual pit area, several rock glaciers at sites slated for sterile rock piles, and one of the exploratory mining roads cutting through an active rock glacier.[6] The mining companies themselves had already produced and published evidence that would put them in direct conflict with the glacier law. Several other mining concessions in the high Central Andes, particularly in San Juan Province, in La Rioja along the Chilean border, in the Famatina Mountains, in Catamarca in the Aconquija Mountains, and possibly in several others farther north, including Jujuy and Salta, were in a similar predicament to Xstrata's El Pachón; they were suddenly illegal.[7]

At first read, top mining executives, even local project managers, did not likely understand the implications of the law. The scenario at companies like Barrick Gold, Xstrata Copper, Minera Andes, Peregrine Metals, NGX Resources, Yamana, Osisko, and several others operating in the Valle del Cura in San Juan and also at lower latitudes along the Chilean border was probably very similar. Managers likely called on legal teams and geologists to determine the damage. Several of these companies probably called in specialists in periglacial environments, since such specialized knowledge is generally not on staff (or wasn't then).

Everyone was trying to figure out if the new glacier law actually created problems for their mining projects. It probably took several days for glacier and periglacial environment specialists to be identified and contracted, and then several more days for them to adequately review project data and the law and make sense of the implications. Once the technical verdicts were in, the respective legal teams would have to examine the law to see where points of conflict occurred. That would take yet a few more days, maybe a week to two weeks to do a preliminary political analysis of the situation.

Although Xstrata Copper had thoroughly mapped out glaciers at El Pachón in San Juan Province and at another high altitude project called Filo Colorado in Catamarca Province, they had not carried out a glacier impact study. The only company that had already gone through the motions of studying and analyzing its glacier vulnerability was Barrick Gold. Barrick Gold had run into problems with glaciers in Chile and had placed in motion several scientific studies that were submitted to the Chilean authorities, some of which were online for public view. A few less detailed glacier studies had also been carried out for the Argentine side of operations, specifically a glacier health study for the Conconta Pass glaciers[8] (along the access road), as well as a glacier mapping exercise near both Pascua Lama and Veladero to determine conflicts with project infrastructure.[9]

So, naturally, Barrick Gold was probably in the best position to evaluate the fallout of the glacier law. It probably also knew by then the full extent of glacier and periglacial relevance and location for both of its mining projects, as well as for the access road to Veladero and Pascua Lama, which had already been the focus of much debate in regards to glacier impacts at the Conconta Pass.

Considering that only a presidential veto could derail the law at this point and that the president only had ten days to veto the law, time was extremely short. The legal team at the Environment Secretariat began silently notching days onto the wall in a countdown to the law's entry into force. They had already started discussing details of the regulatory framework that the Environment Secretariat would develop.

But Environment Secretary Picolotti realized that once the miners figured out the situation, it was just a matter of time before the phone lines started ringing. The mining companies would likely first call provincial authorities to see if they had already identified the evident problems and incompatibility of the mining sector interests with the glacier law. Or they might skip that step to gain time and directly approach the central government executive, which was the shortest and probably the most efficient route, given that time was of essence.

San Juan is a hot and arid province nestled in one of the highest and driest areas of the world. The original inhabitants were the Capayanes (related to the Diaguita indigenous peoples—which today reside just across the border from Argentina, in Chile), the Olongastas, and the Huarpe. They were subsistence farmers and herders of *ñandus* (a smaller Patagonian version of the ostrich), llamas, and guanacos. They cultivated corn, fruits, and other vegetables.

The first Spanish settlers, with help from local indigenous communities, identified precious mineral deposits in the highlands and recruited workers from among the local tribes to work the mines. San Juan City, which serves as the capital of the province, was officially founded in the mid-sixteenth century when the region was still under Spanish colonial rule. In the mid-nineteenth century, and as an independent province of Argentina (no longer under Spanish colonial rule), San Juan's governor and future president, Domingo Faustino Sarmiento, introduced a series of reforms and development projects that established San Juan as a mining province. San Juan's current governor, José Luis Gioja likes to quote one of Sarmiento's phrases from 1869, a favorite of the local mining sector: "Today, mines are the fire that drives people to settle in the desert, and since they require intelligence, they civilize while they settle."[10]

Later economic evolution included a shift from farmland to vineyards to provide grapes to the growing wine sector in neighboring Mendoza Province to the south. Spanish, Italian, and French immigrants settled in the region (including in Mendoza) in the mid-nineteenth and early twentieth centuries to promote the wine industry. Fruit production in mountain valleys also abounds in the province. But San Juan fell behind in the rapid evolution of the region's economy as its neighbor Mendoza became the regional engine for wine and agricultural production, utilizing San Juan grapes incrementally when their own supply was insufficient to meet market demands. Part of the development problem for the province lies in the less attractive, harsh, arid climate in San Juan, which also has less fertile soil than in neighboring Mendoza. So, despite a few development opportunities that appeared during colonial and postindependence years, for the most part, San Juan was and still is considered one of Argentina's poorest provinces and much in need of investment resources.

From indigenous times, water management was always a key factor for the arid province. Strict water laws and concessions always present political difficulties in the distribution of environmental services that are critical for industrial and community development. Even the precolonial inhabitants of modern-day San Juan had to deal with water management challenges to provide for their subsistence. But the hydrological challenge of bringing dozens, and eventually hundreds, of mining projects to these arid lands seemed not to frighten public authorities, convinced that San Juan would find its road to prosperity and development in mining. Nor did they seem concerned about the impacts this industry would have on their principal water sources during the most arid seasons—its glaciers.

As the neo-mining era arrived in San Juan Province in the late 1990s and early 2000s, it centered mostly in the highest regions of the Central Andes, where few Sanjuaninos ever venture, except for local subsistence farmers, cattle ranchers, and llama herders who have lived in the area for numerous generations. They take their cattle to graze in the highland wetland systems (called *vegas*) for weeks on end and have been doing so for many generations. The few remaining indigenous peoples—mostly on the Chilean side of the border (amongst these the Diaguita)—say that their ancestors would cross back and forth between present-day Argentina and Chile through this area. Valle del Cura (the Valley of the Priest), Agua Negra (Black Water), Paso de Flechas (Arrow Pass) were all key passageways through the mountains and served as points of communication between the Atlantic and Pacific Oceans, all of them located in San Juan territory. Indigenous archeological remains are abundant in the highlands, as are mythological tales involving traditional Inca leaders. The lands are considered sacred by the indigenous tribes. There are still many unexplored indigenous remains near both Veladero and Pascua Lama, including burial sites, resting places, and an indigenous cemetery.

In the mid-1990s, then Argentine President Carlos Menem, a native to La Rioja Province (just north of San Juan), built a development plan around the promotion of mining at a national level. Menem's belief was that mining could be the motor for economic development for the poorer provinces of the northwest including San Juan, La Rioja, Catamarca, Salta and Jujuy, as had been the case in neighboring Chile. While prospects for gold, silver, and copper promised bountiful returns for San Juan, La Rioja, and Catamarca Provinces, Salta and Jujuy Provinces had significant lithium deposits, which would be high in demand from the automotive and electronics sectors and could also be converted into wealth to uplift local economies.

Menem introduced a new mining code in the 1990s, offering extensive benefits, including tax write-offs and export credits, to those willing to invest in the extractive industries. Much of the mineral wealth was along the border, and the discovery of gold immediately on the border in the northern areas of San Juan promised to bring significant foreign direct investment. For that to happen, bilateral relations with Chile needed to improve. The border area, the highest terrain that politically and geographically divides Argentina and Chile, had always been contentious, often pushing military forces to the brink of war. Today, however, the prospects for exploiting mineral resources brought the two countries together, and a treaty was eventually signed to promote joint venture mining investments between Chile and Argentina.[11]

The once-poor provinces of San Juan, La Rioja, and Catamarca, now potentially rich in mineral resources, could benefit substantially from the mining reforms and promotional investment policy crafted specifically for the extractive sector. In each case, large mining projects were planned (Veladero, Pascua Lama, and Gualcamayo in San Juan; Famatina in La Rioja; and La Alumbrera in Catamarca). All of these were gold projects. At least two of these were at very high altitudes, above 3,000 meters (9,840 ft) in San Juan and La Rioja. But San Juan took the opportunity much further and ushered in a veritable flood of mining investments, scouring its Andes highlands with exploratory investments.

Barrick Gold centered its search for precious metals in the region known as the El Indio Gold Belt.[12] Whether by luck, exploratory skill, or because its geologists listened to local Inca myths that tell tales of fabulous amounts of golden treasure in the area, Barrick bet on finding gold between the 29th and 30th parallels. Several local projects were already working the area, and Barrick Gold began acquiring ownership of smaller ventures, including the El Indio and Tambo mines just across the border in Chile, near the 30th parallel. El Indio was the first mine to center on gold extraction and production in Chile.

In Argentina, as had occurred in the nineteenth century with European immigrants (particularly from Italy), land was up for grabs. Although there were indigenous stories about gold in the Valle del Cura and mining was already under way in Chile (e.g., at the El Indio project), no one was exploiting these cold and frozen lands of impossible access in Argentina. For less than it would cost to change the tires of your car, anyone could get exploratory rights over the terrain. Barrick Gold arrived in the mid-1990s, and, with the help of local folk, herders, and trekking guides who were the only ones who knew the terrain, they set up camps at the base of the tallest peaks. The locals knew their terrain well and could point out the most promising spots, which were already identified in their local culture and mythology. They'd bring their best animals, offer their guests succulent empanadas, a hearty *asado*, and tasty sanjuanino wine, and then ... the foreigners would drill. When they found promising samples, they'd name the reserves on a whim. Barrick Gold's landmark Filo Federico vein, claims Ricardo Vargas, a local trekking tour operator, was named after San Juan Mining Chamber President Ricardo Martinez's son, Federico. Martinez had accompanied a Barrick team on one of those climbs to search for gold.[13]

Barrick's bet would pay off with positive results at El Indio in Chile.[14] Barrick also acquired a stake in the Veladero[15] mine just across the border in Argentina.[16] Deposits found at the nearby Pascua Lama[17] site (just up from Veladero) straddling the border between Argentina and Chile were deemed one of the largest untapped gold reserves in the world.

The discovery of copious amounts of gold in the El Indio Gold Belt in the mid- to late 1990s only further encouraged gold diggers to come to the area from all around the world, from as far away as Australia, the United States, and Europe. Most came from Canada. They arrived in droves with pickup trucks and drilling equipment to look for gold, silver, and copper. The search area focused on a relatively large strip of land mostly above 3,000 meters (approx. 10,000 ft), nearly 500 km (300 mi) long and about 50 km (30 mi) wide, all along the Argentine-Chilean border, from the 28th to the 32nd parallels.

The new Argentine mining code approved in the 1990s bode well for exploration, and the accompanying rise in precious metal prices, which soared in the 2000s following the speculative investment patterns of key stock markets in industrialized countries, provided highly combustible fuel to push mining into unexplored territories of the Central Andes. San Juan gave out some 200 concessions, mostly to junior and large mining companies in the 2000s.

But, ironically, many of these miners actually didn't want to mine *anything*; in fact, they weren't really miners at all. They were stock market players who had taken off their suits and ties and traded them in for pickup trucks and shovels looking for a bonanza-find to catapult their stock value, at which point they could sell to the bigger fish in the market—the real miners, like Barrick Gold, Xstrata, Anglo American, and Newmont. The stock market investors were taking over the mining sector, or at least the exploratory part.

One of the most notable examples was Rob McEwen, a financier turned miner with Goldcorp. McEwen revolutionized the mining industry when he opened up the once highly secretive geological datasets of a defunct mining project in Red Lake, Ontario, on the Internet, offering US$575,000 in prize money to anyone who could strike gold at the dead site. The traditional miners at the company panicked when McEwen went public with geological information because such information was traditionally kept under lock and key. In this case, argued McEwen, what value is there in a data set that's not producing any wealth? Why not challenge the world to find gold at a defunct and worthless mining site? McEwen's bet paid off: within weeks, Goldcorp received an overflow of information from desk-bound prospectors around the world who, with the data in hand and a good Internet connection, had identified more than 100 new spots where a gold strike was likely, 50% of which the company had not even registered. More than 80% of these virtually identified sites yielded substantial quantities of gold, and, since the contest was put in motion, Goldcorp has found 8,000,000 ounces of gold at the previously defunct project![18] Goldcorp went from being a US$100 million company to having a value of more than US$9 billion, thus raising the value of its stock some thirty times. McEwen's maverick instincts soon turned him to Argentina, where the mining horizon was rapidly expanding and gold was to be found. He bought out one of these small stock market prospectors, Minera Andes, who was looking for gold and copper at the Los Azules project[19] in San Juan Province.

The new California Gold Rush happening now in San Juan had a very effective strategy that followed a very simple logic, one built mostly on speculation and the very human obsession of finding a way to make a quick buck. Buy a mining concession in San Juan for peanuts. Think up a good name for a mining company, something like "Gold Diggers Ltd.," then tell the mining world that you're looking for gold near Pascua Lama. That draws everyone's attention because everyone wants to strike it rich—and Barrick Gold had already done the hard work of actually finding gold and identifying the general vicinity where more of it could be located. Go public so you're trading on the Toronto Stock Exchange, where everyone is looking for strike-it-rich mining stock (or at least they once were). Come up with a good logo to put on your pickup truck so media sources can take your picture and you get in the news. Hire some local geologists, buy a couple of 4×4s, rent a few makeshift semipermanent tents, rent some drilling equipment, hire a few local drivers, and then—the hardest part: head up to the Andes (if you can bear the altitude) and start drilling. Oh, yes—and hire some good PR people to contact the mining magazines and tell them weekly stories about your adventurous work in the mountains of South America where El Dorado is waiting to be found and how close you are to finding it. Human nature does the rest.

If you do get lucky and find some gold, silver, or copper, draft a catchy press release announcing a third of the find. Your stock will initially skyrocket from 50 cents to several dollars because stock market analysts and investors are eager to buy in early to a get-rich-quick scheme. Then your stock will go into a holding pattern awaiting more news from you. But you're already now a millionaire, with a multifold increase in the value of your company. Then, a few months later, announce the second third of the same earlier find. Stock goes up again! Easy money! After six months or a year, announce the final third; maybe by then you've found a little more minerals to send out new press releases the next year. Ride the wave if you've got the funds to pay for local explorers, and, with a press release each quarter announcing new finds, your stock rises. All of a sudden, a stock that started at 50 cents is now worth several dollars or more. Some of the more cunning exploration companies even repackage old press releases and reannounce the same finds, which also inches stock values higher. The real objective is actually not to dig up the gold: but rather find it and announce it, thus increasing the value of the company's stock. Then the concession rights are sold to a big miner like Barrick Gold, Glencore-Xstrata, Anglo American, or another likely culprit. It's a high-risk venture, but one with little upfront investment beyond a few pickup trucks, some light drilling equipment, a good geologist, and very good PR.

Using this modus operandi, exploration companies began to crisscross San Juan's mountains in search of gold. But the work was far more difficult than the venture capitalists had bargained for. They had to introduce roads through frozen rocky terrain. Above 3,500 meters (about 11,400 ft), the temperature freezes (and so does the ground). Sometimes ice is present, making work in such areas painstakingly difficult or simply impossible.

Sometimes the concessions are completely covered in snow and ice, as is the case with the El Potro project concession, which derives its name from a glacier called . . . yes, El Potro (see Figure 5.1). Incredibly, this mining concession was given to the Canadian company NGX Resources for the land *underneath* this enormous glacier. Some of these concessions,

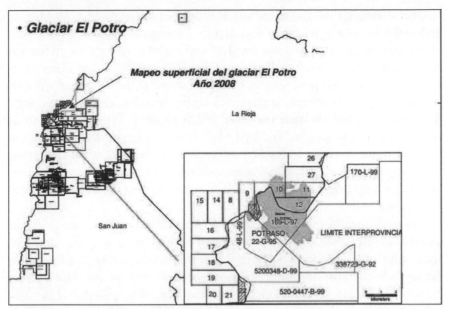

FIGURE 5.1 *Incredibly, this mining concession bears the name "Glaciar El Potro." Nearly all of the El Potro Glacier surface area is concessioned to the company. The El Potro Glacier is visible in the blowout of the diagram as the shaded gray area.*
Source: NGX Resources.

such as Veladero, are nearly 200 km (125 mi) from the nearest town. They are in faraway, remote corners of the planet, where there are no hotels, no supermarkets, no gas stations, and where it's simply impossible to carry out basic human activity. But it is a place where delicate mountain eco systems thrive.

Several problems had to be overcome to make these investments in this part of the world profitable. One of the problems faced by many of these projects was the sheer difficulty of accessing and working in the high-altitude environments where concessions were being granted and where gold was believed to rest under the surface. Winters are harsh in San Juan, particularly above 3,000 meters (nearly 10,000 ft); practically no one lives there because climatic conditions make human activity truly unbearable. With hurricane-force winds, copious snowfall, thin air, and an altitude that makes your heart race and your knees wobbly, not to mention the freezing temperatures for most of the year, it is no wonder that these lands have remained largely unsettled over the course of human history and have gone mostly unused, except for adventurous trekkers, mountain climbers, and herders who dare to come to the high wetland grazing areas when winter eases and the temperatures rise. Springtime weather melts snow and ice, feeding the *vegas* system wetlands that provide excellent grazing for local cattle, llamas, horses, and other livestock.

Transforming this land to industrial usage would not be easy. Mobilizing industrial hardware, trucks, cranes, drilling equipment, and company staff would be a monumental feat, if it could be achieved at all. Most mining companies willing to take up the challenge are forced to adapt to natural climate patterns in San Juan, working with the change of seasons. These companies begin operations in the Southern Hemisphere's springtime (September–November) and work nonstop through the mild to hot summer (December–March). In April, fall is setting in, and by winter (June–August), the snow-packed Central Andes become a treacherous terrain for any sort of human activity. Many companies, particularly the stock-market-investors-turned-miners, simply pack up bags and leave until springtime permits exploratory work to begin anew.

The roads that the miners traced through the Andes are sometimes laid over existing herding paths, etched out through history by indigenous and other local migrants, sometimes on extremely steep ravines zigzagging up mountainsides to the towering summits above 5,000 meters (16,400 ft). Constant treading by llamas or vicuñas along the mountain slopes also etch thin paths across the terrain that are sometimes visible even in satellite images.

Where pickup trucks could fit, they might suffice to get drilling equipment into the area, but many times they did not and explorers were forced to bring bulldozers into the mountains to open up new access roads to their concessions. Sometimes they encountered massive ice bodies, many meters high, and were forced to go around them. Smaller ice formations were broken through to get at their desired destinations. Sometimes even the grounds they cleared contained a rich mix of ice and rock, and that too was moved aside. From season to season, some of these roads that the miners had cleared would disappear under additional ice and rock, as if somehow it was creeping along the surface and returning during the miners' inactive months (they would eventually learn that it *was* moving!). In other cases, the soils they wanted to test were covered with ice so thick and impossible to manage with small machinery that they were forced to drill through the ice to take mineral samples from the earth underneath. No one was telling them where to go, how to get there, or where they should not tread.

Companies introduced exploratory roads indiscriminately through the terrain, wherever they thought they were needed or wherever they thought it was best to channel their conduits through the concessioned terrain. Once drilling was completed, they picked up their equipment and went on to the next site, perhaps never returning to the drill point or expanding activity if their samples came back with positive results.

Many of these exploratory roads (some abandoned, others still in use) carved out in the 1990s and 2000s are still visible today on Google Earth images (Figure 5.2). As the resolution for viewing these images on publicly accessible programs like Google Earth began to improve toward the end of the 2000s, the roads showed extensive impacts to perennial ice, glaciers (large

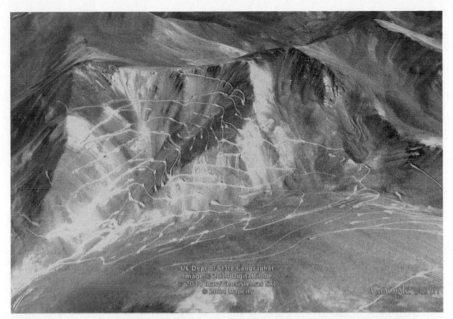

FIGURE 5.2 *Google Earth shows indiscriminate mining exploratory roads and drilling platforms on slopes at the Altar mining project (Stillwater) in San Juan, Argentina, impacting periglacial features. GIS: 31°28'51.57" S 70°28'51.59" W.*

and small), and periglacial areas, particularly to both active and inactive rock glaciers.[20] Before this imagery became available, except for those carving out the roads, little was known about the consequences of introducing roads into these areas. Local folk might have advised drillers or even accompanied and assisted them because this was common practice in the exploratory phases. Where ice was visible and the bulldozers could advance, the damage was done knowingly, with complete disregard for the ice. Where the impacts were to frozen grounds of the periglacial environment, which are ice-saturated earth and rocks, the drivers and prospectors probably had no idea of the consequences of their actions.

When drilling season ended in the late fall and the winter months arrived, exploration companies would pack up their bags and return home until the next season, leaving behind the tracks they had etched in the scenery. Copious snowfall and ice would cover the roads for the winter months. Then the stock market miners would return in the springtime, bringing back their bulldozers to remove the treacherous icy cover.

In some cases, this intervention took place on small perennial ice patches or on small glaciers; if these were active, the removed ice would return due to the natural evolution of the glacier. In other cases, roads were introduced on the surface of glaciers by removing significant layers of seasonal snow and ice from further below. This removal, if recurring each season, progressively intervened with the natural feeding process of snow and ice accumulation on the glacier and began to have noticeable effects on the glacier's mass and health. In still other cases, the roads traversed active and inactive rock glaciers, cutting into the protective rock mantle and dangerously thinning the glacier's protection for its inner ice core. The mining companies saw the ice as a problem created each year due to the snowfall and ice accumulation. They needed to circulate freely from area to area to carry out their exploratory activity. Glaciers, snow, and perennial ice patches were an operational problem resolved by avoiding them, traversing them, moving them, or if needed, destroying them.

But suddenly, with the passage of the national glacier law on October 22, 2008, the mining sector's free and indiscriminate interventions of ice, snow, and the frozen grounds that they were exploring became illegal activity because these icy resources were now protected by law.

Buenos Aires, Argentina—end of October 2008. Nearly a full week had gone by in absolute silence. If in ten working days (taking the date to November 6) the president did not veto the law, the glacier law would stand firm. The Environment Secretary needed to know if something was brewing against the glacier law; the silence was deafening. She had a plan.

The president was a fan of ribbon-cutting at public meetings announcing project inaugurations, public policy, or anything worthy of public mention to gain political favor from her constituency. The Environment Secretariat, which was actively cleaning up waste dumps, forcing the revamping of contaminating industries, and carrying out other environmental works around the country, had hosted numerous such events, sometimes several per month, and this was a perfect opportunity to show strong presidential support for such an important legal triumph. Plus, this particular administration was commonly referred to as the "Penguin Presidency" because the presidential couple (Cristina Fernandez and Nestor Kirchner [the former president]) maintained a weekend home in Patagonia, in the heart of glacier country. Their home was, in fact, in Calafate, a small town used as a hub for tourists visiting Argentina's National Glacier Park.

What better place and occasion to hold an official ribbon-cutting ceremony and have the president announce the world's first glacier protection law? The announcement could be made at Argentina's favorite glacier, the Perito Moreno. The Perito Moreno, a 30 km (19 mi) long massive ice body, is one of the world's few glaciers that is growing. It would be a fitting symbol for the announcement of the glacier law, adding further to the force and visibility of this glacier conservation initiative. It would be a momentous event, one that would go down in political and environmental history as the moment when politics favored ice.

Picolotti sent the president a memo on October 27, 2008, proposing nine different potential public announcement ceremonies that week, among them a sewage works facility in greater Buenos Aires, military land turned into an environmental reserve in Cordoba Province, a river cleanup program in La Rioja Province, a forestry project near the international airport—the list went on. Nestled inconspicuously as the seventh of nine options was a proposed announcement of the recently approved glacier law.

On Friday, October 31, seven business days after the law was voted and four days after her memo to the president, sure that the mining sector would be contacting the president to voice their opposition to the law, Picolotti picked up the phone and called the president's office. She spoke directly to Oscar Parrilli, the president's official secretary, usually the contact point for deciding which political events the president would attend. She reminded him of her memo of October 27 proposing several political public announcement meetings. She called his attention to the seventh item, the announcement of the glacier protection law. "Why don't we fly the president to Calafate and announce the entry into force of the glacier law at the Perito Moreno?" she said inquiringly and matter-of-factly, "it's the world's first glacier law, she's from Calafate, it'll go over very well with the environmental community, it merits loud bells and whistles." She waited for a reaction. Hesitation would be a bad sign, and opposition would signal that the trouble she expected was brewing.

"Yes, good idea . . . she'd like that," Parrilli replied. "What's the date?"

"November fifth is the threshold date when the law goes into force. We should do it on that day, or the next day, the sixth, or we can wait till the weekend, and do it on Saturday the eighth of November."

"Hmm, yes, I like it. I'll check the schedule. She'll like a chance to go home. We'll do it in Calafate. Good idea."

Parrilli sounded energetic, as if he was on board with the idea. There was no opposition from the mining companies or from the provinces, she thought. He made some small talk about other issues and, while on the call, walked over to the president's office, to inquire about that weekend.

"That week is bad, we'll have to do it later in the month. Maybe nearer to the end of the month. I'll get back to you, you'll have to be there." He was already on to the next issue. But Picolotti's mind was clear. There was no visible political opposition to the glacier law. Incredible!

The call went on to cover other issues. Picolotti hung up convinced that no one, *at least not yet*, was protesting. They were just days away from having a fully promulgated glacier law in place. But in politics, a few days is still a long time, and anything could happen.

On November 5, well after close of business, Picolotti spoke with Valls, her legal coordinator. They celebrated. The glacier law would be in full force by the next day. Or so they thought.

Buenos Aires, Argentina—November 6, 2008. The following day, nothing changed. All was silent on the glacier front. As far as the Environment Secretariat was concerned, the glacier law was in full force, but no one was talking about it—yet. Picolotti ordered her legal team to get started on preparing a regulatory framework for the law.

Major media sources had not even reported on the passage of the law. It went unnoticed. Everything was go. But Picolotti was still worried that something would come up. She had a bad feeling: it was impossible that the government of San Juan Province had not reacted. What was happening? Picolotti had expected an explosion of opposition to most likely come from San Juan. But the ten days had transpired, what could go wrong?

Those involved expected that once the provincial authorities fully analyzed the glacier law, they would realize the prohibitive implications for the majority of the exploratory projects under way and also for the three projects in the pipeline and closest to launching: Pascua Lama, El Pachón, and Los Azules. Among these three projects there was well over US$10 billion in investments (an amount that turned out to be even larger). They were all situated squarely in periglacial environments and would probably be deemed illegal under the new law.

When the mining interests finally did figure it out, San Juan's governor would contact the president directly, and the whole thing would explode. Gioja presided over one of the biggest—if not *the* biggest—mining investment promotion initiative to ever come from the public sector, and much of San Juan is right in the middle of glacier and periglacial terrain.

The central problem with the law for the mining sector, particularly of projects under way in San Juan, was Article 6, specifically paragraph (c), which read:

> Article 6. *Prohibited Activities. The following activities are hereby prohibited on glaciers as they could affect their natural condition or the functions cited in Article 1, or as they would imply their destruction, moving, or interference with their movement, in particular:*
>
> a) *The liberation, dispersion or deposit of contaminating substances or elements, chemical products or residue of any nature or volume.*
> b) *The construction of architectural works or infrastructure with the exception of those necessary for scientific research.*
> c) *Mining or hydrocarbon exploration or exploitation. This restriction includes activities in periglacial areas saturated in ice.*
> d) *Emplacement of industries.*

Picolotti was scheduled to participate in a MERCOSUR Environmental Ministers meeting in Rio de Janeiro. On November 7, she left for Brazil. Sixteen days had gone by with not a word. She was to meet with Marina da Silva, her Brazilian counterpart and friend. They were working together on a global agreement to phase out short-life climate pollutants (under the Montreal Protocol) found in many refrigerants and other industrial chemicals.

On November 6 or 7, the call came.[21] But it was not to Picolotti, nor would she find out about it until much later. Most likely, it came from Canada, in English, in a very accented voice asking to speak to Argentina's president. Peter Munk, Barrick Gold's founder and president, had direct access to the presidents of Argentina ever since Barrick Gold first set foot in Argentina in the 1990s to invest in what was slated to be the world's largest gold mine. A problem had occurred, and a solution was needed immediately.

Barrick's legal team and its geologists had already met and reached their conclusions. A report produced by the consulting group BGC Engineering (no relationship to Barrick Gold Corporation) from 2009 entitled *Pascua Lama Permafrost Characterization Study*, which had already been circulated widely to the public and formed part of the environmental impact studies for the Pascua Lama project, summarized in a simple table format that, in regards to the Lama portion of the project (the portion on Argentine soil): "the area of the waste dump was possibly a majority permafrost . . . [that] the area of the pit was all permafrost . . . [that] the plant site had small portions of permafrost."[22]

With the glacier law adopted, the relevance of this summary table for Barrick Gold was profoundly problematic. Most Argentines, including the politicians making decisions about glacier protection, had no idea at the time what the law really meant when it referred not only to glacier protection but also to the protection of periglacial areas, and specifically that mining exploration and extraction was prohibited anywhere in periglacial environments (Article 6). Permafrost is one of the elements of the periglacial environment. Pascua Lama, which had not yet been authorized to begin operations, was, as of October 22, 2008, illegal.

The Argentine glacier law was in direct conflict not only with Pascua Lama, where much of the terrain destined for mining activity was now within periglacial environments protected by the law, but also with projects like El Pachón, Los Azules, Las Flechas, Josemaria, Famatina, Altar, Del Carmen. Pascua Lama was the biggest mining bet in the industry in a long time and too far advanced in its preparatory phase to back out. Argentina's new glacier protection law could not stand if Pascua Lama was to proceed.

Rio de Janeiro, Brazil—November 11, 2008: 8 a.m. It was early morning on November 11. Romina Picolotti, Argentina's Environment Secretary, was on her way to the inaugural session of the MERCOSUR regional meeting of environment ministers in Brazil. She received the news of the veto of the glacier law through a news wire she picked up on her cell phone as her taxi and was pulling into the meeting venue. Moments later, she got a text message from Ernesto Tenembaum, one of Argentina's most respected journalists who already knew of the veto, before the Environment Secretary. He wanted to know if she would resign. She didn't know what he was referring to, but sensed something very serious was happening.

The veto came as a complete surprise, and she found out about it, as did almost everyone else, through the papers. The Argentine daily *Perfil* ran a digital cover story that day headlining "In Cold Blood, Cristina Vetoes the Glacier Law."[23] The story ran with a sinister photograph of the president breaking into laughter in front of Argentina's Perito Moreno Glacier, which was in a region closely associated with the president because she maintained her home base in Calafate, the town closest to Argentina's National Glacier Park.[24]

If things had gone Picolotti's way, the photograph and story should have been of the president announcing the coming into force of the glacier law. Instead, it was a wicked depiction of the unimaginable: the Penguin President vetoing a glacier protection law that had been voted unanimously by both houses of Congress, *a political fiasco*.

Curiously, months later, when the daily *Perfil* became more aligned with the incumbent political machinery, this photograph was changed on the digital version of the story to the one presently shown on the website, taken at the same moment, but showing the president peering through binoculars at the Perito Moreno Glacier.[25]

Still in her taxi, and realizing that all was quickly coming undone, Picolotti immediately thought of the ten-day window for the veto. Valls, her legal advisor, had suggested that the veto window had expired, but there was surely a legal loop hole to consider. Perhaps Congress had not communicated the law to the president's office. She presumed that Carlos Zannini, legal advisor to the president, had done his homework and ensured the validity of the veto. But ultimately any doubts over the temporal details of the veto would soon vanish. The real issue was the president's categorical rejection of a law that Picolotti had carefully and quietly accompanied through the channels of Congress. Her options narrowed immediately and concisely to a likely quick departure from government. The Environment Secretary drafted a resignation text message to President Cristina Fernandez de Kirchner, but she held off sending it while she thought carefully about how to handle the situation. She tried to sort out priorities, attempting to place everything that was happening into a broader perspective, and she tried to think through what the best decision would be for the greater environmental objectives she was working to achieve. Should she stay to try and save the law (if that was even possible at this stage), or should she make a moral statement by resigning and walking away? If she resigned, should she go quietly or with a public statement about her reasons for doing so?

So much was at stake: not only glaciers, but much of Argentina's environmental policy that was only in embryonic form, caught up in a brief era of favorable environmental awareness that had taken over much of the planet but which she thought would surely not last (it didn't). Whatever her decision in those next minutes and hours, Picolotti knew she must get back to Argentina and deal with the veto of the glacier law.

She made several calls to her closest advisors, to her family, and she even made a failed attempt to call the president. That last call, to the president, which she dialed over and over again, would not go through.

She knew that the president would blame her. Her Environment Secretary should have seen this explosive situation earlier, and she should have averted the unfolding political crisis. Picolotti knew that this would be how the president saw the situation. Instead of protecting the president, Picolotti had worked to ensure the bill's safe passage through Congress; now, a firm glacier protection law that was counter to a priority sector (mining), was in place, and it would take a presidential veto to derail it, a veto that was at the heart of a new and potentially explosive political conflict.

Picolotti's brief text message to the president was direct and to the point, void of sentiment or depth:

Dear Mrs. President, I thank you for the opportunity to honorably serve you and my country as Secretary of Environment of Argentina. I hereby present my resignation, effective immediately. Respectfully, RP.

She hit the Send button, but the message came back immediately with a "send error," a common occurrence with telephone communications between South American countries, and particularly between Brazil and Argentina. The worst thing is that the "error" message is not always a reliable indicator of a true error. Did it go through? Had the president received her resignation? She couldn't be sure.

She tried calling Argentina again, to speak to the president herself, but the dialing code prefixes did not put the call through. Picolotti would never know if the message got through to the president, but the events of the next few hours would suggest that indeed the president *had* received her message and knew of her intent to resign.

Her taxi pulled up to Rio's Guarulhos International Airport, and, as she made her way to the check-in lane, a million thoughts raced through her mind. Her heart was pounding, and her phone was ringing incessantly from press agents (*ironically calls came in fine from abroad*), from friends and family who wanted to know if the media rumors were true, if she had resigned, and why the president had vetoed the glacier law. She felt a deep anguish in her chest as she worked her way through the airport. Just over three hours, which seemed like three years, had gone by since the news of the glacier law veto broke as she boarded her flight back to Buenos Aires.

As she sat in her seat, awaiting departure, she read and reread the news pouring out in just about every online newspaper in Argentina, Picolotti was certain that San Juan's governor had intervened. She was sure that the call had come from him.

She decided not to talk to the media, but when she noted one of the calls came from Ernesto Tenenbaum, whom she trusted to convey her message without a spin, the same journalist who had been the first to contact her that morning, she decided to take the call. He must have heard first. Maybe she could tame the rumors. Tenenbaum assumed Picolotti would resign, but wanted to get the scoop directly from the source and get more information as the glacier law veto crisis unfolded.

"Are you resigning?" he said to her straight off, immediately after she answered. Tenenbaum had no idea that Picolotti was in Rio or that she knew less than he about the veto.

She thought back to the send error. Maybe the president already knew of her resignation, and the media team had already spun the story in an attempt to take off some heat from the president. She wasn't sure. No one had called her. From what the media were saying, the veto had been published that same day. Carlos Zannini, the president's chief legal advisor, would have drafted the text the day before or even earlier. Someone could have called her. Zannini himself? The chief of cabinet? The president—why not? It was important enough. The fact that they didn't was a message in and of itself. It was a bad sign and probably meant that her time as Environment Secretary was clearly over.

The send error bothered her as she tried to decide what to tell the press. The president might not yet know of her resignation, and she didn't want to inform her through the media. She felt that the correct thing to do was to ensure that the president knew of her decision before the press did, and so she decided not to say anything definitive until she could get to Buenos Aires and meet with the president or with her chief of cabinet, Sergio Massa.

"I can't answer yet, I'm on my way to meet with the chief of cabinet now," she replied.

"But you're going to resign?" insisted Tenembaum.

"Well, that's a possibility, I can't say anything until I've spoken with Cristina. You know I'm fully behind the glacier law. I saw it through Congress. It got universal and unanimous support. It wasn't even debated. I'm sorry the president vetoed the law. What I can say to you on the record is that I am going to speak to her to see what chance we have of bringing it back," she replied, steering his question to more secure grounds.

"Why did she veto the law, was it Barrick?" retorted Tenembaum, eager to have an inside story over the motives of the president.

"This is off the record Ernesto . . . frankly, I don't know . . . your guess is as good as mine at this point, but I can't imagine that anyone else has the power to muster up a presidential veto, except maybe Governor Gioja of San Juan.

Have you talked to him?" she asked, fully convinced that surely Barrick Gold and Gioja must have come together to pressure the president to veto the law. "Try Barrick, they're obviously behind this! Sorry Ernesto, I have to go, I just boarded a plane and have to turn off my phone. Call me later." She hung up without waiting for a reply.

Picolotti had not heard any news whatsoever at that stage as to who might be exerting political pressure to derail the glacier law. She knew the attack would come shortly, but she didn't know exactly from what corner. Clearly, the issue had been kept secret over the past days and had likely remained only within the most inner circles of the president's closest advisors. She had been in that circle for a time, but now the news of the veto and the fact that she hadn't heard of it before the media showed that the president and her inner circle had left Picolotti outside of the confidence circle.

She turned off her phone, closed her eyes, and tried to collect and sort her thoughts as the plane started down the runway.

Brazilian/Argentine Airspace—November 11, 2008: 2 p.m. During the two and a half hours of flight between Rio and Buenos Aires, Argentina's Environment Secretary tried fruitlessly to figure out what exactly had happened. In order to defend the law, she needed to know where the opposition was coming from.

Provincial opposition to the glacier law was a likely reaction. She expected the opposition from the beginning and was surprised when they didn't derail the glacier law much earlier, when the political cost to do so was extremely low and a derailment in Congress fairly easy.

Picolotti figured that they had surely called the president requesting the veto. It was a logical scenario that she had already considered. She did not know it at the time, but the events of the next few days and weeks would reveal that the provincial authorities were just as uncertain of the source of this decision as she was. They had no clue what was going on behind closed doors. Even worse, they had no clue what the law really meant for the provinces' investments. The provinces had not called the president, nor had they been consulted about the matter, neither by Barrick Gold nor by the president's office. They would find out, like Picolotti, from the papers.

But now, with the veto a reality, the implications for the president were publicly devastating. She was the Penguin President and lived in glacier country, yet instead of announcing that her administration had forged into law the world's first glacier protection act, she would go down in history as being the president *who vetoed the glacier law*, an incredible and avoidable reversal of circumstance. The political costs would be high.

Picolotti shuffled thoughts in her head about what might be happening at the president's quarters. Why didn't the president, her chief of cabinet, or Zannini, the president's legal advisor, call her? The avoidance was obviously intentional. She figured that the president would blame her and Daniel Filmus for the oversight. Filmus was the Peronist senator serving as President of the Environmental Commission in the Senate at the time.

But Picolotti was the first to blame. She knew the issues, Filmus didn't, and, furthermore, she was part of the executive. The president's anger would come directly back to her. Picolotti was Argentina's Environment Secretary, and she had to know about the glacier bill's evolution. It was her responsibility to measure and consider the political implications of adopting this law and—most of all—*the risk* for the incumbent government of having to repeal a legislative vote in favor of the law. She had willfully avoided taking any steps to vet the issue in the executive. Had she done so, the pro-mining factions in the government would have been alerted and would have reacted accordingly, and the bill would never have survived the Environmental Commission and become law. The exact scenario had played out in Chile, in 2006, when Senator Horvath's Glacier Protection Bill simply stalled and died in Congress due to mining-sector opposition.

The Environment Secretary tried to put herself in the president's shoes. What was she thinking? How could she expose the president to the rage of the mining sector? She should have known better—the president would presume (correctly) that she did this intentionally, an indicator that she'd become defiant, working for the benefit of her own agenda, and not for the president's political project. The president also knew Picolotti well, and it would be clear to her that the circumstances were not by chance. This was a coldly calculated decision, and, at this point, the Environment Secretary was a political liability. Consultation was useless. Whichever way she might analyze the situation, the environmental activist Secretary would have to go.

But allowing her to quit or firing her just as the glacier law veto went public was bad timing. It would be too politically costly, making a martyr of Picolotti and drawing attention to the president's outrageous decision to veto a unanimously approved glacier law to protect mining investments. Picolotti had to stay, at least for a while, until the dust settled from the blast.

The president needed a solution to make the veto seem like a rational decision, something that she had done for a greater good. The veto should be seen as a last resort act to favor the will and well-being of the residents of the mining provinces, which would stand to lose much in the way of jobs and development. She would also have to figure out a way to overcome the bad PR of having vetoed the glacier protection law. She needed a solution that would allow her to regain face in the public eye after reneging on the defense of the region most associated with her—glacier country.

Coming back from this faux pas would be tough, particularly in a political arena in which the president was quickly losing political leverage and where she already had a strong media campaign working against her administration.

As the plane began to descend into Buenos Aires airspace, Picolotti considered various ideas from outright confrontation to working with the mining sector and the provinces to develop a workable glacier law. She didn't know whether the president had received her resignation. She had to assume that she had, or at least wait to speak to someone close to the president to see what was happening. What was certain for her was that she had reached the end of the line as Argentina's Environment Secretary.

Ezeiza International Airport, Buenos Aires—November 11, 2008: 3 p.m.
As Romina Picolotti stepped onto the tarmac, she didn't know whether she was still in government, but, even worse, if she was, she wasn't sure that she *wanted* to be in government at that point. Her feelings were mixed. As she looked toward the terminal, she was surprised to see Chief of Cabinet Sergio Massa waiting for her with a crew of journalists eager to get a word from the Environment Secretary.

"The text message went through," she thought—otherwise Massa wouldn't be here with the press. She was still the Environment Secretary. The president had sent her chief of cabinet to derail Picolotti's resignation ensuring that he would speak to her before the media could. The president didn't want to pay the political price of the veto, and the best way to do this was to keep her Environment Secretary aboard, at least for the moment. A resignation now was not good politically for the president.

Massa was not of the inner circle. He was appointed chief of cabinet by the president when Alberto Fernandez resigned from the post following the loss of the soy tax vote in Congress earlier in the year. Fernandez had brought Picolotti aboard as Environment Secretary during the previous administration (under Nestor Kirchner), and he had been her protector and most trusted ally in government; now he was gone, and she had no one to turn to for inside advice—no one was looking out for her. Picolotti would later realize that Massa was not part of the discussion over the glacier law veto, and, like her and despite his role as communicator, he had found out after the fact. But, for the moment, he was the emissary to Picolotti to ensure she didn't leave government just yet and that she was committed to work on a new glacier law.

"What's going on?" said a surprised Picolotti. "What is all this?" she said, looking out to the reporters.

Massa took Picolotti aside and confided in a low voice, inaudible to those present, "Listen, the media wants to talk to you, they think you've resigned; I've already told them that everything is fine and that you haven't resigned," he said nervously.

"But I have resigned," Picolotti responded firmly. "I sent Cristina my resignation by telephone just before boarding the plane. I'm done."

"Don't do this Romina, you can't resign now. You can help us bring back the glacier law. We can draft another law, one that can protect glaciers and allow for mining. The law that was voted was irrational, the miners don't want it nor do the provinces, and you know that under those circumstances, it can't work," Massa pleaded. But it was clear that he held a firm conviction in defense of the presidential veto; his words implied that Picolotti had erred by letting this contentious law slip through. In the end, it was her fault, and now she had to help fix the problem.

"There's nothing for me to do. If Cristina vetoed the law, she has withdrawn her confidence in me. She didn't even call me to let me know it was coming. Under those terms, I can't continue. You know that," she insisted.

"Listen, do you want a glacier law or not? You know that if you don't stick this out with us, there will be no law." Massa was a clever politician and knew very well that Picolotti was not a career politician and would not defend her continuity merely for the sake of holding office; she was, however, a core environmentalist and she could be driven to stay if she saw an opportunity to save the glacier law. He knew exactly what chords to play to the choir.

She listened to his words trying to read through the political message he seemed to be hiding, not quite sure what he really thought of the whole situation. Was there really a chance to bring back the law? Or was this was simply a smoke screen to take the pressure off the president? She couldn't read him and didn't trust him. There was no time to think, as is common in the whirlwind of politics. The journalists were just a few meters away demanding to speak to Picolotti about the rumors of her resignation and also to get her opinion on the president's veto of the glacier law.

Massa could see in the Environment Secretary's gaze that she would stay on to bring back the law. His eyes turned back toward the media crowd, his eyes sparkling with political energy. He put his arm around Picolotti's shoulders for the cameras, posing for a political picture as he called over the journalists to reconfirm that the Environment Secretary had not resigned. For the moment, he had met his objective.

Ezeiza International Airport, Buenos Aires, Argentina: 4 p.m. Picolotti tried to clarify her decision. If she sustained her resignation, all hope would be gone for bringing back the glacier law. There wouldn't be a single environmental voice in the room, and the law would never work. If she stayed on, the likelihood that the law would be any good was slim, and the political implications of her continuing in her post would be demoralizing. Losing this battle to the mining sector would greatly weaken her as Environment Secretary, also undermining many of the other environmental objectives she was pursuing on issues ranging from the Riachuelo River Basin cleanup, pulp mills, and forests to environmental insurance and corporate accountability.

As the cameras went live and the microphones moved in front of her, she wasn't sure exactly what the next few days or weeks would bring, but she was sure it would not be a pleasant journey. For the moment, she continued to be Argentina's Environment Secretary.

Did Barrick Gold Leverage the Barrick Veto?

The presidential veto of the world's first glacier protection law has become euphemistically referred to in Argentina as "the Barrick Veto," on the assumption that it was Peter Munk, Barrick Gold's president and founder who called Argentina's president and leveraged the veto.

Probably only a handful of people in Argentina (not counting Barrick's team) know for certain if this was truly the case, but there is extensive evidence indicating that this was indeed so. I certainly believe so.

It became widely assumed that the mining provinces were behind the veto because these provinces had most to lose from the glacier law and because, if the law stood, specific mining investments would be stalled or canceled. Projects in San Juan, La Rioja, Catamarca, and possibly Salta and Jujuy would be in serious conflict with the new law. Furthermore, the glacier law also banned hydrocarbon exploration and extraction. Noise about the law was buzzing also from more southern provinces such as Neuquén, Rio Negro, Santa Cruz, and Tierra del Fuego, where it was perhaps more mainstream to think about glacier presence. These provinces are rich in petroleum and gas resources and could also stand to lose by limiting oil and gas prospecting areas in the name of glacier conservation. But the real noise was not coming from gas and oil. It was clearly coming from mining. *But did the provinces make the call?*

One revealing bit of evidence came from the president herself, when she later spoke before a group of mining actors and media representatives in San Juan Province. Referring to the presidential veto, and irritatingly aware of the fact that she would pay a high political price for the veto, Fernandez de Kirchner said: "They [the mining province congressional representatives] approved a proposal unanimously, even if it was affecting their interests ... The Congressmen and Senators—amongst them the locals—they lifted their hands lightly without knowing what they were voting for."[26]

This remark is curious in that the president is referring to the same group that supposedly, according to her veto text, had complained to her that the law worked against them. Why would they suddenly have a turn of heart, just days after voting overwhelmingly and unanimously in favor of the law? Clearly, those who complained to the president and those who voted in Congress were not the same people.

News of the veto spread like wildfire on November 11, 2008, as the press picked up the announcement in an official bulletin published that same day.[27] The veto created a very large backlash throughout Argentine society and particularly through the political world. No one saw it coming, but everyone reacted. The cryoactivism born from a small group of advocates in San Juan Province immediately went national.

It was exactly what the president's political opposition was looking for, a gratuitous reason to slam the president—and it hit at the heart of her glacier country homeland. The media was sensitive to and constantly fishing for controversial issues; they immediately picked up the glacier law veto as the new governmental and political crisis. The recent congressional victory overturning the president's decree on the soy tax had empowered the opposition. The glacier law veto could be a deadly blow to the president, following this greatly weakening soy tax debacle. Whatever the reason, the veto of the glacier law would become one of the more controversial decisions of Cristina Fernandez de Kirchner's presidency. Environmental groups reacted instantaneously and ferociously.[28] Whereas most groups had no idea there was even a glacier bill in Congress up for consideration, the reactionary nature of the environmental movement was not slow to attack the veto and lift its banners in defense of the country's glaciers.

In the days and weeks that followed, and as the issues underlying the glacier law and the presidential veto of that law began to surface, the reasons for the veto seemed clear. Society didn't need to speculate or guess at the mining lobby influence in the decision—the reasons were spelled out in black and white in the veto text (salient/relevant sections are translated unofficially):

Buenos Aires November 11, 2008 (from the official bulletin)[29]

Considering the bill 26.418 [the Glacier Protection Bill]

"as the Mining Secretariat has stated, the establishment of minimum standards cannot simply prohibit activities, but rather set minimum parameters . . .

". . . that at the provincial level only activities that can actually be carried out in a sustainable development context and with the protection of the environment, shall be authorized . . .

"that the prohibition of activities described in Article 6 of the bill, if in force, could affect the economic development of the involved provinces, making it impossible to develop any type of activity or work in mountain areas. In this regard, the prohibition of construction of infrastructure works does not take into account that many of these are of public character and of common use, such as international passes; and that the prohibition of mining or hydrocarbon exploration and extraction, including those taking place in ice-saturated periglacial environments, would place environmental aspects above activities that could be authorized and developed in perfect protection of the environment.

"that the requirement of a prior environmental impact assessment before the authorization of all works or activity that could degrade the environment, . . . is excessive.

"that the prohibited activities cited in Article 6, . . . do not take into account that the provinces, by their institutions, and by national and local norms, have in place their sufficient controls to evaluate and authorize activities in infrastructure, industry, mining, hydrocarbons, etc. in perfect harmony, equilibrium and environmental protection.

"and that for this reason, the Governors of the mountainous regions have manifested their concern over the sanctioned norm, as it would impact negatively in the economic development and investments that are taking place in the mentioned provinces." [30]

The veto was purposefully crafted in defense of the mining industry. The president vetoed the glacier protection law because protecting glaciers would impede mining (and possibly hydrocarbon) exploration and extraction. It made reference to mining projects under way, a clear pointer to Veladero and Pascua Lama, the only projects with significant extraction or project preparation already completed and located in glacier territory. They were Barrick Gold's projects in San Juan. The glacier law could also impede other public works that were planned or that could be carried out in high mountain areas. Immediately in mind was San Juan's hopes to finally build a transnational highway across the Andes into Chile at the Agua Negra crossing, which was squarely in periglacial environment. It would later be revealed that preparatory works for the Agua Negra pass had already carved into rock glaciers near the Chilean border.[31]

The text of the veto explicitly referred to provincial government complaints about the glacier law and how it was hurting investments. The text also referred to an ongoing debate regarding other federal laws that encroach on provincial jurisdiction and discretion about what happens (or not) on provincial soil. The tone of the veto suggested that glacier law was yet another example of this controversial encroachment of federal authorities on provincial lands.

These arguments would become central to follow-up discussions that would take place over the next two years on the floor of both houses of Congress. But, even though this was largely the focus and argument submitted by those congressional representatives most highly opposed to the glacier law and even though it was in black-and-white in the text of the veto, *was it really the underlying reason for the veto*?

If it was the mining provinces that were complaining, why did they vote in favor of the law? Why didn't they speak up earlier? Did they all suddenly sit down to read the glacier law immediately after it was passed for some specific reason? And did the implications of the law just suddenly dawn on them?

There is no doubt that a call had come. Someone complained of the glacier law and its implications for mining investments. But did it come from the mining provinces? Evidence emerging after the veto and as the debates against the law moved forward suggests that it was not the mining provinces but rather Barrick Gold that reacted to the law and obtained a direct presidential veto.

What is noteworthy as we consider who might have made this call to the president is that, in the immediate days after Congress approved the law, Argentina's Environment Secretary had several conversations with the president's secretary, Oscar Parrilli, and there was absolutely no indication that the law was being questioned by the mining provinces (or anyone else for that matter), particularly not San Juan Province, which, as would be revealed in time, had the most pressing and problematic issues with the new law. In retrospect, if the mining provinces *had* been complaining, the Environment Secretariat would have been the first to hear of it because the president's office would have deferred to Picolotti on the matter. That is how it usually would play out.

Indeed, the mining provinces had much to worry about, and that is probably why most assumed that it was these same provinces that raised the issue. One of the country's largest gold mining projects, Veladero, and the upcoming Pascua Lama project (both Barrick Gold) were in glacier and periglacial areas.[32] The mega-copper project El Pachón (Xstrata Copper) was next in line and was also in a periglacial environment.[33] The pit area had a prominent, active rock glacier smack in the middle of the terrain to be removed, which would now make El Pachón illegal under the glacier law. Los Azules, then run by Minera Andes and later purchased by the Canadian company McEwen, also had periglacial areas throughout the project influence area that show evidence of impact from exploratory activity.[34] Altar by Peregrine Metals (later Stillwater) also had periglacial issues in the extraction zone. Del Carmen, owned by Malbex and which later would receive investments from Barrick Gold, was in periglacial terrain. These were but a few of the projects that would be affected by the glacier law. It was logical that San Juan would jump at the law, and it eventually did. But oddly, during the first few days and weeks after the law was voted and until after the veto, the provinces remained silent.

Picolotti and Parrilli had discussed and agreed to holding a launching ceremony in Calafate announcing the law as late as October 28 or 29, six or seven days after the law had been approved by Congress. The veto that finally arrived on November 10 had taken some time to review the issues and edit the text, which probably took at least a day or two to complete, maybe more. That places deliberation on the veto, once the president's inner circle was considering, in the first week of November.

Immediately after the veto, Picolotti called Gioja to ask about the province's position regarding the law, presuming he would be furious. He was not one to hold back or hide his temper. But Gioja denied the province had any problems with the law. This surprised Picolotti and threw a curve ball into her analysis. Gioja's Mining Minister, Felipe Saavedra, was confident, with the IANIGLA glacier impact report in hand, that neither Veladero nor Pascua Lama had any impacts whatsoever on glaciers. In fact, a week later, in personal meetings at the Environment Secretariat in Buenos Aires, Gioja and Saavedra spoke to Picolotti and showed IANIGLA geologist Espizua's report to her as evidence that there was no problem between mining and glaciers in San Juan (or so they thought).

In preparation for a meeting planned that week to begin discussions on a new glacier law, Picolotti's team began calling the provinces, sending out feelers on their positions, and ensuring that they would engage in the upcoming meeting. To her great surprise, province after province— San Juan, La Rioja, Jujuy, Salta, Rio Negro, Mendoza, Chubut, Santa Cruz, Tucuman, all with glaciers, and several with mining *and* glaciers—confirmed that they had not intervened with the president, that they had no fundamental problems with the new glacier law, and that they were all happy to engage on a new bill. A review of the president's meeting agenda, which is publicly available online, shows no meetings on mining issues after the approval of the glacier law. Curiously, one meeting that did take place on November 3, 2008, and which included two governors from glacier provinces (Salta and Jujuy), was convened to discuss a tobacco law, but nothing on glaciers.[35]

Something didn't gel for Picolotti and her team. The information in the veto text was very specific, very clear. No one of the president's inner circle knew anything at all about glaciers and much less about periglacial environments. If they had a question, they would have gone to the Environment Secretary, and they didn't. If San Juan hadn't brought the issue to the president's attention, it obviously came from someone who understood and had very clear and specific information about mining and glaciers. It was someone who knew exactly what they were talking about, and it had to be someone with direct access to the president.

The most interested stakeholder adversely affected by Argentina's glacier protection law was Barrick Gold. Gold and silver prices were already on a steady and historic surge in value. With declared reserves of 15 million ounces of gold and 675 million ounces of silver, Pascua Lama alone was worth a whopping US$18 billion in 2006 and had already climbed to US$23 billion in 2008. By 2011, the peak year of the gold and silver surge, Barrick's bet on Pascua Lama had already topped US$50 billion[36] (Table 5.1).

Barrick certainly had a lot at stake at both Veladero and Pascua Lama. It doesn't take much imagination to presume that Peter Munk, who had always had direct access to Argentina's presidents, would not easily walk away from US$50 billion of potential value, particularly considering that the company had already spent probably between US$3 and 5 billion in project preparation. Nor would Munk, known for his very aggressive personal tactics with his board of directors and generally as individual leader of the world's largest gold mining company, turn over the fate of Pascua Lama to the negotiating

TABLE 5.1 Estimated Value of Pascua Lama as Gold and Silver Prices Escalated

Year	Ave. Gold Price (US$/ounce)	Value (billions US)	Ave Silver Price (US$/ounce)	Value (b./QM. US$)	Total Value Gold + Silver (billions US$)
2006	650	9.8	12	8.1	17.9
2007	750	11.3	14	9.5	20.8
2008	850	12.8	15	10.1	22.9
2009	1,000	15	15	10.1	25.1
2010	1,250	18.8	22	14.9	33.7
2011	1,650	24.8	35	23.6	48.4
2012	1,650	24.8	31	20.9	45.7
2013	1,400	21	24	16.2	37.2
2014	1,300	19.5	20	13.5	33.5

skills of San Juan's political leaders in hopes that they could convince the president to veto the glacier law. The stakes were simply too high.

Munk had a one-on-one relationship with the Kirchners. It was closer even than the relationship he had with San Juan's Governor José Luis Gioja. It was not the first time he spoke directly to the most powerful couple in Argentina. It is more than likely that on November 5 or 6, Peter Munk, president of Barrick Gold, convinced Cristina Fernandez de Kirchner, Argentina's president, to veto the world's first glacier protection law in order to save Pascua Lama.

On November 10, 2008, sometime in the late afternoon, three weeks after the glacier law received unanimous approval from both houses of Congress with no opposition on record, Carlos Zannini, the president's legal advisor, finished drafting the presidential veto, to be published the next day in an official bulletin. Cristina Fernandez de Kirchner had vetoed the glacier protection law for Peter Munk.[37,38]

The news was held in complete secrecy that evening, known only to the president, her husband (the former president), and Zannini while they planned out their next steps. Picolotti was in Brazil for an environmental ministers' meeting and had not been informed by the president; in fact, the two would never speak again. By morning, the news of the veto would be out and fallout would ensue.

Notes

1. For the full comment by Argentina's president on glaciers: https://www.youtube.com/watch?v=88gsZssOx4s.

2. For the original 2008 version of the Argentine national glacier protection law (this is not the law currently in effect, but rather the vetoed law), see http://wp.cedha.net/wp-content/uploads/2012/06/4777-D-2007-Glaciares.pdf.

3. See Clarín (October 23, 2008): http://edant.clarin.com/diario/2008/10/23/index_diario.html. and La Nación (October 23, 2008): http://servicios.lanacion.com.ar/archivo-f23/10/2008-c30.

4. *Diario de Cuyo*: http://www.diariodecuyo.com.ar/home/new_index.php?fecha=2008-10-23.

5. For the El Pachón mining project, see 31°44′54.37″ S 70°25′40.99″ W.

6. For Xstrata Copper's Geomorphological Map of El Pachón, see http://wp.cedha.net/wp-content/uploads/2011/09/MAPA-2.6.1-AM-GEOMORFOLOGIA.jpg.

7. See CEDHA's report on El Pachón's impacts to glaciers and periglacial areas at http://wp.cedha.net/wp-content/uploads/2011/08/Glaciar-Impact-Report-el-pachon-xstrata-FINAL-english-version-may-23-2011.pdf.
Google Earth images impacts:

> Road impact to active rock glacier: 31°44′50.39″ S 70°27′07.71″ W
>
> Rock glacier at pit area: 31°44′57.46″ S 70°25′40.32″ W.

8. For the Study of the Conconta Glaciers: http://mineria.sanjuan.gov.ar/pascual-ama/Glaciarconconta0708.pdf.

9. For the glacier mapping and infrastructure review, see http://wp.cedha.net/wp-content/uploads/2011/11/Informe-Glaciares-Lama-Veladero-Espizua-2006.pdf.

10. For Gioja on Sarmiento, see http://mineria.sanjuan.gov.ar/politica/politica.php.

11. For bilateral mining investment treaty, see http://www1.hcdn.gov.ar/dependencias/cmineria/ley_25_243.htm

12. See http://en.wikipedia.org/wiki/El_Indio_Gold_Belt.

13. See Bonasso, 2011, p. 19.

14. The El Indio mining project is about 28 km (17 mi) south of present-day Pascua Lama and is owned and was operated by Barrick Gold in the 1990s; see 29°34′53.80″ S 69°57′04.57″ W.

15. See 29°22′45.83″ S 69°56′51.78″ W.

16. See Barrick Gold History (1994): http://www.barrick.com/company/history/default.aspx.

17. Pascua Lama is literally "on the border," about 5 km northwest of the Veladero mine site. See 29°19′18.35″ S 70°01′04.22″ W.

18. Tapscott, 2007, pp. 8–9.

19. Los Azules (McEwen): 31°03′47.42″ S 70°13′57.21″ W.

20. For Google Earth-visible road impacts to terrain in glacier areas, see

> El Pachón (Xstrata-Glencore): 31°44′53.36″ S 70°25′51.24″ W
>
> Altar (Stillwater): 31°28′41.99″ S 70°28′58.56″ W
>
> Los Azules (McEwen): 31°03′47.42″ S 70°13′57.21″ W
>
> Del Carmen (Malbex): 30°00′45.69″ S 69°54′47.64″ W
>
> Pascua Lama (Barrick Gold): 29°19′08.80″ S 70°00′37.31″ W
>
> Las Flechas (NGX Resources): 28°42′53.09″ S 69°39′24.82″ W
>
> Vicuña (NGX Resources): 28°26′46.45″ S 69°36′31.91″ W
>
> Josemaria (NGX Resources): 28°25′20.34″ S 69°33′01.71″ W
>
> Filo del Sol (NGX Resources): 28°26′57.89″ S 69°38′41.46″ W
>
> Cerro Verde (Anglo American): 28°19′18.93″ S 69°27′52.28″ W
>
> Filo Colorado (Xstrata-Glencore): 27°20′04.71″ S 66°13′22.36″ W
>
> Unidentified: 29°09′36.50″ S 69°51′59.73″ W.

21. Considering that the most obvious limit date for the veto was November 5, had serious contention arisen before that date, the president's legal advisory would have

ensured that the veto was published by the close of business that day. Given that the veto occurred on November 11, a Tuesday, we can presume that the call came some time between Wednesday, November 6 and Monday, November 10.

22. BGC, 2009, p. 32.

23. For the *Perfil* cover story, see http://www.perfil.com/politica/Con-sangre-fria-Cristina-veto-la-proteccion-de-glaciares-20081113-0025.html.

24. For the original picture run with the *Perfil* story, see http://1.bp.blogspot.com/_J8JcMqP3uWk/SR1yehM32eI/AAAAAAAAA2k/rschj2dysKA/s400/Glaciares+Cris.jpg.

25. See http://www.perfil.com/politica/Con-sangre-fria-Cristina-veto-la-proteccion-de-glaciares-20081113-0025.html.

26. See http://www.diariodecuyo.com.ar/home/new_noticia.php?noticia_id=414581.

27. See http://www.boletinoficial.gov.ar.

28. See http://www.greenpeace.org.ar/blog/veto-presidencial-a-la-ley-de-proteccion-de-glaciares/66/.

29. See http://www.boletinoficial.gov.ar.

30. See   http://www.infoleg.gov.ar/infolegInternet/anexos/145000-149999/146980/norma.htm.

31. Rock glacier impacted by preparatory road for the Agua Negra international pass: 30°14′44.20″ S 69°50′20.08″ W.

32. For a report on glacier and periglacial areas surrounding Barrick Gold's Veladero and Pascua Lama projects, see http://wp.cedha.net/?p=12411&lang=en.

33. For a report on glacier and periglacial areas surrounding Xstrata Glencore's El Pachón project, see http://wp.cedha.net/?p=6713.

34. For a report on glaciers and periglacial areas in the Los Azules project (McEwen Mining), see http://wp.cedha.net/?p=9889&lang=en.

35. The president's meeting agenda can be searched by date at http://audiencias.jgm.gob.ar.

36. Value is calculated by approximate averages and Barrick Gold's declared reserves; see

(1) http://www.kitco.com/charts/historicalgold.html; for gold prices and
(2) http://www.barrick.com/operations/argentina-chile/pascua-lama/default.
aspx; for Barrick Gold's declared Pascua Lama reserves.

37. For the original text of the veto, see http://www.infoleg.gov.ar/infolegInternet/anexos/145000-149999/146980/norma.htm.

38. One point of contention that circulated at the time of the veto was in regards to whether the veto itself came late, surpassing the ten-day limit the president had to veto the law, in which case the veto was itself unconstitutional. Congress passed the law on Wednesday October 22, 2008. Ten working days from that date takes us to the end of the business day on Wednesday November 5, 2008. On November 6, 2008, the glacier law would be officially in force and the president could no longer veto it. The official bulletin published Presidential Decree 1837/2008, officially sending the law back to Congress on November 11, 2008, three working days late. Whether this point could be argued is today a matter of irrelevant speculation because the glacier law was deemed vetoed and Congress later passed a revised version in 2010. At the time, however, the issue provoked heated debate within the Environment Secretariat.

{ 6 }

Life Without Glaciers

Climate change is accelerating glacier melt. In the same month that this book first went to the editors, scientists reported the irreversible collapse of a massive portion of the West Antarctic ice sheet at Thwaites Glacier.[1] Thwaites Glacier had already been news years earlier when a massive piece of ice 50 km (31 mi) wide, nearly 150 km (93 mi) long, and 3 km (1.8 mi) thick—that's more than thirty city blocks of ice stacked on top of each other—broke off into the ocean and became Thwaites *iceberg*. Imagine an ice cube about seventy-five times the size of Manhattan Island floating away into the ocean. With the new reported collapse, the entire West Antarctic ice sheet has now entered into a rapid and irreversible melting phase (Figure 6.1).

Thwaites Glacier, as well as others in the Amundsen Bay sector, such as the Pine Island Glacier,[2] form part of a massive ice sheet on Antarctica that is falling to pieces. This is an ice sheet larger than France, Spain, Germany, and Italy combined, and it contains nearly 30 million cubic kilometers of ice (that's about seven million cubic miles; Gosnell, 2005, p. 109). As these colossal ice bodies fall into the warmer ocean, they will begin to melt away, eventually raising global sea levels by about 1.2 meters (4 ft) (Figure 6.2).

The breakdown has come much more quickly than expected and has now entered into an irreversible "runaway process." What should have taken thousands of years in the natural evolution of things will now be complete in just centuries or less. The Pine Island Glacier is a long, flowing ice stream in the northeastern part of Amundsen Bay, and it is the world's greatest contributor of ice to the oceans through melting and calving processes. It is also another of the glaciers at risk of collapsing entirely into the ocean. Thwaites Glacier's collapse is an indicator that the whole ice sheet may be in imminent danger. The Pine Island Glacier, which is 160,000 km^2 (62,000 mi^2), or two-thirds the size of the United Kingdom, has displayed a massive crack for a number of years, and its collapse is also imminent (Figure 6.3).

DROUGHT BREAKING UP
GLACIERS IN THE ALPS

Peaks Present Appearance as in
August—Water in Highlands
Lacking for Cattle.

Copyright, 1921, by The New York Times Company.

Special Cable to THE NEW YORK TIMES.

BERNE, March 16.—The remarkable drought in the Alps and throughout Central Europe continues. There is a poor prospect of Spring grass for cattle, the ground being too hard for grass to grow.

Navigation on the lakes is almost suspended, and the water supply in the highlands is so scarce that cattle.must be driven down long distances to drink.

Orders have been issued for the public to be careful about lighting fires in the open country, especially near woods, for with the present dryness the danger of fires spreading is great.

Many glaciers have become broken up owing to the comparative lack of snow throughout the Winter to cement the ice together. Some glaciers which formerly were almost flat, and over which cattle were driven, are now more like moraines, with huge, yawning crevasses and massive blocks of ice.

High peaks have their August rather than their March aspect, and some venturesome climbers who have ascended mountains recently found there was less snow on the summits than in a normal Summer.

FIGURE 6.1 New York Times *article on climate emergency from 1921.*
Source: New York Times, March 16, 1921.

Christopher White, author of *The Melting World*, summarizes some salient and startling statistics:

> as temperatures climb, ice is vanishing at rates far exceeding natural attrition. The North Pole and surrounding waters will be free of summer pack ice for the first time in human history, as early as 2020. The snows of Kilimanjaro will die by 2033. What is most startling is the timescale of the melting: What normally happens over thousands of years is now happening in less than a century. It's as if—after a thousand years—the Great Pyramids crumbled to dust overnight. (White, 2013, p. 14)

In addition to the rising temperatures that are melting glaciers, the emissions of black carbon (soot) into our atmosphere from industry, diesel engines, and bad combustion from furnaces or stoves also affects glaciers. Soot from such emissions can cause significant changes to a glacier's reflective capacity (its albedo), causing the ice to absorb heat from the atmosphere instead of reflecting it. Just a few parts per billion of soot can reduce the reflective capacity of pure white glaciers by up to 15%. According to a NASA scientist, sooty smoke and industrial pollution from South Asia, for example, is the main reason for the retreat of Tibetan glaciers over the past decade.[3]

FIGURE 6.2 *Thwaites Glacier in Antarctica shows crevasses from a disintegration process already under way. Melt from the Western Antarctic Ice Sheet will cause sea level to rise 1.2 meters (or about 4 feet).*
Source: James Yungel, NASA. GIS: 73°55′56.95″ S 108°01′38.03″ W.

FIGURE 6.3 *Pine Island Glacier in Antarctica breaks open (the crack is shown in a smaller version by satellite image) due to climate change.*
Source: NASA. GIS: 74°33′38.08″ S 100°40′45.13″ W.

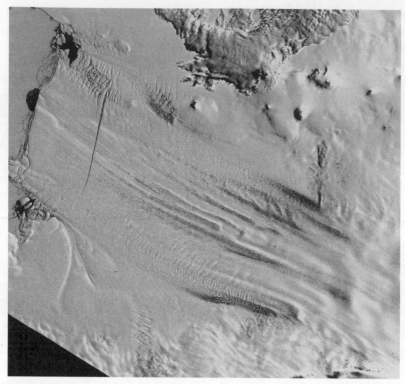

FIGURE 6.3 *(Continued)*

Debris on glaciers, including salts, soot, pollen, and other matter, sometimes dissolve into the glacier in the form of cavities called *cryoconites* (holes), which begin to drill into the ice in the form of vertical tunnels many meters deep. These form vertical water flows that contribute to glacier melt, but that also spawn bacteria, fungi, and other microbes that stain the glacier and also alter albedo, which in turn further accelerates melt.[4]

We've all become almost desensitized to this sort of climate "event." There are even a few climate change skeptics out there who point to a handful of glaciers that are actually growing to suggest that this is proof *that there is no proof* that our climate is changing and that glaciers are receding quickly. And there are those who simply say that it's a natural phenomenon. But, for the most part, we are all finally realizing that we are witnessing an irreversible human-induced deterioration of our climate and the collapse of our glaciers far more quickly than Mother Nature would have it.

We now know beyond any reasonable doubt that human activity is causing serious global ecological imbalances. We hear, see images of, or have to face recurring climatic emergencies and disasters, including droughts, heat waves, hurricanes, and other severely inclement weather. Everyone has seen the comparative pictures of glaciers in the Alps, Greenland, Norway, or Patagonia receding, decade after decade, sometimes by many dozens of meters or even by several kilometers. These are now amplified by the sort of events we are seeing in Antarctica.

As depicted in the recent documentary *Chasing Ice* by James Balog, millions of tons of glacier ice washes up along the world's shores like dead fauna, melting away into the shoreline. The melting of the polar ice caps is sending icebergs of all sizes, including massive ice sheets, into the oceans. These slowly melt away, causing our seas to rise, our freshwater supply to diminish, and our ecological systems to enter into profound disequilibrium.

The collapse of Thwaites Glacier and other ice bodies in the Amundsen Bay region in the Western Antarctic Ice Sheet will cause the sea level to rise by 1.2 m (4 ft). That may not sound like much, but it is, and for many people around the world on every continent living on coastlines, it could mean relocation or imminent flooding. Not only will coastlines around the world move inward simply because the sea is rising, but more frequent raging storms will cause surge events that magnify this rise, reaching far higher than the mere 1.2 m (4 ft) of general rise. Furthermore, if the entire Western Antarctic Ice Sheet completely destabilizes—which, at this point, is not a theory but a reality in progress, says Ian Joughin, a University of Washington glaciologist and one of the scientists studying the Western Antarctic Ice Sheet[5]—then the impacts could be catastrophic, raising sea levels 3–4 m (10–13 ft) with surges of even higher levels.

Eventually, we could be looking at a melting of the entire Antarctic Ice Cover, which would be a cataclysmic scenario, with sea level rises of more than 60 m (200 ft).[6] That's higher than a twenty-story building. According to a recently published hypothetical sea level rise map by National Geographic,

the entire Atlantic seaboard would vanish, along with Florida and the Gulf Coast. In California, San Francisco's hills would become a cluster of islands and the Central Valley a giant bay. The Gulf of California would stretch north past the latitude of San Diego—not that there'd be a San Diego. The Amazon basin in the north and the Paraguay River Basin in the south would become Atlantic inlets, wiping out Buenos Aires, coastal Uruguay and most of Paraguay, . . . in Egypt, Alexandria and Cairo would be swamped by the intruding Mediterranean. . . . London? A memory. Venice? Reclaimed by the Adriatic Sea. . . . the Netherlands will have long since surrendered to the sea and most of Denmark will be gone too. . . . Land now inhabited by 600 million Chinese would flood, as would all of Bangladesh . . . and much of coastal India. . . . [Australia] would gain an inland sea—but it would lose much of the narrow coastal strip where four out of five Australians now live.[7]

Would your home (or that of a family member or friend) be covered by seawater if this occurred? Or would your hometown have to harbor environmental climate refugees from such events?

We all have heard of the melting and retreating mountain glaciers. These are glaciers that are more immediately dear to us, even if we may know little about them. Many of us, knowingly or not, depend on glacier meltwater for our daily water consumption, for the agriculture that feeds us, and for the industries that put money into our pockets and maintain and improve our quality of life.

Melting mountain glaciers, because of their smaller volume, contribute less to sea level rise than do the melting polar ice caps, but they are putting an enormous strain on our ecosystems. And yet, incredibly, we have left this critical source of our most important natural resource completely unprotected.

The changing climate for glaciers and periglacial areas means that warmer temperatures are slowly creeping up mountainsides. If once glaciers could be found at 4,500 m (14,700 ft) in a place like the Central Andes, soon, they'll be at 5,000 m (16,400 ft), and then 5,500 m (18,000 ft), and if the mountains don't go any higher than that, neither will the glaciers. That's what we see occurring in Papua Indonesia, for example, where some of the last remaining equatorial glaciers are scurrying up the mountainside seeking cooler elevations. "Scurrying" is a euphemism in this case, since what is really happening is that the lower portions of the glaciers are melting off. They are reducing in size and only their highest portions are surviving.

In the Indonesian case, the melt fringe is at 4,500 m (i.e., any snow/ice below that melts away), but the mountain is only 4,900 m (16,000 ft) high. These glaciers of the Puncak Jaya Range[8] are desperately creeping uphill in search of a colder environment, but there is nowhere left to go; very soon, they will simply disappear. In the Spanish and French Pyrenees, glacier cover is practically gone. In places like the Sierra Nevada in California, we see many glaciers that are small in size but very important to summer glacier melt. These are nestled in an ever-decreasing glacial belt ranging from 3,500 to 4,500 m (11,500 to 14,700 ft).[9] One of the better-known glaciers in California's Sierra Nevada, the Lyell Glacier, has only about 300 m (980 ft) left to live. In Montana, home to famous Glacier National Park, glaciers have only 400 m (1,300 ft) left of living space, the park will soon become *Glacier-less* National Park.

Christopher White recounts the alarming decrease of Montana's glaciers in his book *The Melting World*, where of 150 glaciers registered in 1850, by 1966, thirty-seven survived. In 2008, twenty-seven were left. Today, only twenty-five survive (White, 2013, p. 7 and inside cover).

Having recently done some work mapping relatively small glaciers of the Aconquija Mountains in Catamarca and Tucumán Provinces of Argentina, utilizing images that dated back to 2008 or so, I was absolutely startled when putting together the following table showing what's left of upward survival crawl space for glaciers. As I revisited the Aconquija Mountains on Google Earth, I noticed that Google has updated images of the area to September of 2013; to my sad surprise, nearly all visible glaciers have fully vanished. Since I started working on glacier advocacy six years ago, these are the first glaciers that I once saw and studied and that have now disappeared.

Table 6.1 shows some elevation survival spaces that are left for a number of glaciers at risk of disappearing around the world. It's evident that they're being cornered into oblivion. With at most 1,000 m (3,300 ft) of upward space to survive for the first examples I've cited, they are all at risk, but, as you go down in the risk list, the severity of the risk is increasingly apparent, with the Mexican glaciers near the country's capital worst off with only 200 m (650 ft) left of survival crawl space before they too disappear.

It's a bit unnerving to realize that we will all see, during our lifetime, areas of the world that completely lose glacier cover, including smaller mountain glaciers in Europe, Africa, Indonesia, and Latin America (including probably full glacier loss in Venezuela and Colombia and certainly in regions of Argentina, Chile, Peru, Ecuador, and Bolivia).

But even the larger mountain glaciers are suffering significant retreat. In the Himalayas, new studies show that glaciers in the Mt. Everest region, for

TABLE 6.1 Survival crawl space for glaciers around the world

Glaciers	Elevation	Survival Space	GPS
Antisana Glacier (Ecuador)	4,500–5,700	1,200 m	0 29′22.31″S 78 08′03.86″W
Huayna Potosi (Bolivia)	5,000–6,000	1,000 m	16 15′48.98″S 68 09′18.93″W
Mt. Ararat (Turkey)	4,100–5,100	1,000 m	39 42′21.34″N 44 18′06.31″E
Tian Shan Mts. (Kyrgyzstan)	3,700–4,700	1,000 m	41 52′40.72″N 78 16′37.63″E
Mt. Damavand (Iran)	4,700–5,600	600 m	35 57′17.51″N 52 06′37.02″E
Cocuy Mts. Glaciers (Colombia)	4,800–5,300	500 m	6 29′44.69″N 72 18′11.74″W
Rwenzawi Glaciers (Uganda)	4,800–5,100	500 m	0 23′04.36″N 29 52′22.58″E
Mt. Kacker (Turkey)	3,200–3,700	500 m	40 50′08.52″N 41 09′07.41″E
Estrecho, Andes Mts. (Chile)	5,050–5,500	450 m	29 17′46.78″S 70 00′55.42″W
Puncak National Park (Indonesia)	4,500–4,900	400 m	4 05′09.36″S 137 10′56.82″E
Glacier National Park (USA)	2,500–2,900	400 m	48 50′57.09″N 113 47′05.96″W
Kilimanjaro (Tanzania)	5,400–5,800	400 m	3 03′15.08″S 37 20′43.36″E
Mt. Famatina Glaciers (Argentina)	5,700–6,100	400 m	29 01′13.24″S 67 50′02.60″W
Mt. Kenya Glaciers (Kenya)	4,700–5,100	400 m	0 09′06.49″S 37 18′33.71″E
Pyrenees Glaciers (Spain)	3,700–4,100	400 m	42 40′51.61″N 0 02′05.25″E
Lyell Glacier (California, USA)	3,600–3,900	300 m	37 44′37.88″ N 119 16′37.62″W
Vignemale Glacier, Pyrenees (France)	2,900–3,200	300 m	42 46′14.03″N 0 08′39.70″W
Iztaccihuatl (Mexico)	5,000–5,200	200 m	19 10′46.10″N 98 38′29.81″W

example, have lost about one-eighth of their mass over the past fifty years, while the snowline has crept up during that time, by about 180 meters (590 ft). Small glaciers have shrunk during this period about 43% since the 1960s.[10] According to the Intergovernmental Panel on Climate Change (IPCC), glaciers have been losing mass in Bhutan, China, India, Nepal, and Pakistan for the past five decades.[11]

Western North America is also suffering rapid glacier and periglacial environment retreat. The past two years, for example, have been especially dry years for California, with very little snowpack to provide water runoff for dry months. Glacier disappearance is impacting forests, drying up terrain, and leading to higher incidences of fires.[12]

In Africa (Uganda, Kenya, and Tanzania; Figure 6.4) and Papua Indonesia (Figure 6.5), we see these countries' last remaining glaciers shriveling up and vanishing before our eyes; these are two areas of the world where glaciers are all but extinct. These glaciers will probably not reappear for many thousands of years, perhaps during the next global glaciation, if ever. Australia, except for some magnificent glaciers in the South Indian Ocean on McDonald and Heard Islands,[13] is already glacier-free.

Latin America is another region that is suffering and will continue to suffer greatly from glacier retreat. We are already seeing the rapid disappearance of glaciers in the Central Andes, particularly in Argentina, Peru, and Bolivia.

FIGURE 6.4 *Some of the last remaining glaciers of the African Continent, in Uganda, at 5,109 m (16,800 ft), on Mt. Stanley Rwenzori Mountains.*
Source: Klaus Thymann. GIS: 0°23′45.94″ N 29°53′25.69″ E.

FIGURE 6.5 *One of the last remaining glaciers in Indonesia, the Carstenz Glacier in Papua.*
Source: Paul Warren. GIS: 4°04'57.83" S 137°10'38.46" E.

In Bolivia, a glacier like the Chacaltaya,[14] which used to house the world's highest ski resort at 5,300 m above sea level (17,400 ft) and which rests above the country's capital, La Paz, home to over 2 million people drinking glacier water, has all but disappeared, with amazingly high melt rates accelerating threefold over the past decade.[15]

For periglacial areas that are frozen year-round, the same predicament holds. For mountains with large frozen swaths of lands above 3,000 m (9,800 ft), the lower limit of this frozen terrain (the limits providing cyclical meltwater year round) will begin to creep upward to 3,500 m (11,500 ft), then to 4,000 m (13,100 ft). Eventually, when the mountains run out of altitude, there won't be any frozen lands to store water during the warmer seasons and during drier years.

The problem with disappearing glaciers and periglacial areas with permafrost is multifold. First, our ecosystems will cease storing water for use in warmer months. Although snow may still fall during the winter, it will not survive beyond spring or early summer, and the mountain environments will dry up during summer and fall, until the cycle begins again the following spring. This will create months of much drier weather during which rivers will dry up, thus disturbing natural ecological balance. Some glacier specialists are forecasting cataclysmic impacts, suggesting that even massive rivers such as the Yellow River, the Yangtze, and the Ganges could dry up in the summertime and become seasonal.[16] Glacier-dependent basins will be less able to rely on glaciers for their water supply during dry seasons or especially dry years or series of years. In South Asia, 500 million people will be impacted by glacier-related problems; in China, 250 million, as reported by the IPCC.[17] In Peru, the melting of the Quelccaya Ice Cap and its outlet glaciers has already prompted water shortages in the famed capital of the Inca empire, Cuzco, a city with some 400,000 people. There, residents have had to resort to periodic water rationing.[18]

Furthermore, as vanishing glaciers uncover once ice-protected lands, exposure of these barren lands to the elements will increase erosion rates, causing further impacts to local hydrological and land systems. Glacier retreat will impact regional species richness, according to the IPCC, as a study focused on three diverse areas of the planet (Ecuador, the European Alps, and Alaska) has shown.[19] All of our ecosystems will have to adapt to these permanently altered climates.

Agricultural lands that once received water year-round will have to ration water because our ecosystem will no longer naturally provide it. Natural lakes and artificial dams downstream from cold mountain environments in places like the Himalaya, the Central Andes, or the Sierra Nevada will have a harder time keeping full, thus placing strains on the drinking water and hydroelectric power supply.

Whether this takes ten years, fifty, or one hundred years, is a matter of geographical and climactic circumstance, but glacier and periglacial systems are vanishing before our eyes, and this will bring inevitable impacts.

James Balog, award-winning photographer and cryoactivist, has documented glacier retreat like no one else has, showing numerous series of still images taken multiple times daily for several years. He has compiled them into film-like footage of dozens of glaciers around the world, documenting progressive and alarming glacier retreat. His fabulous documentary film *Chasing Ice*[20] is a must see for everyone in order to raise our consciousness of a rapidly and dangerously changing planet. It is a chilling depiction of the decline of some of our most valuable hydrological towers.

Many of us have empathetic but short-lived emotional reactions to these images and to stories of the cataclysmic collapse of our cryosphere (our frozen world). We sense a degree of urgency and despair when we hear of this collapse, and we have the ill feeling that we should be doing something to save melting glaciers. But, in the end, we are either skeptical about the veracity of the information, or we see the problem as simply too overwhelming; so, we come up short with answers to address the daunting challenge that we now face. Or we simply don't envision any ideas or actions to actually take steps toward protecting this vital part of our environment. Part of the problem lies in that we probably can't truly understand how the retreating glaciers of Svalbard, Norway,[21] or of National Glacier Park in Montana,[22] or disappearing Bolivian glaciers,[23] or the breakoff of the Thwaites Glacier in Antarctica affect our immediate lives. Even if we could draw that connection, we probably wouldn't know where to begin to take action to save our world's melting ice.

Some questions we should be asking ourselves are:

- Should I care?
- How will glacier and periglacial melt impact my life?
- What are the main risks?
- Can these tendencies be reversed?
- What are other people doing about it?
- And can I do anything about it?

Chapter 8 looks at what people around the world are doing to address some of these impacts.

Notes

1. Thwaites Glacier is at 74°35′19.64″ S 107°50′25.22″ W; news on the collapse of the ice sheet can be found at http://news.nationalgeographic.com/news/2014/05/140512-thwaites-glacier-melting-collapse-west-antarctica-ice-warming/.

2. See Pine Island Glacier at 74°59′01.91″ S 100°20′51.07″ W.

3. See http://earthobservatory.nasa.gov/Features/PaintedGlaciers/page3.php.

4. See http://earthobservatory.nasa.gov/Features/PaintedGlaciers/page3.php.

5. See http://www.theguardian.com/environment/2014/may/12/western-antarctic-ice-sheet-collapse-has-already-begun-scientists-warn.

6. See http://science.howstuffworks.com/environmental/earth/geophysics/question473.htm.

7. See http://ngm.nationalgeographic.com/2013/09/rising-seas/if-ice-melted-map.

8. See 4°05′09.36″ S 137°10′56.82″ E.

9. See, for example, Lyell Glacier at 37°44′37.88″ N 119°16′37.62″ W.

10. See http://moa.agu.org/2013/media-center/press-item/scientists-find-extensive-glacial-retreat-in-mount-everest-region/.

11. IPCC, 2014, chapter 3, Freshwater Resources, p. 13.

12. IPCC, 2014, chapter 4, Terrestrial and Inland Water Systems, p. 47.

13. For Australia's glaciers at Heard and McDonalds Islands, see 53°07′36.98″ S 73°31′54.62″ E.

14. See 16°20′44.93″ S 68°07′24.35″ W.

15. Seehttp://www.treehugger.com/natural-sciences/iconic-bolivian-glacier-disappears-melting-increased-three-fold-in-past-10-years.html.

16. See http://www.treehugger.com/clean-technology/global-warming-melting-glaciers-shrinking-harvests-in-china-and-india.html.

17. See http://www.ft.com/intl/cms/s/0/10e65b88-220a-11e3-9b55-00144feab7de.html#axzz31BGDhOW9.

18. See White, 2005, p. 133.
19. IPCC, 2014, chapter 4, p. 68.
20. See http://www.chasingice.com.
21. See 78°36.876′ N 19°8.159′ E.
22. See 48°45.242′ N 113°43.709′ W.
23. See 16°24.746′ S 67°57.117′ W.

Resurgence

In the days following the president's veto of the glacier protection law on November 11, 2008, the executive power was again in political turmoil. Congress was in an uproar with the veto in part because the congressional majority held by the administration during the earlier months of the presidency was fledgling, and President Cristina Fernandez de Kirchner was slowly losing her political capital to an increasingly empowered opposition, one eager to bring controversial issues to the forefront of the political arena to further weaken her presidency. The glacier law, and specifically the Barrick Veto, played well into this objective.

Following the veto, the Natural Resource Commission of the Lower House of Congress convened a public meeting for November 18 to discuss how to respond to the veto.[1] Only seven of the thirty-one members of the commission (mostly opposition members) showed, shy of the minimum quorum necessary to take official action. The meeting was held anyway. Several environmental organizations were invited. Marta Maffei, the original author of the glacier law, contributed to the discussion, as did Ricardo Villalba, the director of the Argentine Institute for Snow Research, Glaciology and Environmental Sciences (IANIGLA). Villalba made unusually strong public statements in defense of the glacier law, indicating that the scientific community was "shocked and saddened" by the president's decision to veto it.[2] Despite not having a quorum, the commission set out an action plan to bring back the law. They were aware of the official party's intention to develop a new version, one that would appease the mining sector. Clearly, it would be a utilitarian version similar to what had been proposed in Chile. However, those present that day insisted that the same law, just as it had been passed, should be resubmitted, using the exact same text.[3]

In the meantime, the president's office was deliberating on what to do about the fallout. Cristina Fernandez vetoed the glacier protection law because it was incompatible with Barrick Gold's flagship project, Pascua Lama, valued by conservative estimates of the time at upward of US$20 billion. As gold prices continued to rise over the next several years, the value of Pascua Lama would surpass US$50 billion by 2011. Had the glacier law survived, Pascua Lama would run into serious legal conflict. Other large mining projects in the pipeline such as El Pachón, Los Azules, Altar, Famatina, Josemaría, Las Flechas, Vicuña, Filo del Sol, Del Carmen, and a slew of others in exploration in the highest sections of the Andes were also incompatible with the vetoed law.

Mining investments were a top priority for the administration. Not only would they provide job creation for impoverished provinces, but, more importantly, they were a very important source of much-needed tax revenues for the national coffers.

For the president, and for the mining sector more generally, an environ-mental services view of the glacier and periglacial environment would have made more sense, instead of the strictly conservationist view put forth in the just-vetoed glacier law. But this was not how the bill came together, and the veto was directed precisely at the conservationist nature of the law.

The president already had a plan in mind to follow up the veto, one aimed at regaining some political capital while at the same time showing that she did indeed want to protect glaciers. The plan was spelled out in Article 3 of the veto:

> an invitation is extended to the Governors, and to the national and provin-
> cial Senators and Deputies of the mountainous provinces, to construct an
> interdisciplinary forum to discuss the measures to adopt for the protection
> of glaciers and of the periglacial environment.[4]

To do this, she needed a credible emissary to carry negotiations forward. Public opinion was against her. In the public mindset, she had vetoed the gla-cier law to favor the mining industry. As a result, her Environment Secretary was on the verge of resigning. Environmental organizations, the political opposition in Congress, and the media came out strongly against the veto and against the president. She couldn't prop up a mining industry figure (such as the Mining Secretary) or anyone else who would seem aligned with the sec-tor that had leveraged the veto. She and Carlos Zannini, her top legal advisor, discussed the options and both agreed that the best person to lead the effort would be Romina Picolotti.

Picolotti was the only executive branch staff member who had accompa-nied the law through the congressional process. Her imminent resignation was publicly associated with the fact that she stood firmly behind the glacier law and opposed the veto. If she could be convinced to stay on, precisely to defend the glacier law, it would be a gesture from the president's office that the president did in fact support glacier conservation and was trying to seek balanced legislation that would both protect glaciers and, at the same time, allow for industrial activity already planned near glaciers and periglacial areas. This would be a more balanced approach to sustain the mining-driven development plans of provinces such as La Rioja, Salta, Jujuy, Catamarca, and San Juan.

In retrospect, the choice of Picolotti to convene and lead the "Glacier Bill 2.0" process could also be seen as an effort to fully shift the political failure of the veto to her. Given that the president and Zannini presumed she was on her way out anyway, it would be better that Picolotti be pushed out by the president for failing to come up with a good new glacier protection law than to have her resign due to the veto of the first law.

A third interpretation could be simply that the president wanted to rid herself of this controversial issue as soon as possible and that a definitive transfer of discretion over the outcome of the glacier law debate to Picolotti (or to anyone) made sense. Cristina Fernandez paid a price for the veto. We presume she vetoed the glacier law to save a multibillion dollar investment that would provide significant revenues to the national and provincial governments (although, at the time, there were many accusations from the opposition that Barrick Gold paid for the veto). But her defense of the project would not be unwavering, particularly if it expended too much of her progressively diminishing political capital. The veto went with the indication that she would work toward another drafting of the law. And, in subsequent comments she made to the media, she also indicated that she would not veto the law a second time if it were passed again.

In this regard, she was saying to the mining sector and to the provinces, "I had to step in to clean up your mess because you didn't see this law coming, but I won't do it again!" Everyone was now at the table, they could all contribute to the debate and leverage their influence accordingly. The new glacier law debate, unlike earlier, was completely in the open before a sensitized audience; whatever outcome resulted, the president promised to adhere to the congressional vote, even if it were against mining interests. Two years later, that is exactly what occurred. The president kept her word, allowing the Glacier Bill 2.0 to go into force.

Daniel Filmus, former Education Minister and now senator, was also brought into the government sphere of negotiations on the Glacier Bill 2.0. Filmus was Nestor Kirchner's minister, and, like Picolotti, he aligned politically with the former chief of cabinet, Alberto Fernandez. However, Cristina Fernandez de Kirchner had not called him to her cabinet and instead, he had secured a senatorial seat for the term.

Filmus was a Peronist, of the president's party, and headed the Environmental Commission of the Senate. He was the likely candidate to take the process forward as representative for the executive. Some speculation exists as to whether he did this of his own volition, perhaps because he had some affinity with the issue, or because he was specifically asked to do so by Zannini, the president's chief legal advisor. It may have been a combination of the two. He saw his role as one of trying to bring some rationality to the discussion and debate, attempting to harmonize glacier protection with the interests of the provinces and with the mining activity and investments that were the priority of several of the involved provinces. For the mining sector, Filmus's presence was a guarantee that their concerns would be heard and that they would have direct input into the wording of the new text. This is precisely what occurred. The environmental movement, however, saw Filmus as the enemy, standing against the more conservationist Maffei Law. That dichotomy would reign throughout the glacier law debates and would carry on through to the congressional vote that would take place nearly two years later.

What is less clear is to what degree Filmus was actually coerced by the mining sector or by the president's office in the drafting of the text. Filmus did consult with a variety of actors on how the text should read to best accommodate mining interests while still providing a workable framework for glacier protection. And, as the text commandeered by Filmus evolved, it clearly took on an "environmental services" character, which was to be expected if it were to accommodate a broader set of stakeholder interests.

Counterintuitively, however, and despite what most of the public believed, an analysis of the events that were to follow suggests that, very likely, Filmus did not receive instruction from, or even consult with anyone from the president's office on the evolution of the glacier protection bill. It is not clear either that he received any direct pressure from mining interests. The version of the new bill, the one associated with the official government position as well as with that of the mining sector, would be called "the Filmus Law."[5]

Filmus had a strong and constructive working relationship with the Environment Secretary. They had worked together when he was Minister of Education to include environmental material in the public school curriculum. He trusted Picolotti and had identified the "environment" as a political space where he could build capital for himself. In fact, before the whole glacier affair, he had approached Picolotti to suggest that he could become president of the Senate's Environment Commission and that, if she was interested in working with him, he could help her from his position in Congress. Filmus wanted to become mayor of Buenos Aires and needed good political visibility to woo voters. The environment, in the mid-2000s, was an in-vogue issue for the metropolitan *porteños* of Buenos Aires, just the type of issue that could catapult Filmus into the favor of the electorate. This belief may have been one of the deciding factors that saw him take the new glacier bill through the difficult negotiations that were to follow.

The Environment Secretariat was one of the most politically active ministries. Working with Picolotti, Filmus envisioned, would generate media attention in his direction on a topic his constituency cared about. Picolotti needed the Senate's Environment Commission to help her get several bills passed she wanted to become law, so building a relationship with the Commission's president also made perfect sense for her.

One question that remained in the minds of many as discussions were getting under way on a Glacier Bill 2.0 was how the first version of the law slipped passed the scrutiny of the mining sector. Provincial representatives should have picked it out as a contentious issue. *They did not.* The federal Mining Secretary (Jorge Mayoral) and his provincial counterparts (such as Felipe Saavedra of San Juan), could have, or should have, seen it coming. *They did not.* Several provincial governors, including Gioja of San Juan, would later complain that the bill was not circulated as widely as it should have been in Congress and that this was the reason they could not block it earlier. This does not seem to be a valid reason; if they had flagged the bill earlier, they could have easily muscled their influence to have it stopped up until the very last minute. Furthermore, if the provinces were aware of the contentious nature of the law, there would never have been a unanimous vote in favor of it because representatives of the mining provinces in both houses would have voted against it. *But they did not.*

The president, in her veto, makes mention of provincial concern over the glacier law and the alleged negative impacts it would have on the provincial economies, but there is no evidence to date that there was ever any communication from the provinces to the president against the bill prior to the veto. Later, in comments made to the media, the president, irritated by having to step in and pay the political price of a veto, referred to the error of provincial representatives when they voted for the law, claiming that they didn't know what they were voting for.[6] Clearly, she was referring to the fact that no provincial authorities flagged the potential risks of the glacier protection bill for the mining projects already under way in the Central Andes.

What this apparent lack of communication reveals is a serious lack of articulation between government branches, between provincial and federal public officials, and between the corporate and public sectors.

But, as we look back at what transpired, what is perhaps even more surprising is the continued failed and haphazard communication between these actors after the veto and for the two years that would pass before the vote went up again in Congress. Even then, opposition to the glacier law from nonpolitical actors (the mining or hydrocarbon sectors, for example) did not succeed in channeling their concerns and positions through the government. In retrospect, it is perhaps because, like most of us at the time, they simply didn't understand the issues when it came to glaciers. It is hard to believe that they did not have a true interest in the matter.

Picolotti, a lawyer by profession and an environmental legal advocate, was absolutely attuned to the links between public policy and the law, and she had made it a point in her administration of the Environment Secretariat to establish and nurture a fluid relationship between the executive and Congress. She had already taken several laws by the hand through the congressional bodies, and she was versed in the administrative process and political leverage points that needed to be activated for a bill to make it through the voting process successfully.

Although many presume that there is a strong corporatist dynamic governing mining investments and activity, with strong collusion between companies and state, what the glacier law experience showed was just how *unconnected* the primary actors of the sector were with governing officials, beginning with the federal Mining Secretary, who was largely disconnected from other agencies of government and from congressional actors. Mining Secretary Jorge Mayoral, despite his formal role of stewarding national mining policy, as it turns out had very little affinity with the provincial mining authorities, who, in turn, mostly distrusted and avoided federal agencies whenever possible. These disconnects had already worked against mining interests in the run-up to the little publicized vote on the first glacier law, but would also and more surprisingly work against mining interests in the far more publicized vote on the second glacier protection bill.

Whatever the reasons, the corporate sector was not talking to the mining agencies, nor to Congress, and the mining agencies and congressional representatives were not talking to the federal agencies. Picolotti was perhaps the only person actively engaging on nearly all fronts. This gave her a unique vantage point from which to analyze and engage with the discussion while she was at the helm of the evolution of the glacier law. But this situation would create a dangerously headless and errant void if she abandoned the coordinating role she played after the veto and presented her resignation.

For the moment, the Environment Secretary, whose continuity in her executive post was at this point dangling on a thread, had been convinced by the chief of cabinet to stay on to try to save the glacier law. She would be the chosen emissary from the executive to carry forth the planned meeting to discuss the next steps with a possible Glacier Bill 2.0. That favor wouldn't last long.

The Glacier Bill 2.0

This time, unlike the first time around, when no one noticed the glacier protection bill working its way through Congress, the mining sector, the provincial and national legislatures, executive branch agencies, the mining companies, the oil and gas companies, and a slew of other actors that had problems with the law were all called to attention to examine, review, and comment on the future protection of glaciers and periglacial environments. It was expected that the various actors who had most to lose from a strict glacier law, specifically large multinational mining companies operating in San Juan, La Rioja, and Catamarca at above 3,000 m (9,800 ft), as well as the governments of the provinces, would have ample time to contribute to watering down the former bill, working toward a law that would be more tolerant of industry (particularly the mining industry) and that would be less conservationist than its predecessor. Curiously, that's not exactly what happened.

With all issues out in the open, with strong executive support for the mining sector, with the governors and the president on the side of the miners, everything seemed aligned so that the mining sector would get a law it could live with. The end result, which would come two years later, would surprise everyone.

Picolotti knew that her days were numbered at the helm of the Environment Secretariat. Rumors were already spreading in local media that the glacier law conflict would mark the end of her tenure as Environment Secretary—and that the president was waiting for the right moment to dismiss her.[7] This would probably be her last effort at salvaging the glacier law and maybe her last initiative as the head of the Environment Secretariat. She was not wrong, but she didn't imagine that the end was so near—in fact, just hours away.

The opposition parties in Congress were furious with the veto. They couldn't fathom that the president had vetoed a law that had been voted unanimously by all parties. They saw it as an executive affront to Congress— some suggested it was a coup against democracy. The House of Deputies' Environmental Commission met on November 18, and, even though they did not unite the necessary quorum to officially hold and register the meeting, those members of the opposition who did attend proposed to resubmit the original vetoed glacier bill. They confided to Picolotti their doubts that the president's proposed plan to negotiate a new text would guarantee a good glacier law. Furthermore, they were against the presidential effort to come up with a new law, since the vetoed law had the full support of both Houses, and hence, they argued, the original text was the most legitimate. They felt that creating a space for dialogue led by the executive (even if Picolotti were leading it), bringing in industry and pro-mining factions that were in fact already destroying glaciers, was an affront not only to the environment but also to the democratic process. Several congressional representatives vowed to boycott the presidential initiative if it moved forward.

At this point, strong Lower House congressional support for the glacier law was ideological. It was conservationist by definition, and it had taken on a new dimension, one diametrically opposed to large-scale mining. No one in Congress really knew much about glaciers or periglacial environments, or what the law really implied for the mining sector, but the opposition knew that it did not want a glacier law that was at the "service" of industry, and particularly not the mining industry. Glaciers should not be protected merely to serve mining. Glaciers had to be protected *against* the impacts of mining. They sensed that the very wording of the definition of what a glacier was and its importance was at the heart of the debate and that a conservationist definition should prevail. And although not very many congressional representatives in the opposition knew very much about the subject matter, they could obviously see that the mining sector had a serious problem with the law. This, in and of itself, was grounds upon which to assume that mining projects were indeed impacting glacier resources. What they didn't quite understand (and this also held true for the pro-mining voices in Congress) was just how big the problem for mining really was.

The president's proposed meeting to discuss the next steps to take with the glacier law was convened for December 1, 2008. Since the Environment Secretary was the convener, it would be held at the Environment Secretariat's main building in downtown Buenos Aires, at 451 San Martin St.

This ensured that the Environment Secretariat would decide on who would be invited and, more importantly, *who would not*—a key determinant in how political agreements would be mapped out and next steps defined. It also determined on whose territory the negotiations would occur. This may seem a minor issue from the outside, but, on the inside track, the logistics of the meeting, the agenda setting, the invitations (and the exclusions), and the behind-the-scenes bilateral negotiations between the various political actors who attended the meeting would all have their influence on the future glacier law. Picolotti's choice of excluding one of the principal actors in the debate may have determined the destiny of the glacier law. More on that later.

Just prior to the meeting, Picolotti met with Daniel Filmus to discuss preparations and the importance of his taking up leadership in the follow-up to the meeting. Filmus deferred to Picolotti on the substance of the discussion, preferring to follow her lead. But Picolotti stressed to him the likelihood that she would soon leave government and that if he did not carry the agenda forward the glacier law would be swallowed up by controversy and political infighting. Despite the risk that Filmus might steer the glacier law toward a service-oriented dimension, she realized that, at this point, simply keeping the law alive was the priority. There would still be time to revive the power and leverage of the conservationist voices. What was important was that the glacier protection bill not die a quick death after the veto. The issue needed a leader, and, without her at the helm, the risks of it collapsing unless someone with a clear objective took over (whatever the person's objective might be) were high.

Marta Maffei, the House of Deputy member who had originally proposed and drafted the vetoed glacier bill, was no longer in office. In her place appeared Miguel Bonasso, a journalist turned congressman, who saw the glacier law as his chance to leverage political presence in his glacierless city of Buenos Aires, where public opinion was very much pro-green and very inclined to favor glacier protection. Bonasso had already commandeered the Lower House on the voting of Argentina's Forestry Law, which Picolotti had negotiated with both Houses. Bonasso was her point person in the House of Deputies and now appeared as the most likely champion for the glacier law in the Lower House. Bonasso's anti-mega-mining advocacy had already made a splash in media circles, and general anti-mining sentiment was growing in key constituencies. This crusade to hold up a glacier protection law against mining interests fit perfectly with his political strategy. The new glacier bill had not changed a comma or period, but Maffei was no longer around to fight for it; thus, the alternative to the "Filmus Law" became simply "the Bonasso Law."[8]

Bonasso approached the Environment Secretary shortly after the veto. He wanted the inside scoop on what had happened with the veto because he wanted to carry the glacier law baton forward in Congress, and he needed to understand the state of play in the executive. He met with Picolotti at the Environment Secretariat on November 13, 2008, two days after the publication of the veto in an official bulletin. Bonasso recounts his version of the meeting in his book *El Mal*,[9] dramatizing the discussion. He suggests that Picolotti confided to him that the president had vetoed the law as a favor to the provincial governors in exchange for their support for the nationalization of foreign-held retirement investments trading on Wall Street.[10] However, this is erroneous because this was simply not Picolotti's belief at the time. She was convinced, although she did not confide this to Bonasso, that the provincial governors had likely received a call from Barrick Gold and that they, in turn, were exerting pressure on the president to veto the law to protect the upcoming Pascua Lama investment, as well as others that were already in the pipeline.

What is true about Bonasso's recount in *El Mal* is his warning that Picolotti's time in the administration was coming to an end. He recommended to her that she resign immediately. But Picolotti already knew her days were counted, and her main concern before she left was to find a successor to defend the glacier law. In this case, both Filmus and Bonasso stood out as likely candidates to carry the glacier law discussion forward in Congress.

The first point of conflict in the lead-up to the glacier law reconciliatory meeting was what participation would be given (or not) to the Mining Secretariat and to mining interests more generally. Picolotti presumed that the Mining Secretary, Jorge Mayoral, was already fully engaged in the opposition to the glacier law and exerting his influence both on the president and on Congress. Whether it was because Picolotti did not invite Mayoral to the glacier debate, or simply because he simply didn't carry weight among the provincial mining agencies or among the miners, this presumption would turn out to be largely erroneous; in the end, the influence exerted by the federal mining agency was negligible. The veto had curiously and expressly mentioned an opinion of the Mining Secretariat, but it is doubtful that Zannini, the president's legal advisor, contacted the Mining Secretariat at all in his drafting of the veto. It is more likely that the mention was drawn from some previous statement or document and was simply coincidental. The veto read:

> As indicated by the Mining Secretariat, the establishment of minimum standards cannot limit itself to the absolute prohibition of activities, but rather on the contrary, to the establishment of minimum parameters that provinces should assure, more strict even, according to their special environmental circumstances. [unofficial translation][11]

The presidential veto explicitly stated its disconformity with the absolute prohibition of mining in glacier and periglacial environments. As the evolution of the debate went forward, what became very clear was that the mining sector, and more specifically, mining companies, had a real problem with the ban on mining in periglacial areas, which are much more extensive in size and reach than glaciated areas. The nuances of the definitions of what was protected (and what was not protected) would be central to the debate, even if most of those behind the debate could not and probably still cannot clearly define what a periglacial environment is.

Because the veto was grounded on definitional problems for the mining sector, it seemed reasonable that the federal government's top mining official, as well as the provincial mining authorities, should and would want to participate in the discussions. Jorge Mayoral, the federal Mining Secretary, certainly believed so. Picolotti, however, was not at all on good terms with Mayoral. In fact, he had shown outright opposition to her team's efforts to propose federal environmental oversight of the mining industry.

Discussions between the environmental agency and federal mining authorities had become so abrasive on this issue that Picolotti selected her staff to deal with mining relations partially on their experience in dealing with tough political negotiations and conflict. Her team leader was a kamikaze human rights activist from Colombia named Juan Pablo Ordoñez who had experience working under strong pressure in places like Bosnia and Cambodia. Picolotti had met Ordoñez in Cambodia, after the Khmer Rouge collapse, in an international development project financed by USAID to rebuild Cambodia's justice system and thwart human rights violations in Cambodian prisons. Ordoñez, a die-hard activist, was placed in charge of managing relations with Mayoral.

Picolotti decided finally not to invite the federal Mining Secretary to the glacier law meeting. She sustained her decision by arguing that glacier protection was the political incumbency of the Environment Secretariat, and she did not need *federal* mining agency inputs. Provincial mining authorities, however, were in many cases charged with environmental monitoring of mining projects; hence, their presence at the discussions made more sense.

Mayoral was furious. He called her personally on the days leading up to the meeting to demand he be invited. She stoically refused, but, on his insistence, she finally agreed to hold a second meeting at which he could participate (a meeting that never took place). The exclusion of the Mining Secretary at this political meeting sent a strong political message to those present about who was in charge (and who was not in charge) of the discussion, and it set the tone and content of discussion over the revised version of the glacier law. It also had profound implications for the drafting of the new bill and the consideration (or not) of those issues that could be sensitive to miners. But, more than a discussion about substance, this was a meeting about politics and about who would drive the discussion—and, for the moment, mining was out.

The morning of Monday, December 1, 2008—Buenos Aires, at the Environment Secretariat of Argentina. As Picolotti reviewed her briefing notes for the glacier bill 2.0 meeting, congressional and provincial representatives, including several governors, filed into the room. She did not know it at the time, but she was beginning her last forty-eight hours as Environment Secretary of Argentina.

She had already spoken to several provincial governors, including José Luis Gioja of San Juan, about the glacier law veto and possible next steps. Much to her surprise, the sense she was getting from the governors of the mining provinces was that *none* of the provinces had been consulted before the veto; at least, that is what they were telling her. Not only that, but none of the provinces had raised any concern whatsoever to the president about the law. There was no provincial reaction to the glacier law that would have pushed the president to veto the law. They had taken the passage of the world's first glacier law in stride with no apparent concerns. The veto had come as a surprise to most of the provinces, just as it had surprised Picolotti and the rest of Argentine society. What these officials were telling her was that the mining taking place in their provinces was in fact *not* in conflict with glaciers or with periglacial environments and that they had the evidence to prove it.

The Governor of San Juan came to the meeting with his Mining Secretary, Felipe Saavedra. They held a private meeting with Picolotti before the larger meeting with the other provinces. San Juan was the key conflict province for the moment, and the Environment Secretary wanted to get their take on the matter before she went in to talk to the others. If she could resolve San Juan's problem, the others would be easier and would fall in line, or so she thought at the time.

Gioja insisted that glaciers and periglacial environments in San Juan were not in conflict with mining.

"What about Pascua Lama and Veladero," asked Picolotti, "I've seen the maps, they're in glacier and periglacial areas."

Gioja and his mining Minister seemed to be waiting for this comment. Saavedra pulled out the report prepared by Lydia Espizua, a geologist from Argentina's glacier institute, the IANIGLA.

"Yes, but there's no impact," answered Saavedra triumphantly, as he leafed through the report seemingly searching for a key paragraph. "The impacts are negligible. Barrick carried out the studies and they're clear. Here's the report," he added, as he handed Picolotti the Espizua report.

Picolotti hadn't seen this report and was not sure what to say. Saavedra guided her to page 44. She looked briefly at the cover, seeing the reference to the IANIGLA and the report title, "Impact of Mining Infrastructure of Pascua Lama on Glacier and Periglacial Environments." Saavedra had under-lined a concluding phrase by Espizua near the bottom of the page, ready to show to anyone doubting the province's claims:

> In conclusion, the works to be carried out at Pascua Lama, in the Argentine sector, will not directly affect glaciers, snow fields, or rock glaciers, and the impact on discontinuous permafrost will not be significant.[12] [unofficial translation]

Picolotti read and reread the highlighted sentence and thought about the phrase as she continued to leaf through the study, "How could this be?" she thought to herself. Her gut told her not to believe the Espizua report or, at the least, to look at it more closely to understand just how Espizua could reach this conclusion. Picolotti knew that there was definitely conflict between min-ing and glaciers; she had reviewed the information provided by Juan Pablo Milana showing clear impacts at Barrick Gold's Pascua Lama and Veladero projects. The title was suggestive. It pointed to *infrastructure* and perhaps was only looking at hardware placement of the project. It was definitely not an environmental impact study, and this was probably the reason for such a cat-egorical statement.

Many thoughts raced through her head as she contemplated the report and decided on her response to the governor and to the minister. Here they were, holding up a report suggesting that Barrick Gold was not impacting glaciers in San Juan Province. She knew that something must be wrong with the report and its conclusions. The activist in her wanted to refute the gover-nor and the provincial Mining Minister, but the more seasoned Environment Secretary had learned that conflict was not the route to convince the politi-cians. Ultimately, she realized that whether or not the information was true, what was more important was that the provincial mining minister and the governor of San Juan *thought* it was true. Counterintuitively, this was the key to engaging the provinces! Let them go with this idea. The Espizua report was actually just what she needed to rope them in. Her job was not to convince the provinces that mining projects were impacting glaciers and that they thus needed a strong law to protect against this (which in their minds would work against mining and against provincial interests). Instead, just the contrary: if they truly believed that mining was not impacting glaciers, a strong glacier protection law would not affect mining in their provinces.

A clean bill of health for Barrick Gold at Pascua Lama could save the glacier law. She looked at Gioja and then back at Saavedra. "Well then, if this is true, then Pascua Lama will have no problem with the glacier law," said Picolotti. Both the minister and the governor smiled, presumably confident that they had convinced the Environment Secretary that there was no conflict with glaciers and mining in San Juan.

As the meeting with San Juan came to a close, her thoughts returned to the comments made by several governors she had recently spoken to. They *all* believed more or less the same thing: that their mining projects were not in conflict with glaciers or with periglacial areas. Whether this was true or not was irrelevant; in fact, she thought, if they all thought this to be true, all the better. Getting the glacier law through Congress wouldn't be as difficult as she imagined.

And if San Juan's Mining Minister was so convinced that Pascua Lama was clear of problems, then it was indeed possible that they had not leveraged the veto. Thus, the likelihood that Barrick Gold (which knew better than anyone the true implications of the glacier law) had been the reason for the veto became increasingly more likely.

But she also knew that, in time, the provinces *would* come to realize that there indeed were huge conflicts between the glacier protection law and their mining investment intentions. This would present serious challenges if the law was not approved soon.

San Juan, Mendoza, Santa Cruz, Tierra del Fuego, Chubut, Neuquén, Rio Negro, Catamarca, La Rioja, Salta, and Jujuy Provinces, all provinces with glaciers and all with mining activities under way, were invited to the glacier law debate at the Environment Secretariat and all sent either their governors or very high-level representatives to the meeting.[13]

Picolotti kept thinking to herself, "Could it be that the provinces really had not intervened in the veto? Could the president have acted solely on pressure from Barrick Gold alone?" Her other concern was with how negotiations would evolve once the provinces figured out the true conflict between the law and mining operations. Could she quickly come up with a new glacier law, one that could provide significant real protection and still leave room for mining investments, and get it through Congress?

Picolotti had placed one condition on Zannini to stay on as Environment Secretary: that the executive firmly stand behind the renegotiation of the glacier law and that a Glacier Bill 2.0 would be presented to Congress in the near future. Zannini assured Picolotti that the president was in full agreement with revising and resubmitting a new glacier law, but one that would be in harmony with the interests of the provinces and of the commercial investments that were under way and planned for the future. Picolotti could work with that. It was with this objective in mind that she entered the packed meeting room at 451 San Martín. She was surprised to see so many governors present. The Environment Secretariat had never been able to muster so much executive political weight in its history. But this leverage would soon be entirely unraveled. This would be one of her last meetings as head of Argentina's Environment Secretariat and one of the last meetings since that day in which the Environment Secretariat carried any real political weight.

Her message to the group would be simple: the will of the president was to come up with an improved glacier protection law. The president had vetoed the glacier law because she considered it was not in the best interests of the provinces, and she, Picolotti, was there as a presidential emissary, ready to work through a law that everyone could agree to. The message was strong for three reasons: (1) it laid out the president's full commitment to negotiate and reach a new law; (2) it specifically showed presidential priority for mining investments, which would be key to get the mining provinces to support the law; and (3) it reaffirmed the president's intention to protect glacier resources. Everyone could identify something they could hold on to legitimize the negotiation process. A fourth underlying message could also be understood: that there was hope for salvaging the glacier law particularly because Picolotti was chosen to lead the discussion because everyone in the room knew that she was a staunch supporter of the conservationist vetoed law—this was further buttressed by the explicit exclusion of the Mining Secretary (a fact that surprised everyone in the room). This would appease some of the more hardline critics of the veto while at the same time tell the mining sector that it had to work with the Environment Secretariat toward a law that would truly and effectively protect glaciers.

This set the tone for the next two hours of debate, one that would subsequently drive discussions on the Glacier Bill 2.0 for the next two years.

To Picolotti's surprise, but also as a reaffirmation of what she had heard from the provinces, none of the provinces or other actors present in the room aggressively attacked the content of the vetoed glacier law. Either they didn't really understand it, or they didn't have fundamental problems with the text. This reconfirmed that the real enemy of the glacier law was not in the room. Some of the contentious issues that emerged as provincial authorities began to understand the implications of the glacier law—issues that would carry forward in the negotiations of the new glacier protection bill—included:

- The absolute prohibition of industrial activity in glacier and periglacial areas, which now several actors opposed

- The discretion and jurisdictional authority over the implementation of the law, which pitted provincial jurisdictions against federal agencies
- The role of the provinces in the implementation of the law, which provinces preferred rest entirely in the provincial domain
- The budgetary responsibilities of the implementation of the law, which no one except the federal Environment Secretariat wanted to assume.

All these were reasonable concerns and issues, but none of them was a deal breaker. The discussion around a new glacier protection law went forward amicably and constructively for the next few days.

What also became clear to Picolotti and to those on her team was that no one in the room knew much about glaciers. They didn't know where Argentina's glaciers were and even less about what a periglacial environment was or where it was. Most believed that the periglacial environment was the immediate area surrounding a glacier. In fact, throughout the next two years of negotiations, few people really ever understood the periglacial environment and that protecting it in absolute terms (as the new glacier law would do) would effectively exclude mining activity in a vast mountainous area.

In the minds of the congressional representatives, and even in the opinion of environmental actors and the general public, it made sense to protect glaciers and also to establish a buffer zone to ensure they were not affected. So, the idea that the periglacial environment was that buffer zone was also an understandable mistake, conveniently—albeit falsely—providing a sense of security for pro-conservationists that the law protected the glaciosystem when, in fact, it didn't. But the periglacial area (as we learned in Chapter 4 of this book) is not the area surrounding a glacier or even a glacier buffer zone. It is instead a much more extensive swath of land lying below the glaciated regions of a mountain. There can even be periglacial areas without any visible glaciers on a mountain, as was the case in the El Pachón and Los Azules projects that were coming into the pipeline. Protecting the periglacial environment practically placed mining activity off limits along an entire strip of land between 3,500 and 5,000 m (11,500 and 16,400 ft) in the Central Andes, where most of Argentina's new mining exploration activity was taking place.

This erroneous thinking about periglacial areas was a fortuitous occurrence for those espousing the conservationist thinking that dominated the Maffei version of the glacier protection law, and it would help ensure that those components of the bill that protected the most valuable hydrological resources in the vetoed law would survive in successive reincarnations of the bill. If the general public and the Congress realized that periglacial environments were far more extensive than merely the glacier's immediate surroundings there might have been much more debate and conflict about including the periglacial areas under the protection of the law, and this could have killed their inclusion in the bill. Fortunately, that did not happen.

The Environment Secretary and the President of the Senate's Environment Commission, Daniel Filmus met briefly after the meeting to discuss a communication strategy for the follow-up to the meeting. Filmus needed to be at the helm of the discussion to ensure continuity, since Picolotti realized she was on her way out. Filmus later met with the press and sent several public messages about his view of the upcoming Glacier Bill 2.0. His idea was to come up with a new version of the law, one that could reconcile mining activity with glacier protection. Probably confused himself about the periglacial dimension of the discussion, he also indicated that the "periglacial environment" would have to be looked at more closely so as to better define the region and to reach an acceptable definition that would serve everyone's interest.[14] Filmus was hearing complaints from the mining sector and although he was just starting to discover glacier issues, he could already sense from his discussions with the mining sector that the devil was in the frozen grounds of the periglacial area.

Filmus, knowingly or not, revealed the real essence of the conflict for the miners. The periglacial area was simply too large and too extensive to reconcile a conservationist law with mining activity. At the time, even for glacier experts, defining the periglacial environment was extremely difficult, and this in the end would provide stalling opportunities for the implementation of the law. Even Argentina's national glacier institute, the IANIGLA, was complaining that the vetoed law unrealistically called for carrying out an inventory of Argentina's periglacial environment within five years, and a reinventory of it every subsequent five years, which they viewed as impossible. They simply didn't have the staff, technology, or capacity to carry out such a labor-intensive process, one that involved reviewing satellite images of thousands of glaciers, conducting field visits to areas stretching across thousands of kilometers in many unreachable areas, and taking ample temperature readings at hundreds of thousands of points to determine ground temperature. It simply could not be done in five years, and maybe never. The technical limitations of identifying periglacial areas would soon change, but no one could envision this in November 2008.

Even Dario Trombotto, the periglacial specialist who had introduced the term *periglacial* into the bill, had originally intended the law to protect *certain periglacial forms* within the periglacial environment, such as rock glaciers saturated in ice. He was not thinking of the entire periglacial area when he drafted these key sections of the glacier protection bill.

Deputy Maffei, however, had the Colombian *páramos* in mind (tropical mountain wetlands), and she wanted to set aside a vast area for protection. She envisioned that a much larger water source domain should be protected by law; so, for her, the inclusion of the greater "periglacial environment" made perfect sense. It was this conservationist position and interpretation of the value of glaciers and periglacial environments that would survive congressional discussions over the bill.

December 2, 2008—Buenos Aires. The next morning, Picolotti received a call from Carlos Zannini, the president's legal advisor, who, in the relatively calm media aftermath of the glacier law meeting, wanted the Environment Secretariat to formally contribute a legal position justifying the veto of the glacier law. He was worried about what was brewing in Congress because opposition deputies were orchestrating a comeback of the vetoed glacier law text. It was an ultimatum that Picolotti could not refuse if she were to stay on with the administration.

Zannini knew what he was asking and guessed that Picolotti would bend to the request in her desire to stay on to work through the law. He also realized that Picolotti was on the edge and might walk away at any time; this request, if accepted, would not only bring her back to the president's side of the fence, but would also have her take some portion of the blame for the veto off the president's plate.

"Sorry Carlos, I'm not participating in this," she replied by telephone.

As she refused Zannini's order, Picolotti realized she was effectively resigning as Environment Secretary. She remembered the day two years earlier when representatives from communities in San Juan appeared at her office with evidence that mining activity was destroying glaciers. It was clear from the position of the governors that Barrick Gold had leveraged the veto, *not the provinces*. If the president wanted to protect Barrick's interests, whatever her reasons, that was her prerogative, but Picolotti would not be part of it.

As she listened to Zannini screaming on the other end of the line in reaction to her refusal to draft the veto justification, she realized that this was the end of the road for her political career. She would have to step aside. Zannini represented the president, and his order to write the justification of the veto was the president's order. Whether the president was actually in on the demand didn't matter; in fact, she probably wasn't. This was probably Zannini's move to gain an upper hand with Picolotti and bring the rogue secretary under his control. If she refused, he'd convey her decision to the president, and the president would demand her resignation. Picolotti did not know if the president or even Nestor Kirchner (the former president and husband of the incumbent president) had devised the idea to corner her in this manner, and she would never know, but her refusal was an adamant confession that she was no longer willing to support the president. In the book of politics, that meant the end of the line; even though Picolotti was not a career politician, she knew it. She sensed that her time to step down had come.

At about 5 p.m., on December 2, 2008, Romina Picolotti, Argentina's
Environment Secretary, received a call from the president's chief of cabinet,
Sergio Massa asking her to come to his office. She knew what he wanted to
discuss, but her decision had already been taken hours earlier and she was
already preparing for her exit. The glacier law was now adrift in the corri-
dors of Congress and on a parallel track within the sphere of the executive. It
seemed a long shot that it would ever survive.

It took only a few minutes for Picolotti to get to Massa's office in the
presidential Palace in the Plaza de Mayo, just a few blocks away from the
Environment Secretariat. The discussion was short and to the point. Massa
indicated that the president was not happy with the handling of the glacier
law and made reference to the request that the Environment Secretariat draft
a justification of the veto. Picolotti reaffirmed what she had said earlier to
Zannini: "I'm not going to support the veto. I don't want any part in it. I supported the glacier law and the president vetoed it. There's not much more to
say than that. I think my constructive contribution to this government has
come to an end, and it's time for me to go," she said as she stood up and prepared to leave the room. "My team will collaborate with you in the transition
to the new Secretary, but I am leaving."

Massa made no effort to hold her back. He knew her well by then and realized that there was nothing at that point to keep her from leaving. Romina
Picolotti, environmental advocate and Environment Secretary, at age thirty-
eight, left the presidential palace, relieved on the one hand that she was holding
to her convictions, but extremely worried that her job to protect glaciers had
not been finished. As she returned to the Environment Secretariat, she made a
passing comment to a journalist standing outside the building waiting for an
interview, indicating that she was through. She could hear some of the journalists commenting that the chief of cabinet had just called a press conference.

Moments later, digital headlines on the Argentine dailies ran that Picolotti
had resigned as Secretary of Environment in disagreement with the president's veto of the glacier law. Thirty minutes later, the headlines changed,
suggesting that the president had let her go alluding to an underexecution of
the Environment Secretariat's budget, which was far from the truth. Massa
had held his press conference, and the president's office was already spinning
the news.

The press began calling Picolotti incessantly. She took only a few of the
calls, from a few journalists she trusted most and only to comment that she
had supported the glacier law and that, following the president's veto, she
considered that she could no longer contribute constructively to the administration, repeating the exact words she had said to the chief of cabinet. She
had already drafted a letter of resignation the day before, sensing that the end
was near, and now she sent it to the president. As she walked into the
Environment Secretariat for the last time, she took a deep breath, called her

closest staff, and confirmed what they were already hearing in the press: that she had resigned as Secretary of Environment of Argentina. Some were tearful to realize the road had come to end, but most understood that it was the right thing to do. She called her family and a few of her closest environmental advocate colleagues to give them the news. She returned to her desk and collected her personal belongings. Before leaving, she took down a few personal photos from the wall, but decided to leave behind the large photo in the meeting room of the populous march against the pulp mill factories at Gualeguaychú. For several weeks, maybe longer, it continued to hang as a reminder of where the Environment Secretariat had once stood.

December 3, the National House of Deputies, the Argentine Congress—Buenos Aires. On December 3, 2008, two days after the first meeting of the executive branch's initiative to come up with a new glacier protection law, congressional Deputies Miguel Bonasso and Eduardo Macaluse presented a motion before the Environmental Commission of the Lower House of Congress to resubmit the original text of the vetoed glacier law. They did not trust the executive effort to come up with a new consensus law and insisted that the glacier law should come back by the hand of Congress.

If they could come up with two-thirds of the vote from both houses, they could override the veto; it was very unlikely that they could do it, even though the Congress had voted unanimously in favor of the law the first time around. No one had voted fully understanding the issues and now, with mining interests clearly in the open, numerous Deputies would turn their vote. The presidential majority party knew the implications of the law and would invariably side with the president, no matter how they had voted previously. The new divide in favor and against the glacier law turned out to be about even. The motion could not muster support to override the veto.

The Glacier Bill 2.0 Debates 2008–2010

With all interested actors now alerted to the evolving political and legislative conflict over the veto of the glacier law, a period of heated debate ensued in both houses of Argentina's National Congress. The debate also carried over to provincial legislatures, as provinces considered introducing their own glacier protection laws to remain in the driver's seat on a very conflictive issue.

This was a fascinating process to watch, as politicians who one day didn't even know there were glaciers in their own province were all of a sudden ferociously defending their ice while others more in line with industry interests in high-altitude environments attempted to contain the conservationist tone of the discussion and opt for a more services-oriented approach to glacier protection. Now everyone seemed to be versed in glacier knowledge!

At the National Congress, the camps were divided into those behind Senator Daniel Filmus, a stance seen as supporting the political position of the executive branch and the mining sector, *the services-oriented faction*; and those behind Deputy Representative Miguel Bonasso, *the conservationists*, who simply resubmitted the same glacier protection bill that Marta Maffei had submitted in 2007.

Everyone jumped into the discussion.

Environmental groups, politicians, governors, mining sector representatives, representatives of companies, opposition media, congressional representatives of all parties, environmental agencies, academics (particularly geologists and geographers), school teachers, radio personalities, actors, talk show hosts—it seemed that everyone suddenly cared about glaciers.

The debates were for the most part informed by the IANIGLA experts. The IANIGLA is Argentina's glacier institute, one of the few public agencies in all of Latin America (and of the world) specifically focused on glacier studies. This national agency, located in Mendoza, Argentina, where most of the country's glaciers are found, was completely off the radar screen for most Argentines, but now the IANIGLA was in the center of the storm. Ricardo Villalba, the IANIGLA director who had participated in the crafting of the law with Dario Trombotto, a specialist in geocryology (an academic mix focusing on the dynamics of rocks and ice), was catapulted into the spotlight and interviewed by dozens of media sources following the presidential veto.

Villalba was visibly disturbed by the veto and said as much in the many interviews he gave in the days that followed.[15] Villalba spoke broadly to Argentine society for the first time about glaciers—a rarity for the IANIGLA, which kept most of its work among academics. He spoke of glaciers in San Juan, in La Rioja, in Catamarca, and even in the extremely arid provinces of Salta and Jujuy, where practically no one knew there were any glaciers. He spoke of debris-covered glaciers, which are glaciers *covered* with rocks, and of rock glaciers, which are glaciers *mixed* with rocks, and of periglacial environments and permafrost, which are frozen earth mixed with ice-saturated rocks. This was all very confusing to most who were hearing for the first time of thousands of glaciers in places they couldn't imagine. There might be ice up in the mountains, and there were even glaciers underneath the Earth that we couldn't see. He mentioned that glaciers were feeding much of Argentina's agriculture and that glacier water was helping power hydroelectric dams. This was all news to most.

It was a sudden period of massive and instantaneous glacier education for Argentine society, which had never really heard much about the role glaciers play in daily life. Villalba highlighted the merits of the vetoed glacier protection law, communicating a few basic concepts underpinning why a glacier law was important, concepts that the general public and the media would grasp and carry forward in their newly acquired appreciation for glaciers. Some of these messages were that:

- Glaciers were important because they *stored* water.
- Glaciers were important because they *regulated* water flow.
- A law would establish glaciers as the *public interest*.
- Glaciers were important to hot and dry provinces with high mountain environments.
- There were *hidden glaciers* and *frozen environments* that also contained ice.
- The law would oblige Argentina to take stock of its glaciers.
- Certain activities could harm glaciers.
- Argentina needed to evaluate glacier impacts.

These basic messages, repeated widely to most of Argentina's media outlets during the days following the glacier law veto, would become the buzz phrases for most actors engaged in the debate and certainly for the congressional representatives who would take the legislative discussions forward. The media was also quick to take up these ideas, draft articles, and develop news programs around them.

But the issues underlying each of the messages were not at all clear to those listening and, more importantly, to those who most needed to understand. Defining a glacier was not as straightforward as it might seem. Understanding what the periglacial environment is was not at all simple. Trying to explain to a novice that the "peri" glacier was not the "peri-meter" of the glacier, as the word might suggest, would prove to be an almost impossible task. To this day, most believe erroneously (but understandably) that the *peri*-glacier is the area around a glacier. Understanding the difference between a *rock glacier* and a *debris-covered glacier*, which look similar and sound similar in definition but that are actually very different, was even more complicated. The fact that no environmental organizations were focused on glacier protection or on glacier education also made things difficult. And, as is generally the case, there were no interlocutors between the scientific community and the nonscientific public. Villalba was trying to make that bridge, but, as is common in such circumstances, the communication between science and the vernacular was lost in translation.

The intense interest of congressional representatives to move this debate forward rapidly, combined with the complexities of the glacier debate at a technical and academic level, resulted in the serious gaps, misunderstandings, and confusion that would dominate the congressional glacier debates for two years. This frustrated scientists like Villalba, Trombotto, and Milana who had dedicated much time and effort into coming up with definitions and text for Maffei in the first run of the glacier bill. They were further frustrated when they were called to Congress to explain to representatives what glaciers and periglacial environments were—only to find that much of the congressional body had already pretty much defined their position based on political ideology, irrespective of what the scientists might share from a technical standpoint.

Although an academic glacier expert might find the nuances between a glacier, a debris-covered glacier, a glacieret, a perennial snow patch, or a rock glacier critical for scientific discussions, policy makers were on a very different wavelength, generally taking all types of glaciers together and focusing more on the general hydrological importance of the many different cryogenic forms. That made sense from a conservationist perspective, even if the scientific community cringed when politicians called glacierets *glaciers* or a debris-covered glacier a *rock glacier*. Reconciling policy with science became an almost impossible task during the glacier law debates. There was a clear disconnect between that which drives science and that which drives policy makers.

Non-glacier experts formed their own conceptualization of what they understood was a glacier, a rock glacier, or a periglacial environment and forced their personal conceptualization to fit their policy and political objectives. For many, this was more about containing mega-mining or opposing the incumbent administration than it was about protecting glaciers. Many times, the concepts used were incorrect or misinterpreted, but they served the purpose of building a platform upon which to defend a conservationist approach to glacier protection—which is what prevailed in the end, much to the dislike of some members of the scientific community and especially to the displeasure of mining interests.

The congressional glacier law debates over the revamping of the vetoed glacier law centered on:

- Whether the periglacial environment (and its various components) would be protected
- The components of the periglacial environment and which ones would be protected (e.g., active vs. inactive rock glaciers)
- Whether periglacial features actually contribute water to ecosystems
- What gets inventoried and by whom
- What is prohibited in glacier areas or in periglacial environments
- What happens when an activity (such as a mining project) is found to be in a periglacial environment and whether the activity should stopped, redesigned, or relocated
- What agency would implement the law

Most of these issues involved intricate scientific detail and debates that hardly anyone was able to (or had the interest to) engage. Even scientists had debates about many of these questions and could not come to definitive and conclusive positions; policy makers with little knowledge of glaciers before this discussion could not be expected to do any better.

One of the biggest controversies that ensued in the scientific community (as well as at the political level) regarding the Argentine and Chilean glacier debates is whether the periglacial environment contributes water to the ecosystem. The mining sector sent a private consulting firm (BGC, which works for Barrick Gold) to tour Argentina and give seminars on the periglacial environment. One of the key messages delivered by BGC to audiences in San Juan, La Rioja, and in Chile was that the hydrological contribution of rock glaciers to the environment was negligible. Although a few glaciologists strongly associated to the mining industry seconded this idea, more mining-critical glaciologists with extensive experience in periglacial areas suggested that the hydrological contribution of rock glaciers and the periglacial environment more generally was not only significant, but that it was also largely unknown because very few academics had studied it. Not knowing how much water the periglacial environment contributes, said these latter experts, is not the same as saying that they don't contribute. We need to study their hydrology.

If we stand aside from the debate and think rationally about the ecosystem, the idea that periglacial environments do not contribute significant amounts of water to the ecosystem seems preposterous. In part, those sustaining this theory point to rock glaciers and argue that they contribute very little water because they are in equilibrium. This presumption is already skewed because it is clearly false.

First, climate change is warming our planet; hence, rock glaciers, as part of an ecosystem that is in disequilibria, are suffering from increased heat. For this reason, we can presume that, on the whole, just as with normal, white, uncovered glaciers, rock glaciers are also melting more than they are acquiring ice.

Second, the entire periglacial environment (the frozen swath of land) is also moving upward on mountainsides in search of cooler grounds. This means that the lower fringes of these frozen terrain are also thawing and melting, thereby releasing water into the environment.

And third, the very definition of the periglacial environment implies that, even if these frozen terrains are in equilibrium (which is unlikely), they are permanently freezing and melting at the movable fringes—that's precisely their function. Taken in their entirety, it is physically impossible that periglacial environments do not contribute significant amounts of water to downstream ecosystems. The fact that few or no scientists have studied (or measured) this contribution is not grounds to suggest that these frozen environments do not make a substantial hydrological contribution to our water supply.

Fourth, the periglacial water skeptics usually speak of "net" contribution. This is another falsity used to suggest that periglacial environment, even glaciers, do not contribute water, but that they simply receive and release equal amounts of water. This position suggests that if a glacier or periglacial feature receives and releases the same amount of water/snow each cycle, then, in the end, it is hydrologically irrelevant. This is ridiculous! A glacier is a cold body that captures snow and releases meltwater. The fact that the glacier exists is significant, since the "capturing" of this hydrological resources is aided and prolonged because the ice mass is cold. The glacier or periglacial feature holds the water for a longer period of time, much longer than would occur if the glacier or periglacial environment were not there. So, even if it were true that the net hydrological contribution of a glacier or periglacial feature is zero, what is of critical importance is the "delayed" factor of the water release, which is precisely why these cryogenic features are so important to our ecosystems.

Although it is clear that the contribution of cyclical thaw of fringe regions cannot contribute to downstream ecosystems in the same magnitude of what seasonal snow melt contributes in the first months after the winter season, periglacial environments should not be compared to glaciated or snowed covered mountain regions. On the contrary, we're talking about hydrologically important terrains *after the snow has melted*. These features have no visible ice. It is land *below the visible ice* that is storing water, protected from the warmer temperatures. We're talking about lands that contribute much less water than that of snowmelt after the winter, but still release significant amounts of water to keep rivers running when other seasonal sources have vanished.

It seems absolutely reasonable, logical, and more than likely that the collective hydrological contribution not only of rock glaciers but of all periglacial features to downstream ecosystems during spring and summer is fundamentally important. It is the melt from glaciers, rock glaciers (active and inactive), ice-saturated permafrost, and other periglacial features that keeps the water flowing throughout the year.

The fact that no one has studied this phenomenon to give academic certainty and provide data to this theory should not in any way serve to suggest or to presume (as numerous glaciologists and hydrologists hired by the mining industry have done) that the hydrological contribution of this frozen swath of land, which is permanently melting and contributing water to downstream ecosystems, is negligible. On the contrary, the lack of data forces us to make use of reason, and, to this end, we must employ the *precautionary principle*. We should presume by logic and reasonability that the hydrological contribution of periglacial environments is extremely significant after postwinter snowmelt. If the mining sector's glaciologists and hydrologists want to counter this theory, then it is their responsibility to collect data to prove the contrary. Until now, they have not been able to substantiate this claim.

Although glacier and periglacial specialists had participated in the drafting of the Maffei Law, their input to the debates grew less and less relevant and lost significance as the debates became more heated and more political, and particularly as congressional representatives became increasingly entrenched in their own views of what should be protected and what should not.

Dario Trombotto, a periglacial expert and Argentina's representative before the International Permafrost Association, grew increasingly irritated as his original definitions and focus on ice-saturated features of the periglacial environment were distorted to accommodate politicized objectives of the congressional representatives.

Pro-conservationist factions in Congress utilized at will selective parts of experts' declarations to serve a conservationist leaning in the discussions, whereas others cherry-picked the academic input for nuances of glaciology and periglacial terminology that would create distinct categories of what should and should not fall under the protective aegis of the law. One example was the distinction between *active, inactive,* and *fossil* rock glaciers. The first two (active/inactive rock glaciers) contain ice. Fossil rock glaciers do not contain ice. The first, active rock glaciers, are moving ice and rock bodies, regenerating and theoretically in equilibrium with the environment, maintaining their size and volume. The second, inactive rock glaciers, are also ice and rock bodies, but they have no forward movement; they're slowly losing their ice mass, and they are in disequilibrium with the environment, a sign of a warming climate. The third category, fossil rock glaciers, are relict, with no ice; they merely indicate past ice presence and, as such, have no hydrological value. Various actors in the debate stood for the protection of all types of glaciers, while others suggested that only active rock glaciers should be protected, not inactive ones. But these distinctions were not very rationally based, nor did they make much sense if the idea was to protect water reserves, because inactive rock glaciers may contain colossal amounts of stored water and could take decades, even centuries, to fully melt away—meanwhile, they are still providing meltwater.

The discussions around frozen grounds and permafrost also resulted in great debate and confusion. Academics distinguish between permanently frozen grounds and cyclically frozen grounds. They explained to many science-averse congressional representatives that *permafrost* (which means *perma*-nently frozen) may or may not be "permanently" frozen. Go figure! Confusion abounded.

In the end, no one was happy with the discussions, and, for the most part, everyone was confused over what was really being debated.

Accusations between factions took on categorical, ideological, and sometimes completely inflexible positions, whether founded on science or not. Bonasso and Filmus, whether they were actually on distinct sides or not, became heads of the two accepted sides of a heated debate that no one fully understood. The public, which was not all versed in glacier and periglacial terminology, knew much less than either of the two congressional representatives, but nonetheless went along with this false dichotomy and took sides in the debate according to which faction they most associated with or which they most trusted (or against the faction they least trusted). If you were *pro-mining* or wanted to accommodate mining interests into the glacier law, you went with the Filmus Law; if you were pro-glaciers, anti-mining, or skeptical of mining, you went with the Bonasso Law. There was little rational middle ground to choose.

Daniel Filmus was associated with the position of the national government. True to the facts or not, the Filmus Law was seen as the will of the mining companies or of the mining sector and the mining provinces more generally. Perhaps at the early stages this criticism had some legitimacy because Filmus had agreed to follow the president's lead and convene the actors (originally along with Romina Picolotti—after her exit he was alone in this mission). He talked to the provincial representatives, to mining companies, to academics, and to mining agencies and came up with a law that, in spirit and in theory, would be accepted by the mining interests but that still could protect glaciers. At least that was the idea.

The Filmus Law incorporated a few adjustments to the Maffei Law that made it, as could be expected, more aligned with the "services-oriented" approach desired by the mining sector. This mirrored what had occurred in the discussion in Chile, balancing environmental protection with industrial usage of glacier and periglacial resources. The Filmus Law was stricter and not as definitionally inclusive of what qualified as a protected glacier. It was not as categorical with prohibitions, and it did not have a retroactive character, thus making it friendlier to the mining sector and the mining provinces.

The definitional article of the Filmus Law reads:

Article 1. Object: The present law establishes the minimum standards for the protection of glaciers and of the periglacial environment with the objective of preserving them as hydrological resources of strategic reserves, for human consumption, for agriculture and industrial activity, as providers of recharge water for hydrological basins and for the generation of hydroelectric energy, as a source of scientific information and as a tourist attraction.[16] [unofficial translation]

Furthermore, the Filmus Law limited protection of the periglacial environment only to rock glaciers, which are but one of many elements of the periglacial area. The Maffei Law reached further on this point, protecting the entire periglacial area as long as it was contributing water to the basins. The Filmus Law did not require inventorying periglacial areas, as did the Maffei Law. Additionally, the Filmus Law did not establish that a competent authority could order the suspension of activities that were in violation of the glacier law at the time of the passage (a retroactive characteristic of the Maffei Law that would effectively give the Environment Secretariat veto power over existing mining projects).

Miguel Bonasso, a former Kirchner ally but now firmly in opposition to the administration, came out strongly against the veto. He was also against the president's initiative to come up with a new glacier law through executive convening of the interested actors; that is, he opposed the idea that Picolotti and Filmus gather all of the actors to redraft the law, a task he believed could only be done by the Congress. As such, he refused to participate in the meeting called for December 1 to lay out the next steps following the veto. Bonasso mobilized opposition members of Congress, civil society actors, and academics working on glacier protection against the presidential veto and kicked off his own effort within Congress to save the glacier law and oppose what was perceived by the more conservationist stand to be a mining-friendly surrogate devised by the president's clan and the mining companies.[17]

Bonasso took up Maffei's flag and initiative in what became a personal quest, and, although he readily accepted the assumed authorship of a bill he never participated in drafting, to his credit, he defended Maffei's bill stoically for two years and held up a conservationist platform that would, in the end, shape the final glacier protection law approved in 2010, which was improved substantially from the previous Maffei bill.

Maffei's law called for, and Bonasso's main negotiating points were:

- Glaciers and periglacial areas had an intrinsic natural reserve value, which should not be seen as service oriented, as in the Filmus Law.
- Glaciers must be broadly and amply defined, to provide maximum protection for *all types and all sizes* of glaciers.
- The periglacial environment must be broadly defined in order to protect all of it (not just rock glaciers, as in the Filmus Law).
- Glaciers *and* periglacial areas must be inventoried.
- Activities must be banned that would impact glaciers, specifically mining and hydrocarbon extraction.
- An implementation authority must be empowered to decide on the cancelation of activities that might affect glaciers or periglacial areas.[18]

The Provincial Glacier Laws

At least half of this book is about the birth of the world's first national glacier protection law, but the fact is that Argentina's national glacier law (the one finally adopted in 2010) was actually not the first law on the books to focus solely on glacier protection. In fact, as the glacier law debates moved forward in Congress, it became clear to those provinces with glaciers that the issue merited further local consideration. Glaciers were important water reserves, and protection of these reserves was necessary, but it could also hamper certain economic activities in some provinces. Furthermore, leaving this issue in the hands of the federal government, for many provinces, was not desirable.

The national debate about the need for glacier protection continued to be very real, and the risk that the debate would derail local interests seemed increasingly more plausible as the discussions became more heated and conflictive. For this reason, provinces decided to move forward independently to introduce *their own* glacier legislation.

Two reasons could underlie this decision, the first being that if there were provincial glacier protection laws, the debate in Congress might ease up and there would be less urgency to pass a national glacier protection law. Second, the provinces preferred to be in the driver's seat on the glacier discussion, and, more importantly, they wanted jurisdictional discretion over glacier protection, particularly if protecting glaciers might mean containing industry. If the provinces could decide on contentious cases, they would ensure complete discretion over the more difficult choices, such as allowing projects like Pascua Lama, El Pachón, Filo Colorado, or others to move forward or not. They were not in that seat at the national debates. With local glacier laws, they could regulate glacier protection however they wanted, and, in this way, they could also determine for themselves how *and if* they protected glacier resources, as well as what exceptions they might make for certain industries. Mining concessions (and in the future possibly hydrocarbons) were also on the mind of several provincial authorities. At the time this book went to press (early 2015), there was incipient exploration for shale gas and oil in Mendoza Province, on the fringes of periglacial areas. Furthermore, an eventual federal law would defer to the provinces if these already had a good glacier protection framework in place. It made sense for the provinces to come up with their own laws first. And that is exactly what they did.

On April 10, 2010, Santa Cruz Province, home to Argentina's National Glacier Park, became the first jurisdiction in the world to adopt a comprehensive glacier protection law.[19]

The Santa Cruz provincial glacier protection law (Law 3123)[20] was a mix of the vetoed glacier law and some elements from the Filmus Law. The first two articles addressed some of the key issues in the discussion:

Article 1. *Object. The present Law establishes the guidelines for the protection of glacier and periglacial environments, with the objective of preserving them as strategic reserves of hydrological resources, and as providers of recharge water for hydrographic basins.*

Article 2. *Definition. To the effects of the present Law, protection comprises, within the glacier environment, uncovered and covered glaciers, and within the periglacial environment, rock glaciers, bodies that comply with one or more of the environmental and social services established in Article 1. By uncovered glaciers the law understands those bodies of perennially exposed ice, formed by the recrystallization of snow, whatever its form and dimension. Covered glaciers are those bodies of perennial ice that have a detritic or sedimentary cover. Finally, rock glaciers are bodies of frozen debris and ice, whose origin is related to cryogenic processes associated with permanently frozen grounds and with subterranean ice, or with ice derived from uncovered and covered glaciers. All of these ice bodies have a clear spatial delimitation to the effect of defining the protected areas. In addition to the ice, the detritic rock material and the internal waterways are a constituent part of each glacier.* [unofficial translation]

In sum, the Santa Cruz Province glacier law established the protection of glaciers and periglacial environments due to their strategic reserve value as hydrological resources and as providers of water recharge for hydrological basins, while Article 2, the definitional article, provided a broad definition for glaciers, one that included glaciers of all sizes and shapes. The same article also mentioned the "social and environmental services" value of glaciers. (It does not mention industrial services, however). Like the Filmus Law, the Santa Cruz Province glacier law focused on a reduced interpretation of periglacial areas centered solely on rock glaciers. And, like the vetoed national glacier law and other versions circulating in congressional debate, it banned mining and hydrocarbon exploration and extraction activities in glacier areas (Article 6).

The following month, in May 2010, the Province of Chubut introduced a glacier protection bill.[21] Chubut's legislators led the text with a firm statement indicating that glacier resources belonged to the province:

Article 1. *The present law custodies the property of glaciers and periglacial environments saturated in ice as belonging to the Province of Chubut, with the objective of preserving them as strategic hydrological resources reserves and as providers of recharge water of hydrographic basins. Glaciers and periglacial zones, saturated in ice are the property of the Provincial State (as per Article 124 of the National Constitution) and they are considered non-commercial in nature, and as such are inalienable.*

This paragraph differs from all other laws that would surface in its very direct initial emphasis firmly establishing glaciers and periglacial areas as "state property," as well as defining them as *non-commercial*. The provinces were clearly beginning to stake out their glacier territory vis-á-vis the federal government as they realized that the glacier debate would be one of jurisdiction, perhaps more than of glacier protection per se. They were not wrong.

The other provinces with glaciers, the more controversial ones, however, took a different approach to the legislative evolution of glacier protection and came together in a collective plan to delineate how glaciers should be protected and under whose jurisdiction. Whether a legitimate jurisdictional prerogative or not, the way in which this strategy was executed further fueled speculation that mining interests, and more specifically Barrick Gold, were behind the mining provinces' lobby against a national glacier law.

In late June 2010, several provincial governors traveled to Canada to meet with Peter Munk, the president of Barrick Gold. The meeting was held in Toronto, Barrick Gold's headquarters, on June 26, 2010.[22] President Cristina Fernandez was also in Canada at the time for a G20 meeting and would also meet with the gold mining company, with José Luis Gioja, Governor of San Juan at her side. It seemed obvious at the time that there was no coincidence to this public meeting of the Argentine president, the head of Barrick Gold, and the governors of the mining provinces. The encounter sent a clear political message about where executive politics stood with regards to the glacier law. Their coalescing at such a key political moment on Canadian soil was obviously planned to send a signal to the financial markets showing that the executive power and the provinces were aligning to ensure glacier policy would not harm mining investments. What was most suggestive, however, and what really irked the conservationists in the glacier law debates, was that the very moment the governors returned to Argentina, even before returning to their home provinces, they called a meeting in Buenos Aires on July 6 to announce they would introduce their own provincial glacier laws and that a model law of common agreement was already drafted.[23] They had drafted the model law in Canada, presumably following their meeting with Barrick Gold. The obvious accusation from the more hardline environmentalists was that Barrick Gold, in fact, had drafted the model law.

Whoever the author, eight governors would propose to their respective legislatures in the coming days—and in some cases, *in the coming hours*—the adoption of these glaciers protection laws. "Propose" is actually a euphemism, since there was no proposal to speak of. The bills were submitted directly to vote with no legislative discussion whatsoever.

National Mining Secretary Jorge Mayoral (who had been excluded from the meeting with Picolotti immediately following the veto) was present at the meeting, and the strategy was openly presented as having been devised during their visit to Canada. They made no effort to conceal the mining sector origins of the new provincial glacier protection laws. Several other provincial representatives (who had not been in Canada) also joined this meeting, including representatives from Mendoza, Neuquén, Rio Negro, and Catamarca Provinces.[24]

That same, day, July 6, the executive power in Jujuy Province delivered a glacier protection bill through their representatives in the provincial legislature. It became law a mere two days later, on July 8, with no legislative debate.

On July 7, the same occurred in La Rioja Province.[25] In this case, within twenty-four hours: on July 8, the provincial legislature of La Rioja adopted its glacier law with absolutely no legislative discussion.

On July 14, a week after its formal submission, San Juan Province adopted its glacier protection law[26] with no legislative debate whatsoever.

On August 5, only one month later, Salta Province adopted its glacier protection law[27] with no legislative debate whatsoever.

Also on August 5, Catamarca's Provincial Senate approved a glacier bill[28] that would lose momentum and not proceed through the Lower House.

These four laws and one bill differed only slightly in their content and treatment of the central issues. The main unifying characteristics were:

- They protected glaciers as strategic water reserves.
- They insisted on provincial jurisdiction over glaciers.
- They singled out "active" rock glaciers for protection (except for the Catamarca bill), but not inactive ones.
- They prohibited activities that might imply glacier destruction.

The Province of Rio Negro produced a glacier and periglacial protection bill[29] in May 2010, quite independent from the other provinces, which took elements from various models and established provincial jurisdiction over glaciers, as the other provincial laws and bills did. It used a broad definition of glaciers and added a unique feature to the law, one prohibiting the sale of lands containing glaciers. The bill has not moved to final vote as of the publication of this book.

Sometime later, in October 2011, the Province of Neuquén,[30] in a more independent and less controversial process, introduced a homegrown legislative bill to protect glaciers, which is also still under consideration. The Neuquén glacier protection bill, which comes after the eventual adoption of the federal glacier protection law, protects both the glacier and periglacial environments and directly adopts the National Glacier Law definition of a glacier. Overall, it takes a very integrated approach with the National Law, promoting collaboration between provincial and national agencies.

The public understood that the provincial glacier law initiatives, particularly those of San Juan, La Rioja, Catamarca, Jujuy, and Salta, were a direct result of the provincial executives' alignment with mining interests, and, in the case of San Juan, directly with Barrick Gold. In the case of provinces like Mendoza, Chubut, Rio Negro, Neuquén, and Santa Cruz, the subservience to mining interests seems less likely because these laws were derived from more homegrown initiatives that followed broader and more inclusive discussions both at the national and provincial levels.

We should note that these more "independent" provinces seem not to have conflicts between mining activity and glacier resources. Furthermore, the glacier debate in these territories did not involve significant mining actors in these jurisdictions, which is likely the reason for the more transparent and participatory processes that occurred in the presentation of the bills to the local legislatures.

What we can say in favor of the independent evolution of glacier protection legislation at the provincial level is that, as occurs with the management of other natural resources, *all* of the listed provinces would have contentious issues of a jurisdictional nature with the federal glacier law, and, in such a context, they would legitimately want to advance with provincial-level policy and legislation as they addressed glaciers and periglacial protection.

What is absolutely legitimate to critique, however, was how the introduction of the provincial laws was handled in those provinces with clear deference to the mining sector. These bills were presented immediately after meeting with the president of Barrick Gold and received a de facto treatment in their legislatures, with no legislative debate, no community debate or consultation, and with no effort at rational dialogue between any actors whatsoever.

While the provinces placed their own legislative machinery in motion to adopt local glacier laws, the national debate on the federal glacier law was in full swing at the National Congress.

Bonasso and Filmus became two diametrically positioned advocates, each unwaveringly committed to his platform. Both were convinced that they were aiming for glacier protection, with Filmus convinced that his proposal was more rational and inclusive of a legitimate and more diverse set of interests and that it could be politically viable to bring the mining provinces into harmony with glacier protection. Bonasso stood firmly on an inflexible conservationist platform, one that was rabidly anti-mining in nature.

They agreed to hold public debates on the differences. These took on a violent character, particularly on behalf of Bonasso, who repeatedly and publically screamed at Filmus on national television and accused him of being servile to the interests of the president, of the mining provinces, and even directly of Barrick Gold.[31]

Late to the table, but now eager to engage, civil society did not stand aside in the conflict. Greenpeace, which had not registered the glacier protection law before the veto, took up the issue and became one of the leading civil society forces in the glacier debate. Juan Carlos Villalonga, Greenpeace Campaign Director, took on the issue personally, confronting Governor Gioja of San Juan. Gioja was staunchly against the environmental movement that opposed mega mining in his province, and he agreed to debate Greenpeace in a televised press conference, which happened a few days shy of the final glacier law vote in the Senate, in September 2010.[32] Gioja defended San Juan's prerogative to choose mining as a development model for the impoverished province and upheld that mining was not impacting glaciers. As he had two years earlier in his meeting with Picolotti just after the veto, he held up the IANIGLA report by Lydia Espizua, which, according to the Governor of San Juan, proved that Veladero and Pascua Lama were not affecting glaciers.[33] This gesture was quite revealing for those who knew the underlying issues. Nearly two years had passed and still the authorities in San Juan did not see or understand the conflicts between Pascua Lama and the periglacial environment.

Other groups such as the Argentine Association of Environmental Lawyers, Conciencia Solidaria, and the Center for Human Rights and Environment (CEDHA), began producing analytical documents to help congressional representatives and the general public understand the issues underlying the debate.[34] These organizations themselves had to learn about glaciers and periglacial environments, since they had never focused on these natural resources before the congressional debate ensued in 2008.

The most heated televised debate between Filmus and Bonasso took place on July 7, 2010.[35] Following acidic exchanges and the seemingly irreconcilable differences between the two representatives, Filmus contacted Bonasso and proposed a private meeting that would go public only if they reached an agreement. Bonasso agreed to those conditions, and they met at his office on Tuesday July 13, 2010,[36] only a day before the scheduled Lower House hearing of the Glacier Bill 2.0.

That day, Daniel Filmus and Miguel Bonasso hammered out the historic text of a new glacier law, one that would be referred to henceforth simply as the "Bonasso-Filmus Agreement."[37] After their meeting, they decided to distribute the agreement publicly. Considering the implications of the negotiated text, this was probably a key factor in carrying it forward in the public domain and for its survival in Congress.

The Bonasso-Filmus Agreement

Bonasso stated publicly that he was surprised to see Filmus make the concessions that he did in this new document. He commended him for his bravery and independence. Something had happened. A new and improved glacier protection bill was born on July 13, 2010.

Gone was the "environmental services" orientation of the Filmus Law. Gone was the mention that glaciers were preserved for industry or for hydroelectric usage. Glaciers and periglacial areas were back to being a public good, and the agreed upon text that would go to the congressional floor took on the same conservationist approach of the vetoed Maffei Law.

In Article 2 of the agreement, the glacier and periglacial definition was broad and inclusive, accepting glaciers of all shapes and sizes. An important specification was made for the protection of periglacial areas: they were protected if they were contributing as basin regulators (legally freeing up frozen grounds with no hydrological content). Glaciers and periglacial areas would be inventoried. The rest went more or less along the lines of the original Maffei/Bonasso Law.[38]

The one significant concession that Bonasso made to Filmus was in Article 11, on sanctions in the case of violations of the law. Here, the ball rolled in favor of the miners and the mining provinces:

> Art. 11.—*Infractions and Sanctions. The sanctions for non-compliance of the present law and the regulations that shall be introduced, beyond other responsibilities that might apply, shall be those that are established by the jurisdiction according to its corresponding policing power and which shall not be lower than those established here. Jurisdictions that do not have a sanctions regime, shall apply the following sanctions which correspond to the national jurisdiction:*

The text stated basically that if a mining company violated the law, it was the province that would rule on the sanction, not the federal government. This would keep the veto power over mining projects in the hands of provincial authorities.

To this day, a mystery that remains is why Filmus conceded to Bonasso and dropped the services-oriented Filmus Law. Curiously, the Senate would go on to vote between the Bonasso-Filmus Agreement and the Filmus Law—which, ironically, Senator Filmus no longer supported. The agreement dropped all wording important to the mining sector, keeping only Article 11 as a last resort that provided a loophole to give provinces discretion over when and if they would suspend mining activity due to glacier impacts, but the rest of the law favored the conservationists and was an improvement over the original Maffei Law.

Attempts to speak to Filmus on this point before the publication of this book were unsuccessful, but several possible reasons could be put forward to explain his abandonment of his own Filmus Law and his near full concession to Bonasso and the conservationists:

- Filmus arguably had taken on the official position and supported the mining interests, but in reality it is possible that he did not do this as an order from the president, but on his own volition in an earnest and honest effort to produce a balanced law meeting the interests of all stakeholders.
- As social pressure grew against the Filmus Law, and as signs appeared in public spaces and in front of Congress claiming that Filmus was Barrick Gold's employee and that he was bought out by the miners, he may have felt pressure to separate his position from the official stance and from the mining sector stance.

- It is possible that, in fact, no one was supporting Filmus within government and, as such, his crusade was sustained only by himself. As he saw the official position losing ground in Congress and realized that failure was imminent, he preferred to side with the greater general public.
- Filmus came to this debate with a personal agenda seeking the Buenos Aires mayorship, and it was mostly people from Buenos Aires who were criticizing him for siding with conceding text to the mining sector and to the presidential position. In the end, Filmus was not willing to fight someone else's fight and go down with the ship when it went sank, thus losing his own constituency in the fall.
- Filmus knew the official position from the presidency and from the mining provinces would stick to his Filmus Law. He didn't need to defend it at this point and could jump ship to the other bank, in the process taking credit for the Bonasso-Filmus Agreement, which he did.
- The negotiated Article 11 was a last hold-out negotiating point; when the pro-mining lobby saw that they might lose the battle in Congress, they gave him the green light to reach an agreement on the rest of the law, but told him to hold firm on sanctions jurisdiction.

The final irony of the Bonasso-Filmus Agreement was that just when everyone believed that the Glacier Bill 2.0 would end up falling somewhere in between the very conservationist Maffei Law and the environmental services-oriented Filmus Law, it actually ended up being even more conservationist than the original Maffei Law.

July 14, 2010—The Lower House of Congress—Buenos Aires. The Bonasso-Filmus Agreement took everyone by surprise, particularly those who had by then become savvier and better understood the technical issues being discussed. The more hardline environmentalists doubted the agreement. Did we get the wrong version of the text? What was going on? How could the mining lobby accept this? "This can't possibly be okay by Filmus," was the sentiment as the bill went to the congressional floor that evening.

A lengthy debate and fourteen-hour marathon session ensued in the Lower House. The president's lobby was trying to draw out the debate in hopes of losing the quorum necessary to keep the congressional session open. During the session, Agustin Rossi, the president's point man in the House of Deputies, made a motion to temporarily suspend voting until early August because his party (the president's party) of eighty-seven members had not had enough time to read and review the agreement. Critics responded by pointed to the voting in the provincial legislatures where no discussion or reading took place whatsoever and that did not seem to bother governing party representatives at that time.

Bonasso rejected the proposition outright and made the accusation to the floor that the provinces were buying time to pass their own provincial glacier laws before the national glacier protection law went into effect. This was true to some extent, but several provincial laws had already been passed, irrespective of the national congressional outcome. The motion for the delay of the vote went to the floor and was rejected.

The Lower House went on to hear thirty-nine speeches from the floor for more than fourteen hours. The most heated debates centered on historic abuse and the hypocrisy of a wealthy federal jurisdiction over poorer provincial ones. The mining provinces pleaded with Congress that they be allowed to develop industrially and that Congress not use glacier conservation as an excuse for further impediments to such development. Environmentalists argued for the protection of water over multinational interests. Accusations of presidential as well as congressional servitude to Barrick Gold and to foreign interests abounded, as did those pointing to arrogant extremism in the environmental camp.

But as the night dragged on, it became clear that the overwhelming majority favored the agreement ironed out by Filmus and Bonasso. Filmus, a member of the Senate, sent a message during the session of the House to indicate that if they voted for the Bonasso-Filmus Agreement, he would see to it that the Senate voted on the same proposal.

Once it seemed clear that the official position (in favor of the former Filmus Law) was losing ground and that the vote was near, Agustin Rossi asked for the floor to deliver a message from the president. He said that after the suspension vote was rejected earlier that evening, he had called the president to inform her of the outcome and of the evolution of the debate. She had asked him to send this message to the Congress: "whatever the decision today by Congress, I will not veto the glacier law again."[39]

Seeing an imminent failure to override Bonasso and the conservationists, President Cristina Fernandez de Kirchner was sending a warning to the mining provinces to get their act together and ensure that what happened on the floor that evening met with their approval because she was not going to intervene again on their behalf.

The House went first to a general vote. Of the 216 Deputies present (there were 87 supposedly aligned with the president), 129 voted in favor of the Bonasso-Filmus Agreement and 86 voted against. One vote (from the president's soldiers) abstained. But the vote was not over because the Congress now had to hear and approve the bill article by article. Articles 2 (definitions), 6 (prohibitions), and 15 (retroactivity) were all key, controversial, and could still derail the law or end up in revisions that could water down the law significantly. No one was talking about Article 11.

The House of Representatives began listening to the reading of the Bonasso-Filmus Agreement, article-by-article, voting after each. The quorum was maintained, and the articles approved through the reading of Article 5. But as the reading of Article 6 ensued—the article that banned mining and hydrocarbon activity in glacier and periglacial areas—those members of Congress aligned with the president walked out of the room, effectively destroying the required congressional quorum to continue the vote. The full approval of all articles would have to wait. After a failed attempt some days later, on August 11, 2010, Bonasso proposed a full reading of Articles 6 through 16, with the floor voting on all of them together, not one by one. The motion was granted, and the approval vote finally materialized. But, before proceeding to the final vote, Bonasso motioned for the incorporation of a new article, Article 17, giving the IANIGLA discretion to carry out special inventories at sites where new projects were proposed in glacier areas. It was a last ditch effort to keep some discretion over mining activity in the hands of the federal government. The House of Representatives gave their approval to the Bonasso-Filmus Agreement, including the new Article 17.[40] Half the battle was over.

September 29–30, 2010—the Senate of Argentina. A month and a half after the Lower House of Congress voted in favor of the Bonasso-Filmus Agreement, the Senate was scheduled to treat the glacier protection bill. The Senate would be a tougher battle, with the vote too close to call. Enough Senators of the president's political machinery had a favorable opinion of the Bonasso-Filmus Agreement and would not necessarily fall into ranks behind the president's preferred glacier bill the Filmus Law (no longer supported by Senator Filmus), which was more lenient to mining interests. But the president had already announced on more than one occasion that she would accept the Senate vote and would not veto the law whatever version passed. This was an invitation to break ranks, and that made official party members, the mining provinces, and mining interests very nervous.

What nobody anticipated was that the version of the law that was likely to pass, if the opposition party could secure the votes, was the version agreed-upon between Senator Filmus and Deputy Bonasso, which ironically was not a watered down version of the vetoed law but actually a stronger law.

The only modification coming from the Lower House was the elimination of Article 17, which Bonasso had introduced at the last minute before the approval vote at the House of Deputies. This article gave the IANIGLA discretion to call for glacier inventories at new proposed project areas. Article 17 was dropped.

More than thirty Senators spoke that night on the proposed glacier protection bill. The vote didn't come until after 4 a.m. The final vote count was thirty-five in favor and thirty-three against with one abstention.[41] Ironically, the president's closest ally in the Senate, Miguel Pichetto, cast the final vote (at which point the Bonasso-Filmus Agreement was winning 34–33). His vote (presuming it was against the agreement) could have forced a tie at 34 votes. That outcome would have forced the vice president (the president of the Senate) to vote to untie the count.

Julio Cobos, the vice president and member of the Radical Party that was now gaining power in the opposition, had already fallen out of ranks with the administration by untying a vote against the president's position on the soy tax several months earlier (July 2008), and he had gained enormous political leverage in Congress by doing so. This unanticipated vote against the administration propped him up as a rogue vice president, empowered him in the opposition, and earned him the scorn of President Cristina Fernandez de Kirchner and her party; they were not about to give him another opportunity to define a congressional vote against the president. So, Pichetto, against all political logic, voted in favor of the Bonasso-Filmus Agreement. Pichetto explained his vote diplomatically, stating that it showed that the president had freed everyone in her party to vote according to their own conscience and that he was proud of the debate that had taken place in the Lower House and Senate.

And so, on the early morning of September 30, 2010, the world once again had its first glacier protection law; ten days later, as promised by the president, it would enter fully into force.

Articles 1, 2, and 6 read:

Art. 1: *Subject*
The following law establishes the minimum standards for the protection
    of glaciers and the periglacial environment with the objective
    of protecting them as strategic freshwater reserves for human
    consumption; for agriculture and as sources for watershed recharge;
    for the protection of biodiversity; as a source of scientific information
    and as a tourist attraction.
Glaciers constitute goods of public character.
Art. 2: *Definition*
As per the present law, we understand glaciers to be all perennial stable
    or slowly-flowing ice mass, with or without interstitial water, formed
    by the re-crystalization of snow, located in different ecosystems,
    whatever its form, dimension and state of conservation. Detritic
    rock material and internal and superficial water streams are all
    considered constituent parts of each glacier.
Likewise, we understand by the periglacial environment of high mountains
    the area with frozen ground acting as regulator of the freshwater
    resource. In middle and low mountain areas, it is the area that
    functions as regulator of freshwater resources with ice-saturated ground.

Art. 6: *Prohibited Activities*

All activities that could affect the natural condition or the functions listed in Article 1, that could imply their destruction or dislocation or interfere with their advance, are prohibited on glaciers, in particular the following:

a) *The release, dispersion or deposition of contaminating substances or elements, chemical products or residues of any nature or volume. Included in these restrictions are those that occur in the periglacial environment;*

b) *The construction of works or infrastructure with the exception of those necessary for scientific research and to prevent risks;*

c) *Mining and hydrocarbon exploration and exploitation. Included in this restriction are those that take place in the periglacial environment;*

d) *The installation of industries or the building of works or industrial activity.* [unofficial translation]

The conservationists had won.

Notes

1. For a partial video of the session, see https://www.youtube.com/watch?v=4JC8IidW-Lw.

2. For press statement by Villalba against the veto, see https://www.youtube.com/watch?v=Fa56y1IjnvY; see also https://www.youtube.com/watch?v=zxCompsZiFM.

3. See http://walterrojas.wordpress.com/2008/12/09/novedades-sobre-ley-de-proteccion-de-glaciares/.

4. See http://www.infoleg.gov.ar/infolegInternet/anexos/145000-149999/146980/norma.htm.

5. For the Filmus Law version of the glacier protection bill, see http://www.cedha.net/wp-content/uploads/2011/04/v-filmus.pdf.

6. See http://www.diariodecuyo.com.ar/home/new_noticia.php?noticia_id=414581.

7. See http://www.elintransigente.com/notas/2008/11/27/nacionales-7750.asp.

8. See http://eventos.senado.gov.ar:88/8316.pdf.

9. See radio interview with Bonasso on the publication of *El Mal*: https://www.youtube.com/watch?v=Q1LrovsyUMk.

10. See Bonasso, *El Mal*, 2011, p. 307.

11. http://www.infoleg.gov.ar/infolegInternet/anexos/145000-149999/146980/norma.htm.

12. See Espizua, 2006, p. 44.

13. According to records, the following governors were invited: Eduardo Brizuela del Moral (Catamarca), Mario Das Neves (Chubut), Walter Barrionuevo (Jujuy), Luis Beder Herrera (La Rioja), Celso Jaque (Mendoza), Jorge Sapag (Neuquén), Miguel Saiz (Río Negro), Juan Manuel Urturbey (Salta), José Luis Gioja (San Juan); Daniel Peralta

(Santa Cruz), and Fabiana Ríos (Tierra del Fuego). There is no record of the governor of Tucumán Province being invited to the meeting, despite it having numerous rock glaciers and periglacial areas along the Aconquija Range (bordering with Catamarca). This was likely an oversight because very few people associated Tucumán with glaciers.

14. See http://www.miningpress.com.ar/nota/36344/glaciares-picolotti-filmus-y-provincias-harn-nuevo-proyecto; or http://laangosturadigital.com.ar/v3/home/interna.php?id_not=7398&ori=web; or http://www.elintransigente.com/notas/2008/11/27/nacionales-7750.asp.

15. See, for example, https://www.youtube.com/watch?v=zxCompsZiFM.

16. For the original Filmus Law, see http://wp.cedha.net/?attachment_id=14232.

17. For Bonasso's initial reactions, see https://www.youtube.com/watch?v=4JC8IidW-Lw; for later positions, see https://www.youtube.com/watch?v=Tptw4rYIALM.

18. For the original Maffei/Bonasso Law, see http://eventos.senado.gov.ar:88/8316.pdf.

19. The Santa Cruz Glacier Protection Law, see http://wp.cedha.net/?attachment_id=9445.

20. The Santa Cruz Glacier Protection Law: http://www.santacruz.gov.ar/ambiente/leyes_provinciales/ley%20N_3123%20de%20Glaciares.pdf.

21. The Chubut Glacier Protection Bill: http://wp.cedha.net/wp-content/uploads/2011/04/glaciares-docs-ley-glaciares-chubut.doc.

22. Including governors from La Rioja, San Juan, Jujuy, and Salta.

23. See http://wp.cedha.net/wp-content/uploads/2011/04/glaciares-docs-ley-glaciares-declaracion-gobernadores.doc.

24. See http://www.pagina12.com.ar/diario/elpais/1-149331-2010-07-12.html.

25. The La Rioja Glacier Protection Law: http://wp.cedha.net/wp-content/uploads/2011/09/Ley-de-Glaciares-La-Rioja.pdf.

26. For the San Juan Glacier Protection Law: http://wp.cedha.net/?attachment_id=10857.

27. For the Salta Glacier Protection Law: http://wp.cedha.net/?attachment_id=9543.

28. For the Catamarca Glacier Protection Bill: http://wp.cedha.net/wp-content/uploads/2011/04/Ley-de-Glaciares-para-la-Provincia-de-Catamarca.pdf.

29. For the Rio Negro Glacier Protection Bill: http://www.cedha.net/wp-content/uploads/2011/08/Proyecto-de-Ley-Glaciares-Rio-Negro.doc.

30. For the Neuquén Glacier Protection Bill: http://wp.cedha.net/?attachment_id=9538.

31. For the heated Filmus-Bonasso debate, see https://www.youtube.com/watch?v=wBqZoPcPi7Y.

32. For the Gioja/Greenpeace debate, see https://www.youtube.com/watch?v=Qms2_t-oNco.

33. For the Espizua report: http://wp.cedha.net/wp-content/uploads/2011/11/Informe-Glaciares-Lama-Veladero-Espizua-2006.pdf.

34. See http://wp.cedha.net/?page_id=1277.

35. See https://www.youtube.com/watch?v=wBqZoPcPi7Y.

36. Bonasso recounts the secret meeting events in *El Mal*, 2011, pp. 396–397.

37. The Bonasso-Filmus Agreement: http://wp.cedha.net/wp-content/uploads/2011/04/0078-S-2009.pdf.

38. For a full analysis of the Bonasso-Filmus Agreement, see http://www.cedha. net/wp-content/uploads/2011/04/glaciares-docs-analisis-articulo-por-articulo-acuerdo-bonasso-filmus.doc.

39. Quoted in Bonasso, 2011, *El Mal*, p. 411.

40. To see the actual session: https://www.youtube.com/watch?v=yFYp2fhEW3M.

41. For an analysis of the voting, see http://www.perfil.com/politica/El-Senado-volvio-a-proteger-glaciares-20100929-0027.html.

Amazing Glacier Stuff

We are the first generation to feel the sting of climate change, and
we are the last generation that can do something about it.

—*GOVERNOR OF WASHINGTON STATE, JAY INSLEE*

This chapter is about what glaciers—and particularly what glacial and periglacial melt—mean to people and communities around the world. We often don't realize that people interact daily with glaciers. Some go to visit and hike on glaciers or to photograph them for their magnificent beauty. Some ski on glaciers. Others extract water from glaciers for personal and industrial use. Others fear glaciers for their potent fury and destruction.

People and communities are adapting to climate change and its impacts on glaciers, sometimes without even knowing it. Others are very aware of glacier vulnerability and are taking measures to address the changing cryosphere. They are mitigating circumstances and are adapting to impacts. In this chapter, we share stories and facts about glaciers and periglacial environments, which most people are probably unfamiliar with, and we explain how lives in these environments are changing due to climate change.

Glacier Tsunamis: GLOFs

Few people have heard of *glacier tsunamis*,[1] but they exist, they're real, they're ferocious, and they can kill. Scientists call them *glacier lake outburst floods* (GLOFs). And as climate change deepens, more and more GLOF phenomena can be expected.

Imagine you live at the foot of a mountain range like the Rocky Mountains, the Himalayas, or the Central Andes. On a nice sunny day, you can see the snow-capped mountains in the distance, maybe 20 or 30 km (12–18 mi) out, maybe even more. You are sitting at home when all of a sudden you feel shaking and hear a rumble. People start screaming. You look out the window and see people running frantically and erratically about. Then a woman yells, "The mountain! It's coming! Run!"

Imagine a large glacier the size of a dozen or so city blocks, perched atop a mountain. It's 180 meters thick (600 ft), which is as tall as a sixty-story building. Below it, time and climate have formed a lake, a *glacier lake* occupying the same spot where the glacier once rested, pushing rock and earth out and forward as the glacier flowed downhill when it was solidly frozen and healthy. As the climate warmed and much of the glacier melted away, it began leaving a big empty space where it once thrived. The void is now surrounded by walls of rock and earth that the glacier had displaced for hundreds, maybe thousands of years (these are called *moraines*). This space, some 350 meters (1,200 ft: that's as tall as the Empire State Building) deep is now filled with icy meltwater constantly streaming from the remaining ice of the glacier, now dangerously perched above the lake and slowly melting, slowly shifting. Large chucks of ice, abruptly and unannounced, break away from the glacier body as more of the ice melts and the glacier deteriorates as the climate continues to warm. Trekkers come regularly to the glacier to witness its majestic beauty. They take boat rides to the front of the glacier and gaze serendipitously, watching as pieces fall in slow motion into the lake. Anyone that has witnessed this phenomenon can't but recall it in extreme natural awe.

One day, maybe on an especially warm day or maybe on just a normal day, unannounced, a large piece of ice, or more appropriately an enormous piece of ice, the size of a city block, breaks off, plunging hundreds of meters (thousands of feet) down onto the surface of the lake. The plunge instantly forms a giant wave, 20 meters (65 ft) high, racing across the lake toward the moraine. It's so powerful and so large it not only surpasses the natural dam, sending water over the edge and down the canyon below the lake, but it actually breaks through the moraine, leaving a gaping hole 90 meters (300 ft) wide and 120 meters (400 ft) tall and causing a full and instantaneous drainage of the lake. Billions of liters of water instantly come gushing forth, mixed with ice boulders as large as a house, creating a deadly ice and water wave taking out anything in its path, picking up stones, ripping trees from the ground, causing massive mud slides, and gaining strength as the avalanche of ice, water, rock, earth, snow, and vegetation comes rushing down the mountain canyon at more than 100 km per hour (60 mph). That's why your neighbor screamed, "The mountain is coming!" For anyone witnessing this sight, it must be an absolutely terrifying experience, but it is one that has happened many times and will likely happen many more.

Mark Carey has written a book called *In the Shadow of Melting Glaciers: Climate Change and Andean Society* that provides a fantastic review of one region of the world—the Peruvian Andes—where glacier tsunamis are a common occurrence. He tells the chilling tale of Peruvian GLOFs and how climate change is creating deadly glacier lakes waiting to burst down on communities below. He calls the section that describes the phenomenon, "The Killing Ice of the Andes":

Peruvians have suffered the wrath of melting glaciers like no other society on earth. The Cordillera Blanca mountain range in north-central Peru contains more than 600 glaciers, which account for about one quarter of the world's tropical glaciers and half of Peru's glacial ice. These mountains rise up more than 6,000 meters [nearly 20,000 feet], jutting 3,000 meters [nearly 10,000 ft] above surrounding valleys to make not only a stunning landscape but also a place—as in Scandinavia, Alaska, Iceland, Canada, the Alps, and the Himalayas—where people live in close proximity to glaciers. These Cordillera Blanca glaciers, like most others worldwide, have been retreating since the 400-year Little Ice Age ended in the late nineteenth century. Over time, hundreds of thousands of Peruvians inhabiting mountain slopes and river valleys found themselves living directly beneath crumbling glaciers and the swelling glacial lakes that often formed below them, dammed precariously behind weak moraines (piles of rocks). The number of Cordillera Blanca lakes has risen dramatically from 223 in 1953, the year Peruvians conducted the first lake inventory, to more than 400 today.[2]

Carey goes on to recount many of the glacier tsunamis that have killed tens of thousands of people in Peru over the past several decades. Until the nineteenth century, says Carey, death tolls from GLOFs were not that large, but in 1941, things changed. Lake Palcacocha burst through its natural moraine and wiped out the town of Huaraz, killing 5,000 people in its wake and leaving a 200 km (125 mi) path of destruction. A colossal chunk of ice breaking off the glacier, seen in Figure 8.1, smashed into the glacier lake, causing such an enormous wave and impact that it took out a large portion of the natural dam, nearly 100 meters (330 ft) high, left by the retreating glacier. The ice, water, and debris picked up by the flash flood came crashing down the steep mountain gorge, wiping out Huaraz and killing thousands.

In 1945, 500 people died at Chavín de Huantar. In 1950, 200 people died when the Cañon del Pato hydroelectric dam was overrun by a glacier tsunami. In 1962, Mount Huascarán burst into a glacier avalanche and swept through the town of Ranrahirca, killing 4,000. And then, in 1970, came the most deadly glacier disaster in world history when another avalanche from Mount Huascarán expelled a mass of glacial ice about 730 meters (2,400 ft) wide—or about eight city blocks—that came crashing down into the lake below, causing a glacier tsunami carrying 50 million cubic meters of debris and ice nearly 15 km (9 mi) down the valley at more than 200 km/hr

FIGURE 8.1 *Collapsing chunks of ice into glacier lakes at the base of Peruvian glaciers can be deadly for downstream communities. Note the break in the moraine and detail of proximity to communities below. GIS: 9°22'35.52" S 77°23'19.74" W.*

(125 mph), triggering an earthquake killing 15,000, leveling the town of Yungay, and bringing the total death count in Peru due to glacier tsunamis to nearly 25,000.[3]

We can expect similar tragedies to occur as other glaciers upstream from communities around the world melt and become unstable. But glacier tsunamis are not only a risk for Peruvians. In May 2012, in Pokhara, Nepal, Mount Machhapuchure of the Annapurna range produced a GLOF into the Seti River that was documented by local residents with their cell phones. The videos show the instantaneous and unstoppable ferocious force of these phenomena.[4] The Tsho Rolpa Lake in Nepal, a little more than 100 km (62 mi) northeast of Kathmandu, is one the most GLOF-prone and high-risk lakes of the world due to the progressive melting of the Trakarding Glacier.[5] In 1985, a GLOF event at Dig Tsho caused millions of dollars of damage and disrupted the lives of downstream communities wiped out by a GLOF.[6] Another nameless but very deep lake, referred to by Daniel Byers simply as Lake 464 in the Hongu Valley, poses enormous risks to downstream communities. Byers produced a short video useful for understanding glacier lake dynamics and risks.[7] In total, fourteen GLOF events have been documented in Nepal in recent decades.

Terrifying images were also captured of the Island Mountain Glacier (in Icelandic, the Eyjafjallajökull) GLOF.[8] Also in Iceland, in 1996, when the volcano Grimsvotn erupted and caused a glacier lake overflow, the resulting river flow exceeded 50,000 cubic meters per second, with a wall of ice water sludge some 4 meters (13 ft) high destroying everything in its path and carrying ice blocks of some 100–200 tons.

In Bhutan, with thousands of glacier lakes, Himalayan range glacier melt has greatly increased the risk of GLOFs; the government there has launched a GLOF study and is developing emergency action plans for many communities living downstream from glaciers at risk.[9] Several documented GLOFs have occurred in the last several decades including in 1950, 1960, and 1968. In 1994, a lake in the Lunana area of northern Bhutan burst, killing twenty-one people.[10]

In the United States, Grasshopper Glacier in Wyoming caused a GLOF in 2003, when the lake at the head of the glacier burst through the natural glacier dam, sending an estimated 2,460 million liters (650 million gal) of water and debris for more than 32 km (20 mi) downstream.[11] GLOFs and risks of GLOFs have also been documented in Alaska, Canada, Chile, Pakistan, parts of Europe, Scandinavia, Kyrgyzstan, China, and Tibet.

Lakes can also form on *and even inside* glaciers, and, as ice shifts due to temperature changes and melting, a glacier can suddenly release an entire lake's worth of water into rivers and valleys below. Live footage of such a release was captured in 1988, when Argentina's Perito Moreno's natural ice bridge and dam broke suddenly, releasing an entire dammed lake into another lower portion of the lake. This rare video is a must-see in order to comprehend the force of nature in action at glacier sites.[12]

These phenomena are likely to occur more frequently as more glaciers melt away as average ambient temperature rises. Groups like the International Center for Integrated Mountain Development (ICIMOD) have been working for decades to produce knowledge and exchange information about mountain environments and glacier dynamics; as glaciers continue to retreat, more information is being produced about changing glacier environments. A USAID-sponsored program called High Mountain Adaptation Partnership (or HiMAP), for instance, has brought together people of Peru and Nepal to exchange information and experiences on melting glaciers and the risks they pose to local communities and environments.[13]

Glacier Time Machines

Glaciers hold history trapped deep within their ice. Hollywood has been obsessed with the fantastic idea of making a discovery inside of a glacier of a live Neanderthal, a dinosaur, or some other prehistoric creature—or even of an alien spaceship—that suddenly surfaces after tens of thousands or even millions of years of hibernation. Just recently, a very real photograph went viral of supposed pyramids discovered under the Antarctic ice as climate change is melting the South Pole's ice cover. The alleged pyramids were hailed as proof of an ancient civilization predating the Egyptians that inhabited the now desolate continent.[14] As it turns out, the supposed pyramids were actually naturally occurring pyramidal peaks and/or small rock protrusions that also occur naturally in such icy areas, called *nunataks*—it was a fascinating idea while it lasted!

But, in fact, glaciers do hold fascinating secrets, although maybe not of the sci-fi type many of us would like to believe. Animals, rocks, people, trees, and even entire forests can and have been swept up by glaciers and buried for many years, centuries, and even millennia, one day to resurface, largely preserved, sometimes mummified, revealing bits and pieces of an unknown past. The comical trailer animation of Blue Sky Studio's animated film *Ice Age*, of a prehistoric squirrel falling into the ice and emerging thousands of years later to continue pursuing her nut is not so far from the truth: although ice will not preserve complex organisms alive, it can preserve organic and other matter in very good condition, so that, when revealed, we not only learn about prehistoric organisms, but we can also gauge environmental conditions as they were many years before. Depending on where you fall into a glacier— if into actively moving surface ice or into slow and stationary core ice—your resurfacing might take anywhere from a few years to a few centuries, and even a few millennia or more.

Ötzi the Iceman

Ötzi the Iceman was discovered in the depths of glacier ice in the Italian-Austrian Alps in 1991. He was by then 5,300 years old! He was so well preserved that scientists could determine that eight hours before his death (in 3,300 BC), he had eaten a well-rounded meal consisting of deer meat and herb bread, grains, and even fruit for desert! They knew (by calculating when the fruit harvest might have been) that he died in springtime. The scientists also could determine what his diet was like—both during his childhood and also as an adult—and his medical conditions, including illnesses that he suffered and when he suffered them. They could put together some pretty good and very specific theories of how he died, his activity before his death (which included struggle and possible homicides that *he* committed); they could determine that there were people with him when he died who helped him in his final hours and several more rather detailed and incredible bits of information derived from preserved clues in the freezing environment.[15] Glaciers are, in fact, very good storytellers!

Glaciers and Indigenous Peoples

Glacier treasures and remains have also helped indigenous communities learn details about their past. Doug Macdougall, in his book *Frozen Earth*, reports how archeologists and biologists have begun making systematic surveys of melting glaciers in Alaska and northern Canada, not to monitor their retreat, but to retrieve the whole animals, human hunting implements, bones, and even the fresh-frozen animal dung that is disgorged as the glaciers melt back. He cites that, in 1990, melting ice in northern British Columbia yielded a human body of about 550 years accompanied by clothing and various tools and implements, all frozen in glacial ice. Named Kwaday Dan Sinchi (long ago man found), this man has become the focus of intense interest on the part of both native peoples of the region and scientists. Analyzing DNA from humans and other species preserved in glaciers has the potential, says Macdougall, to open up whole new areas of biology and anthropology for investigation (Macdougall, 1994, p. 11).

In the Yukon, a receding perennial ice field exposed a 4,300-year-old dart shaft in caribou dung. Other artifacts, such as such as 2,400-year-old spear throwing tools, a 1,000-year-old ground squirrel snare, and bows and arrows dating back 850 years, have also been found under receding ice. A 1,600-year-old copper implement predating indigenous European contact has been found, suggesting advanced smelting knowledge held by indigenous tribes of the time.[16]

The Llullaillaco Children

In 2007, an Inca mummy of a 15-year-old girl as well as two younger children (named the Llullaillaco children), preserved in ice for more than 500 years, emerged as if they had just passed away a week or so earlier.[17] The natural high mountain deep-freezing process of mummification did a much better job of preserving the children's bodies than have elaborate artificial embalming techniques used in lower altitudes by ancient civilizations. With DNA testing, a descendant of the girl was found in Peru, thus establishing direct lineage to the Inca children.

Gelid Climbers

Glaciers can and often do swallow up people in accidents. Mountain climbers have fallen into glacier crevasses and, through those tragic accidents, we've learned something about glacier movement. As crazy as it may sound today, there was a time when our society didn't realize that glaciers actually moved over the surface of the Earth. Mariana Gosnell in *ICE* (2005, pp. 85–93), recounts one of the first global awakenings to the fact that glaciers can swallow up people and churn them in the depths of the ice for many decades before spitting them back out. It was in the nineteenth century that scientists began to conclude that glaciers advanced, and frozen climbers helped make that leap of knowledge. Gosnell recounts reflections of John Tyndall, a prominent nineteenth-century physicist who studied glacier motion. One of the pillars of evidence Tyndall offered to prove that glaciers actually move was corpse transfer. Gosnell cites Tyndall:

> Here at the head of the Grand Plateau, and at the foot of the final slope of Mont Blanc, I should show you a great crevasse, into which three guides were poured by an avalanche in the year 1820 (he is speaking in 1871). ... The crevasse was so deep in 1820 that the guides could not be rescued, and gradually it filled with snow and closed over them. Decades passed, during which scientists learned more about glaciers, enough so that one of them, English geologist J. D. Forbes, had the temerity to predict not only that the guides would reappear but when—35 or 40 years after they vanished—and where—at the foot of the glacier.

Almost on cue, tells Gosnell, the guides reappeared forty-one years later:

> On August 12, 1861, another (luckier) mountain guide discovered, at the edge of a crevasse at the foot of the Bossons Glacier (according to Stephen d'Arve's Histoire du Mont Blanc, as condensed and quoted by

Twain), several dreary objects: three human skulls, or parts thereof; a jaw "furnished with fine white teeth"; tufts of black and blond hair; a forearm (flexible); a hand, all fingers in tact, blood visible on the ring finger; a left foot, "the flesh white and fresh"; portions of waistcoats, hats (one straw, one felt), and hobnailed shoes; a pigeon's wing (one of the guides had planned to free a cagefull of pigeons on reaching the summit of Mont Blanc); a boiled leg of mutton, which, though odorless, when extracted, soon was not. (Gosnell, 2005, p. 93)

She goes on to recount how friends of the hikers reacted to the find forty years after the accident.

Two guides who had survived the avalanche of 1820, elderly by 1861, identified the objects as belonging to their long-lost colleagues, with one of the old men clasping the resurrected hand, grateful for the chance to touch again the flesh of his brave friend before he too quit this world. A blood-stained green veil emerged from the ice the next year ... as did a second arm, whose "extended fingers seemed to express an eloquent welcome to the long-lost light of day." (Gosnell, 2005, p. 93)

You can also scale up Mt. Everest to find the world's highest glaciers and meet up with a number of famed mountaineers who have frozen in place, some to reappear as climate change begins to thaw ice. As many as 200 frozen bodies of unlucky climbers are purportedly preserved on the world's tallest mountain; by 2011, 216 mountain climbers had fallen in the area above 26,000 feet known as "the death zone," where the air is light (atmospheric pressure is about one-third what it is at sea level) and oxygen tanks are obligatory.[18] Generally, if you fall and break a leg at this altitude, you're a gonner because rescuing you is often impossible.

Perhaps the most famous frozen mountaineer is (or was) Everest hiker George Mallory, who disappeared with his climbing partner Andrew Irvine on the peak in 1924. Some mountaineering historians suppose that Mallory and Irvine were actually the first humans to reach the peak on this tragic ascent. That honor was instead given, in 1953, to Sir Edmund Hillary and his Nepalese partner Tenzing Norgay. The Mallory-Irvine team was last seen 900 feet below the 8,848 meter (29,028 ft) mountain. Mallory's body finally reemerged from the ice in 1999, seventy-three years later, sun bleached, frozen, and mummified.[19]

Such accidents take place around the world in several mountain glacier zones. The life of William Holland, a geologist, would end in tragedy when he fell into a glacier in 1989 while climbing Mt. Snowdome, in the Columbia Ice Field. He would not reappear for more than twenty years as he was slowly transported through the glacier ice core. Hikers came upon his frozen remains at the front of an advancing glacier.[20]

Chilly Flights

Airplanes have crashed into glaciers and disappeared into the ice, their debris frozen in time and churned through the cryogenic dynamics of glacier evolution and movement only to be spit out years or even decades later. The entire crew of a Globemaster C124 US military plane perished when it crashed into Colony Glacier in Alaska in 1952. Debris started surfacing nearly six decades later.[21] The same occurred in Argentina, with British South American Airways Flight CS59 disappearing in August 1947 in the Central Andes. The flight crashed into a glacier on Mt. Tupungato and only surfaced some seventy years later.[22] In 1991, on Chilean glaciers, a Rolls Royce airplane engine was pulled from the ice, presumably from that same crash. In 1983, the remains of another military plane that disappeared in Patagonia in 1950 popped out of a glacier terminus in the Perry Fjord in Chile. The frozen environment had conserved the remains for more than thirty years.[23] In 2003, a Russian plane that disappeared in 1968 with 102 souls on board resurfaced from the Dhakka Glacier in the Himalayas, in India.[24] Although it seems to take glaciers a matter of decades (depending on length) to process debris from entry to exit on active portions of the glacier, perhaps as glacier melt advances and full melt of core ice occurs we will begin to discover what might lie in the profound depths of millenary glaciers. Who knows? We may still find some interesting surprises in these massive time-frozen environments.

Frozen Treasures

How about glacier treasures? In September 2013, a glacier climber in Chamonix, France, stumbled upon a stash of jewels including emeralds, rubies, and sapphires, neatly packed away in the ice, in containers labeled, "Made in India." The origin of the jewels is believed to be from one of two ill-fated plane crashes, one in 1950, when Air India Flight 245, the "Malabar Princess" went down in a storm, or Air India Boeing 707, the "Kanchenjunga," that crashed at nearly the same spot near the Mont Blanc sixteen years later.[25]

Glacier Messages

Glaciers have also been purposefully employed by people to tell us stories. In December 2013, a fifty-three-year-old message was found in a container, emerging from a glacier on Ward Hunt Island in Canada. The note, written in 1959, was from Paul Walker, a geologist studying the glacier; he had recorded measurements of the glacier hoping that his observations would one day resurface and future scientists could use his observations to better understand the glacier's dynamics.[26]

Glaciers and Our Earth's Past

But a glacier's knowledge of history is actually far broader and far more informative than the mere circumstantial bedside tales that we can extract from these recent glacier findings and mysteries. We can use millenary ice to reveal profound bits of information about our planet's past environment.

Much like trees that add bark layers each year, and capturing environmental contaminants seeped up through their roots between years revealing details of what the climate was like in our past, glaciers can also tell us their age and past atmospheric air quality through their stacked layers of fresh deep snow alternating with thin layers of particle contamination deposited on glacier surfaces between snowfalls.

Heavy snow accumulation during intense snowfall is generally stored in thick pristine white layers of compacted snow-turned-to-ice. When the snow stops for several months, the ice surface hardens, and dust particles in the air accumulate on the surface for many days, weeks, or even months on end. These successive layers repeat themselves, capturing atmospheric information neatly divided into very distinguishable strata. Air bubbles and airborne particles get trapped in these layers, suspended in time, and may remain untouched by new atmospheric conditions for many years, decades, or centuries. In this way, glaciers that may be many meters thick, even several kilometers thick, can store a massive amount of layered atmospheric information spanning thousands of years. Scientists remove these cores very carefully in tubular containers and then use fluorescent light to distinguish a series of dark and light bands, similar to tree rings. Each light band represents one summer, whereas dark bands correspond to winter. The difference has to do with compression and air content. Together, a dark and light band form a year of climate information.[27] Ice cores taken from Antarctica and from parts of Greenland have revealed history dating back over 100,000 years and even as far back as 750,000 years (see Macdougall, 1994, pp. 178–179).

We can learn a lot about our past from these layers of contamination on glacier surfaces. We can count ice layers to count years, and we can measure yearly quantities of snowfall and thus know how old the glacier is and how much it snowed each year; we can analyze dust particles trapped between layers and know if, for example, a volcano erupted and when it erupted, or if there was a large forest fire that year, or if man-made carbon emissions resulted in an increase of black carbon pollution in a given year or season.

Using dust analysis, we can even tell where the source contamination was from, even if it was on the opposite side of the planet!

We can also take ice core samples through the ice and analyze air bubbles captured in the ice to determine air quality in past eras. In this way, scientists can and do learn from glaciers quite a lot about our planet's climate history.

Glacier Dust

Kimberley Casey, a NASA glaciologist, is an artist-turned-scientist. She studies glacier dust and can tell you by swiping a finger over a glacier just where the dust comes from, be it a volcano on the other side of the planet, smog from Los Angeles, exhaust from a car driving by, or emissions from a nearby coal-fired plant. Actually, Kimberley uses satellite images to identify dust types and their sources on glacier surfaces, and she has produced glacier debris maps discriminating dirty and clean ice, but also describing the composition and source of the particulates found on glaciers. This could be fundamental research to later determine liability for global climate change impacts.[28]

Extraterrestrial Glaciers?

Hollywood has on occasion attempted to fly earthlings into our solar system to ward off asteroids or land on ice-encrusted comets and meteors. But are there glaciers in outer space? The presumption is quite rational, and the probability that there are glaciers beyond our Earth is very real. In fact, considering that the properties of physics probably hold true throughout our vast universe, as long as we have rocks, ice, and gravity working their magic together, we can expect to find glaciers (particularly rock glaciers) on distant planets. With the discovery of many dozens, even hundreds of planets similar to Earth, the probability of the existence of extraterrestrial glaciers grows quickly with each new find. That's exactly what scientists are studying on the surface of Mars in their quest to find water on the Red Planet. We read of the work by Whalley and Azizi (2003) in Chapter 4 that looks at possible rock glaciers on the Martian surface (see Figure 4.6). You can visit these alleged rock glaciers on Google Earth simply by placing Google Earth in the "Mars" mode and going to the following GPS coordinate: 42°11′03.33″ N, 49°59′11.61 E (rotate the image so that N points down to see the image the same way we portrayed it in Chapter 4). The site shows what seems to be evidence of talus rock glaciers at the base of an incline, suggesting that that there once were, or still may be, ice and glaciers on Mars!

Gas Hydrates: The Ice That Burns

Have you ever heard of ice that can catch on fire? Well, you just may need your fire extinguishers on your next glacier trek, because it exists! As glaciers melt, and once permanently frozen grounds containing rich organic matter in some previous planetary era is uncovered, the organic matter releases captured methane. Methane is a very flammable greenhouse gas, many more times more potent than carbon dioxide (CO_2) in terms of global warming.

Gas hydrates are ice-like crystalline molecular complexes formed from mixtures of water and methane gas, making this form of ice extremely flammable. To be clear, it's the methane in the ice that catches on fire and not the ice itself (Figure 8.2).

The petroleum sector has expressed much interest in gas hydrates because this is the most abundant source of natural gas on our planet. According to the US Geological Survey, global stocks of gas hydrates outnumber conventional deposits 10:1. The problem for the oil and gas sector is that we don't know of an efficient way to capture gas hydrates. They are so dispersed in our

FIGURE 8.2 *The ice that burns: gas hydrates are a mix of frozen water and methane. Light a match to the ice and it catches fire!*
Source: USGS.

environment that bringing them into a manageable centralized stock would be far too expensive to merit the effort.

Of interest to our debate about glaciers and frozen environments, gas hydrates are found in permafrost areas, as well as below the surface of the ocean (the vast majority is in the oceans). Experiments are under way to tap into gas hydrate reserves. In Canada, experiments took place in 2008 to extract gas hydrates from permafrost terrain, but we are still far from making such extractions economically viable.[29] Currently, companies are exploring the Central Andes (in Mendoza Province) near glacier and periglacial environments for shale gas reserves.

Protecting Glaciers?

So, after dedicating the better part of this book to the evolution of a glacier law passed in Argentina in 2010 and to the impacts to glaciers of anthropogenic activity, it's pretty clear that I believe in glacier protection. I cannot accept the fatalist view that we simply need to stand by and watch our climate change, giving in to the fact that glaciers are melting, that they will continue to melt, and that soon they will vanish, and that there is simply nothing we can do about this.

As a global society, we need to work to slow climate change and even attempt to reverse it. This means, in the long run, ending fossil fuel consumption and moving fully to renewable and sustainable energies. But this book is not about this quest, although this goal is something that I am personally dedicated to. This book is about the predicament of glaciers in our present context. It's about their vulnerability, it's about their changing dynamics, and it's about communities awakening to the fact that glaciers are melting and that we need to take action to deal with this phenomenon—in some cases, to brace and prepare for impacts (adaptation) and in others to change our lifestyles to try to slow or reverse that impact (mitigation).

And yes, glaciers can be protected.

ICE HOUSES

Protecting ice used to be an everyday responsibility for lots of people. Back in the old days, before household electrical refrigeration, in addition to milkman visits, *the iceman*[30] would visit, bringing big blocks of ice for family consumption. Walt Disney recently depicted the work of the ice man in its 2013 film, *Frozen*. The character Kristoff (alias "iceman") works as the iceman of Arendelle, harvesting ice, cutting out large blocks of ice during the winter from a frozen lake or other source, and transporting it back to local households.

Some houses even had "ice rooms" or small "ice houses" [31] built into their cellars to keep and preserve ice. In some communities, larger ice houses were constructed to keep ice for a larger population. Without constant refrigeration, how could this ice survive for more than a few hours? Simple: sawdust.

Sawdust is a very poor temperature transmitter, which means it works great as insulation, keeping heat and cold contained. Ice was brought to the ice house in hempen bags, placed in large vats, and then covered with sawdust and/or straw. This kept the ice from being immediately exposed to ambient heat and allowing it to survive for up to several months and even through the summer! We can experiment with this procedure in our own refrigerator, packing away ice cubes in sawdust and placing them in a plastic container. The ice can survive a very long time with this protection, and if the temperature is very cool in the refrigerator, it may survive many days or longer—although it may need a little assistance by recreating natural dynamics. Nighttime temperatures drop, often below zero, and if our preserved ice benefits from such a drop, it has a much better chance of survival. If we recreated dropping nighttime conditions in our refrigerator—for example, by placing the ice in the freezer for a few hours each day and allowing it to fully refreeze—we can make it survive many weeks.

The Aztecs and Incas used frozen soils (periglacial environments) wisely, burying perishables underneath the permanently frozen earth, which worked much like our modern refrigerators![32]

THE ICE TRADE

Back in the nineteenth century, ice was big business, mostly deriving from parts of Scandinavia (Norway) and the United States. The ice trade involved chopping up ice from ponds or lakes, carving it into large cubes or blocks, and then shipping them off by rail or boat around the world. The ice trade revolutionized commercial shipping by allowing for the transport of perishable goods. Ice even found its way to Caribbean locations for wealthy consumers who liked to drink their rum cold. Trade in ice reached the farthest corners of the planet, including locations such as China, South America, India, and Australia. At its peak, the ice trade industry employed more than 90,000 people and was valued at more than US$600 million (in present worth). Norway alone exported a million tons of ice per year. A pound of ice (or about .5 kg) sold for approximately 25 cents (about US$3.70 today). [33]

SAWDUST

Engineer Benjamin Morales Arnao of the Patronato del Museo de las Montañas Andinas, of Peru, probably got a visit from the iceman when he was a kid growing up in Huaraz, Peru. He lived through terrifying GLOF events (glacier tsunamis) that drove him to become a glacier expert. Benjamin, now in his seventies, but still a practicing glaciologist and concerned with melting glaciers in the Cordillera Blanca Mountains near Huaraz, experimented with sawdust for glacier conservation, hauling hundreds of bags of sawdust to the nearby Pastoruri Glacier and onto Mt. Chaupijanca. His experiment consisted of covering an area about the size of a football field with about 15 cm (6 in) of sawdust just as winter was ending. Then he waited to see what happened as spring arrived and everything started to melt. Well, just as he knew from his days with the iceman, covering ice with sawdust makes good thermal sense if you want to preserve the ice. After just three months, the surface covered in sawdust, towered nearly 5 meters (16 ft) above the surface of the uncovered bare ice that had been left exposed to the sun. This was a simple experiment that used an old, lost tradition to conserve ice.

GLACIER BLANKETS

As glaciers of the French and Swiss Alps and in other parts of Europe melt, some of the most visible and concerned stakeholders are the owners of ski resorts. Seasons with no snow and receding ice are indicating that ski season is getting shorter and, in some years, it can be suspended all together. At mountains such as Germany's Zugpitze[34] and France's Val Thorens,[35] ski resort owners have been rolling out white tarps after the snow season to better reflect sun and protect ice.

Making Glaciers?

"The Glacier Man,"[36] as he is called by many in India (aka Chewang Norphel), is a seventy-four-year old retired engineer who got tired of seeing glaciers melt away over many decades. So, in 1987 he decided to make them himself! How does he do it? Easy. He simply builds small embankments at various points along a small stream running through his village. These embankments create intervals of small flooded areas that freeze up in wintertime and allow for the accumulation of ice, which in turn then survives further into the spring and summer, thus acting as mini-glaciers. He calls them simply "artificial glaciers," and he has made about ten of these along rivers near communities that need the water most. One small community, Stakna, loves its Stakna Glacier, which they have learned to keep, maintain, and service. Norphel's glaciers, criticized by some academics as not really true glaciers, provide millions of liters of water to communities that would otherwise not have such abundant water resources for much of the year. This has improved crop yield and even introduced new crop cycles during months where agriculture was impossible without the artificial glaciers. He points to many added benefits of creating your own glaciers, including recharging ground-water; increase of cash crop farming, fuel, livestock fodder, and livelihood incomes; mitigation of climate change; and many benefits to soil quality. He is now looking to replicate his model in water-stricken countries like Kyrgyzstan and Kazakhstan.

MAKING ROCK GLACIERS?

In Chapter 4 we created our own rock glacier in our refrigerator. But can we reproduce natural rock glacier phenomena at a natural large scale? "Yes" says Cedomir Marangunic, a Chilean glacier specialist of Croatian descent who is one of the world's most experienced glacier and rock glacier specialists. He founded and runs a consulting firm called Geoestudios, in Chile, providing cryogenic consulting services mostly to government agencies and mining companies. Cedo (pronounced Chedo), as his friends and colleagues call him, has probably spent more time in the ice than anyone else in Chile, studying, among other things, rock glaciers and debris-covered glaciers. Chile, like other countries around the world, is also facing glacier melt, but Cedo says we shouldn't just stand there and let our glaciers melt away—we can work to protect them and even create them, too! In 2007, he was hired by a company to try to create a real life-sized glacier by moving about 30,000 tons of ice from one location to another.[37] He chose to replicate nature's debris-covered glaciers by placing a thin film of rock cover on the new glacier, effectively helping preserve the ice while the new glacier formed. The results of the artificial glacier were significant. Whereas the natural glacier was receding at about 15 cm (6 in) per year, the artificial glacier brought the number down to about 3 cm (1.2 in). In another experiment, Cedo set up *wind walls* on flat areas at high mountain altitudes. These walls changed the wind pattern and made snow drift up over the wall, which in turn resulted in an artificial vertical drop of snow just past the wall, which would not have occurred if the wall were not in place. This caused an artificial accumulation of more snow at the point of the drop, which in turn led to the creation of a small glacier at the site. The experiment is yet another example of how creative design can help manage snow accumulation and contribute to the creation of artificial glaciers. In yet another experiment, he is using chicken wire laid out on a mountain slope to help retain snow and ice after the winter snowfall. In Cedo's latest glacier studies, he claims to have successfully drilled through an entire rock glacier, taking a core sample for analysis and dropping a video camera through the hole to film the interior. He believes that this is the first time we will be able to see the complete depth of an active rock glacier. We're waiting to see this one, Cedo!

GROW YOUR OWN GLACIERS (GLACIER GRAFTING)

The term *glacier grafting* or *glacier growing*[38] refers to techniques to artificially make glaciers grow by placing seed ice in key areas. This is a traditional practice that has existed for many years, for example, in Northern Pakistan, in the Karakoram Mountains, where tribesmen collect ice at lower elevations and strategically place it in higher frozen areas at altitudes where freezing weather conserves the ice to be used during future warmer months. By mixing the ice with earth, rock, and water, the locals spawn the formation of perennial ice bodies that they count on for future water needs. This is essentially the artificial creation of rock glaciers.[39] This technique has existed for centuries, if not longer. Today, Pakistan has included in its National Climate Change Strategy the recommendation to promote "the use of glacier grafting techniques in high altitude areas," as well as the undertaking of "scientific studies to preserve glaciers and explore [glacier] grafting techniques."[40]

WHITEWASH GLACIERS

In Peru, Eduardo Gold, a self-trained glaciologist, is experimenting with whitewashing a mountain to better reflect light. He won a World Bank prize in 2010 for his attempt at glacier creation by painting a mountain on the Chalon Sombrero peak at 4,756 meters above sea level (15,600 ft) near Ayacucho.[41] He uses environmentally friendly ingredients: lime, industrial egg white, and water. The project is grounded on the idea that by changing the *albedo* (the Earth's reflective capacity) we can cool the ground, create a cooler surface temperature, and thus make it more ice-friendly. He claims that surface temperature is many degrees cooler where he has painted the rocks white. Slowly, ice is accumulating at these sites.

How About Intentionally Melting Glaciers?

Climate change is rapidly and dangerously melting glaciers, so why would you want to melt a glacier? However crazy that might sound, the most obvious reason to melt a glacier may be to regulate water flow to provide agricultural or drinking water for a downstream community. But other very legitimate reasons may also exist to warrant voluntary accelerated ice melt.

For hundreds, even thousands of years, people have known that dark colors absorb heat more effectively; so, if you need water when you don't have it, but you have ice handy, darken it and you'll get meltwater. Farmers in Asia did this over a thousand years ago, and more recent experiments have also attempted to change glacier color to provide drinking water to downstream communities. One effort in Chile attempted to spraypaint a glacier with a small plane to have it provide more meltwater to farmers and communities downstream.

Ice Pits

In mountainous areas of Afghanistan and Iran, some communities collect and pack snow, which they place in watertight pits, thus preserving the ice as water storage for later use—not too different from the ice houses we spoke about earlier in this chapter. The snow is then covered with soil, which acts as an insulator. A storage facility of about 300 m³ (10,600 ft³) can provide a family of ten with water for up to two years.[42]

The artificial glaciers mentioned here are in fact created to be melted on demand, so, here again, we actually *want* these glaciers to melt so we can use their water. We should note that even healthy glaciers in equilibrium with their environment melt and recharge cyclically, so that glacier melt is in fact a very positive thing for our environment. But we must clarify: the problem with glacier melt today is that climate change is hindering the glacier's capacity to recharge. In artificial glacier creation, we must find this ecological balance by trial and error, and melting a glacier in this case is a part of this process.

You may also want to melt a glacier to reduce its risk of collapse. If a glacier is becoming dangerously unstable, accelerating melting of part of the glacier may be a way to ease stress and avoid catastrophic sudden collapse.

Bombing Glaciers with Missiles?

Really? Yes, really! We've mentioned the Perito Moreno Glacier[43] many times in this book, largely because it is Argentina's most famous glacier. It's located in Argentina's Glacier National Park in Santa Cruz Province. The glacier is more than 30 km (19 mi) long and has a massive front terminus that spans some 5 km (3 mi) across the surface of Lake Argentina. The frontal ice of the glacier, which is continuously calving into the lake, has a visible above-water towering front of nearly 60 m (200 ft).

In addition to its majestic beauty and its jaw dropping magnitude, the Perito Moreno Glacier has the novel characteristic of advancing over Lake Argentina toward the shore. The glacier advances so far against the land that it ends up touching up against the peninsula, damming up one side of the lake which begins to exert pressure on the wall of ice pressed up against the lakeside land. When the pressure of the ice advance, plus the pressure of the dammed water reaches a critical point, the glacier front bursts into

FIGURE 8.3 *The spectacular calving of the natural bridge formed by the advancing Perito Moreno Glacier occurs irregularly but provides a breathtaking sight.*
Source: Unknown. GIS. 50°27′59.91″ S 73°01′43.2/″ W.

thousands of pieces in a spectacular event. This occurs on irregular cycles, but sooner or later, the glacier explodes into a massive calving at this point of connection with the continental coast. Tourists from around the world come to see the event and sometimes camp out for days waiting for the glacier to break.[44]

In 1939, the Perito Moreno Glacier in Argentina surged forward, touching land and dangerously raising the lake level. It began flooding nearby areas. Local populations were frightened that a GLOF would ensue and requested government intervention. After some debate about what might be done, the government decided to bomb the glacier, sending in the army with two airplanes to drop explosives on the glacier surface and terminus. The glacier was not in the least affected by the blasts. Eventually the glacier broke of natural causes and alleviated the flooded areas[45] (Figure 8.3).

In the Buff, in the Ice

Greenpeace was behind a campaign in Switzerland and around the world to raise awareness of melting glaciers; in 2007, the organization convinced 600 followers and supporters of the Greenpeace Glacier Protection Campaign to strip down and pose naked on the Altetsch Glacier to draw attention to global warming.[46]

The Glacier Republic

In an effort to draw attention to glacier retreat and global warming, Greenpeace Chile created the Glacier Republic. Anyone can become a citizen of this virtual republic, which is dedicated to promoting glacier protection and trying to get Chile to adopt a glacier protection law. Get your passport before the republic melts away![47] I'm a citizen of the Glacier Republic, are you?

How to Measure Glacier Speed with a Bicycle Wheel

Take a look at this incredibly simple technique devised by a glacier expert in France to measure the advancing speed of a glacier by mounting a bicycle wheel on a cantilever and letting the ice do all the work.[48]

Ice Bars?

In the enormous salt flats of the Central Andes in Bolivia, you can spend a night in a hotel made of salt, but how about having a drink at a bar made of ice? The attraction of ice for bar goers seems to have spurred more than one entrepreneur to offer a completely iced-over environment in which to have a drink. Originating in Sweden in 1994, ice bars have opened doors in places as diverse as Las Vegas, London, Orlando, New York, Amsterdam, Calafate, New Delhi, Madrid, Tokyo, Stockholm, Bariloche, Budapest, Panama City, Mexico City, Munich, Barcelona, Saint Petersburg, Paris, Bogota, Boston, Montevideo, and Dubai. All of these cities have introduced ice bars![49]

How a Whisky on the "Glacier" Rocks Could End You Up in Jail!

But be careful what you order at those ice bars! In 2012, a man was jailed in Chile for stealing 5 tons of glacier ice from the Jorge Montt Glacier in Patagonia (valued at US$6,000) and selling it to local bars so that they could serve whisky on the rocks . . . that is, on glacier rocks.[50] Whisky on glacier ice is a popular beverage in both Chile and Argentina (and probably among scientists working on the Antarctic Ice).

Ice Homes?

We've all heard of the famous Eskimo igloos of the Arctic region of the planet. In the Inuit culture, which has made igloos famous, the word itself simply means home and does not necessarily have to be associated with ice. Igloos in snowed-over frozen areas are used to this day for temporary housing by hunters and mountain climbers, but also by many communities for year-round living. The Inuits have used ice igloos for seasonal and even permanent dwellings. While temperatures outside may drop to 40 or 50 below 0°C (32°F), snow acts as a very good insulator, and walls made of ice can allow the sheltered environments within the igloo to climb to 16°C (61°F) with body heat alone as the source of warmth! Sometimes whalebone and animal hides are used to buttress and line ice igloo constructions. Fresh powder snow is used to craft the walls, and melting snow and ice on the interior refreezes to help solidify the construction upon itself. In this fashion, and counter to what we might imagine, ice igloos harden from the inside out. In this manner, tunnels and connections can be added to expand an igloo that houses one or two people to one that houses an entire family or more; as many as twenty people may be housed in a single igloo. Once solidified into ice, igloos can become so strong that people can walk on top of the dome. Furs and skins can be used to cover cold ice surfaces for comfortable seating and bedding. Until the beginning of the nineteenth century, the Inuits of Northern Canada built large buttressed snow domes for singing, dancing, and wrestling competitions.[51]

House-Invading Glaciers

On May 13, 2013, ice started washing ashore on the coast of Mille Lacs Lake in Minnesota, in the United States, and in Manitoba, Canada. The ice suddenly appeared as a massive slow-motion tsunami ice wave that simply started crawling up the beach. Onlookers thought it a curiously funny incident and started filming the phenomenon until it inched onto the lawns of dozens of houses. Then, to the horror of homeowners, and in slow motion, it moved onto porches, breaking through doors, and into their homes. The mound of ice just kept coming. About 15 km (9 mi) of shoreline were covered, and some of the ice walls that formed ashore reached upward of 10 meters (30 ft) in height. Before the eerie ice wave came to a halt, it had seriously damaged several homes in the Mille Lacs Lake area.[52] The ice wave was caused by loose ice floating on the lake combined with unusually strong winds.

Cities Under Ice

One of the first ice cores ever drilled, according to Mariana Gosnell in her book *ICE*, was at an American military camp built in 1959 beneath Camp Century in northwest Greenland, a veritable "underground" city carved out of snow and ice. The camp was powered by a nuclear reactor, had twenty-one tunnels, a main tunnel "street" 1,200 feet long, a movie theater, chapel, infirmary, skating rink, and dorms where one hundred men were able to live for two years, walking around coatless and taking steam baths while temperatures above ground registered −54°F (-48C) (Gosnell, 2005, p. 124). The site not only served as a military base camp, but also provided some 100,000 years of layered information about the planet's past climate in the ice.

Glaciers Versus Volcanoes

Ice doesn't like fire (except for gas hydrates!). Thus, we might imagine that glaciers don't care much for volcanoes. Nonetheless, many glaciers exist on some very tall volcanoes, simply because it's cold up there and because the sunken craters are great for capturing snow and containing ice. In the case of Mt. Saint Helens, in the western United States, the Crater Glacier began a retreat from the active volcano. A sequenced video of the retreat was captured by the United States Geological Survey (USGS) and shows the incredible evolution, from 2004 to 2014, of how a glacier moves away from a volcano source to find cooler grounds.[53]

Surfing Glaciers?

In the summer of 2007, two world-class surfers, Garrett McNamara and Kealii Mamala, decided they would try the inconceivable idea of surfing glacier waves caused by the calving of large ice boulders falling into glacier lakes. They traveled to Alaska to attempt this crazy feat at Child's Glacier. It must have been a cool wave (sorry, I had to say that!).[54]

Cryokinetic Powers

And finally, to close this chapter, by request of my kids, I draw attention to our social fixation on the power and mystique of ice. Human cultural history and folklore, as well as Hollywood, have helped create some pretty "cool" characters over time, including the most recent—Elsa the "Snow Queen" from Disney's animated movie *Frozen*, who harnesses up her cryokinetic powers to transform ice into castles and toboggans.[55]

Then there was Frozone: remember him from the 2004 animated movie *The Incredibles*? Frozone performs superhero tasks by instantaneously converting water or even humidity into ice. Like melting glaciers, his powers are limited if there is no water in the environment—and one of his weaknesses. ... his eyes are sensitive to the burning sun.[56] Sounds like our glaciers!

Marvel Comics created a less memorable character, the Iceman, a founding member of the X-Men. Iceman can freeze anything around him, including himself.[57]

Folklore, myth, and Hollywood have depicted various version of the *Yeti* or the "Abominable Snowman," which many believe once existed as a hairy Neanderthal-type beast inhabiting the Himalayas; the locals there refer to this creature as *metoh-kangmi* or "wild man of the snow."[58] Such a figure was immortalized in the 1950s through the animated, fearsome-turned-cuddly Abominable Snowman in the short animated film *Rudolph the Red-nosed Reindeer.*

The White Witch of Narnia (in *The Chronicles of Narnia* by C. S. Lewis) also bears cryokinetic power, which she uses to dominate the land of Narnia with her evil intents.

Jack Frost is everyone's favorite ice boy, bringing icy cold weather each winter to our homes and neighborhoods. Frost has also been personified in animated films, such as the *Rise of the Guardians*, where Jack Frost uses his cryokinetic power to help us defend the world from the evil Pitch Black.[59]

And finally, Frosty the Snowman, is, of course, every child's favorite ice hero, bringing fun every winter to those fortunate enough to live in or visit snowy places. Dating back to 1950, first in a children's book and then as a short film in 1954, Frosty really made it big as a sidekick character for Rudolph the Red-Nosed Reindeer. In 1969, he got his own film (Rankin-Bass), which sent him into global stardom.[60]

Notes

1. From Wikipedia: A *tsunami*, also known as a *seismic sea wave*, is a series of water waves caused by the displacement of a large volume of a body of water, generally an ocean or a large lake. Earthquakes, volcanic eruptions and other underwater explosions, land-slides, glacier calvings, meteorite impacts, and other disturbances above or below water all have the potential to generate a tsunami.

2. Carey, 2010, p. 7.

3. Carey, 2010, p. 7.

4. See videos: (1) https://www.youtube.com/watch?v=8ScHkS_cqk4; (2) https://www.youtube.com/watch?v=yCoocd2TzzU.

5. See http://en.wikipedia.org/wiki/Tsho_Rolpa.

6. See ICIMOD, 2011, p. iv.

7. See https://www.youtube.com/watch?v=ZN8a-pP6owk.

8. See https://www.youtube.com/watch?v=fJII-u-41Lg.

9. See https://www.youtube.com/watch?v=Z-1MSW74pmo.

10. See http://www.jstor.org/discover/10.2307/3673897?uid=3737512&uid=2129&uid=2&uid=70&uid=4&sid=21103743179731.

11. See http://en.wikipedia.org/wiki/Grasshopper_Glacier_(Wyoming).

12. See incredible footage of Perito Moreno's bridge calving and sudden lake release at https://www.youtube.com/watch?v=Dfl4DAtHkYQ.

13. See https://www.youtube.com/watch?v=LiZuaCm1Xj8&list=PLVHFdw3YfHyP5U YJ3VQrygg2FvXQwnjGL.

14. See http://beforeitsnews.com/science-and-technology/2012/08/melting-polar-ice-caps-reveal-antarctic-pyramids-2446548.html; the following area is full of nunataks and/ or pyramidal peaks: 78°19′08.10″ S 85°10′11.79″ W.

15. See http://en.wikipedia.org/wiki/Ötzi.

16. See http://tywkiwdbi.blogspot.com.ar/2010/05/retreating-glaciers-and-melting-ice.html; see also http://articles.latimes.com/2003/jan/03/science/sci-mummies3.

17. See http://www.dailymail.co.uk/news/article-480514/Preserved-mummy-500-year-old-Inca-Ice-Maiden-wows-visitors.html.

18. See http://sometimes-interesting.com/2011/06/29/over-200-dead-bodies-on-mount-everest/.

19. See http://edition.cnn.com/WORLD/asiapcf/9905/02/everest/; see also http:// en.wikipedia.org/wiki/George_Mallory.

20. See https://www.youtube.com/watch?v=int6PKPmQ2s.

21. See https://www.youtube.com/watch?v=k3yyxB3ea3s.

22. See http://www.bbc.co.uk/mundo/noticias/2014/03/140312_avion_desaparecido_ andes_stardust_aw.shtml.

23. See http://www.taringa.net/posts/noticias/9846461/Encuentran-avion-argentino-desaparecido-en-1950.html.

24. See http://www.usatoday.com/story/news/world/2013/09/01/newser-body-himalaya-crash/2753019/.

25. See http://edition.cnn.com/2013/09/26/world/europe/france-mountain-jewels/.

26. See http://thechronicleherald.ca/canada/1173466-54-year-old-message-in-bottle-helps-chronicle-death-of-glacier.

27. See Gosnell, 2005, pp. 121–122.

28. For Kimberley Casey's work on remote sensing of glacier dust, see http://earthob-servatory.nasa.gov/Features/PaintedGlaciers/page1.php.

29. See

    (1) http://science.howstuffworks.com/environmental/energy/
    natural-gas-hydrates.htm;
    (2) http://www.usgs.gov/blogs/features/usgs_science_pick/
    gas-hydrates-and-climate-warming/.

30. On "the iceman," see http://en.wikipedia.org/wiki/Iceman_(occupation).

31. See http://en.wikipedia.org/wiki/Ice_house_(building).

32. See Corte, 1983, p. 265.

33. See http://en.wikipedia.org/wiki/Ice_trade.

34. See http://www.dw.de/germanys-largest-glacier-keeps-melting-despite-sun-screen/a-2467214.

35. See http://pistehors.com/news/ski/comments/0737-val-thorens-tarps-glacier/.

36. See http://www.ipsnews.net/2009/11/india-lsquoglacier-manrsquo-vows-to-build-more-artificial-glaciers/; for a slide show of artificial glacier creation, see http://www.scientificamerican.com/slideshow/artificial-glaciers-to-survive-global-warming/;

for a video documentary, see http://edition.cnn.com/video/data/2.0/video/world/2012/06/04/udas-india-glacier-man.cnn.html.

37. See http://www.ipsnews.net/2014/02/like-new-glacier/.

38. See http://en.wikipedia.org/wiki/Glacier_growing.

39. See also Johansson: http://lup.lub.lu.se/luur/download?func=downloadFile&recordOId=2760100&fileOId=2760103.

40. See http://wp.cedha.net/?attachment_id=14236, pp. 14 and 32.

41. See http://www.bbc.co.uk/news/10333304; for a video documentary: http://inhabitat.com/video-painted-mountains-in-peru-are-slowly-bringing-back-lost-glaciers/.

42. See Johansson, 2012, p. 12.

43. The Perito Moreno Glacier is at 50°28′15.67″ S 73°02′20.03″ W

44. The bridge calving of the Perito Moreno on video and incredible footage of the breakage and release of glacier lake damming: https://www.youtube.com/watch?v=Dfl4DAtHkYQ https://www.youtube.com/watch?v=vc6Wvp6J4Aw.

45. See http://www.lanacion.com.ar/1029662-el-dia-que-bombardearon-el-perito-moreno.

46. See http://advertisingforadults.com/2009/10/greenpeace-glacier-nudes/.

47. The Glacier Republic: http://www.republicaglaciar.cl/index1.php.

48. See Wheel and Glacier at https://www.youtube.com/watch?v=njTjfJcAsBg.

49. See http://edition.cnn.com/2010/TRAVEL/12/07/five.to.go.icebar/.

50. See http://pijamasurf.com/2012/02/ladron-arrestado-en-chile-por-extraer-5-toneladas-de-hielo-de-un-glaciar/.

51. On igloos, see (1) http://www.kstrom.net/isk/maps/houses/igloo.html; (2) http://en.wikipedia.org/wiki/Igloo.

52. For ice wave, see http://foxnewsinsider.com/2013/05/13/amazing-creeping-ice-damages-homes-minnesota-canada.

53. Volcano vs. glacier video: https://www.youtube.com/watch?v=sp2wbcrJN6I; see also https://www.youtube.com/watch?v=IuDRD_oPic8.

54. See The Surfer Project:
    https://www.youtube.com/watch?v=QSnKJogYH8w;
    https://www.youtube.com/watch?v=CJkPniK-hMU;
    http://www.zapiks.com/the-glacier-project.html.

55. On Elsa the Snow Queen: http://disney.wikia.com/wiki/Elsa_the_Snow_Queen.

56. On Frozone: http://en.wikipedia.org/wiki/List_of_The_Incredibles_characters#Frozone.

57. On Iceman: http://en.wikipedia.org/wiki/Iceman_(comics).

58. For the yeti: http://en.wikipedia.org/wiki/Yeti.

59. On Jack Frost and the *Rise of the Guardians*: http://en.wikipedia.org/wiki/Rise_of_the_Guardians.

60. On Frosty the Snowman: http://en.wikipedia.org/wiki/Frosty_the_Snowman.

9

Implementation

There are no glaciers in the areas around Pascua Lama nor
around Veladero ... Neither Veladero or Pascua Lama
contemplate impacting glaciers ... As I said, there are no glaciers
near Veladero or Pascua Lama ... Our operations do not impact
glaciers in the area.

—*BARRICK GOLD'S VICE PRESIDENT*
FOR SOUTH AMERICA, RODRIGO JIMÉNEZ (2009)[1]

This chapter looks at the context and circumstances surrounding the
implementation of Argentina's glacier protection law. We also examine the gaps
that exist in the implementation of the law and activities that groups like the Center
for Human Rights and Environment (CEDHA)—an Argentine nonprofit environmental organization—have carried out to push for this implementation. It's
a long chapter with lots of different activity and so I've divided it up into sections
that group sets of issues. The first section will look at the context for implementation, the legal attacks by the mining sector and the provinces against the glacier
law and how, in this case, CEDHA organized to address these challenges. The second section looks at how, in the absence of information from the state, CEDHA
went about carrying out unofficial glacier inventories to draw attention to the risks
glaciers and periglacial areas face from industrial activity. The third section looks
at analytical work to assess mining impacts to glaciers, as well as complaint actions
presented in specific cases where glaciers have been or are being impacted.

I. Context

The passage in the Argentine Senate of the Minimum Standards Regime for
the Preservation of Glaciers and Periglacial Environments (law 26.639) on
September 30, 2010, was an important stepping stone to achieve a framework

and a guiding path for glacier protection in Argentina, but glacier protection was far from a done deal.

The glacier law would still have to be regulated and implemented, the key actors responsible for its implementation would have to carry out their responsibilities effectively, and the law would also have to confront systemic legal and political attacks from key detractors, the first two of which had publicly declared themselves strongly against the law and were ready to wage battle: namely Barrick Gold, the mining company that had the most to lose from the implementation of the glacier protection law, and the executive branch of the Province of San Juan who had bet heavily on a development model based on the promotion of mining activity much of which happened to be in glacier and periglacial environments. Both were already planning (together) the glacier law's derailment.

Although their initial hope was that the president would again veto the law, this was highly unlikely. President Cristina Fernandez de Kirchner had already indicated that she would not pay the political price of a second veto, and, on this point, she kept her word. Whereas generally the executive power actively promulgates laws that it favors before the period allotted for a presidential veto expires, in this case, the law quietly and simply went through self-executing motions, was officially entered into the books, and was announced in an official bulletin on October 28, 2010.[2]

So, for the moment, the glacier protection law stood.

A QUESTIONABLE CONTEXT FOR IMPLEMENTATION

Public concern and perception at that time regarding glacier risks and impacts, whether founded on tangible evidence or not, was that mining was destroying glaciers. Quite a lot had been said about Barrick Gold's Veladero and Pascua Lama projects, but there was more mining on the way in the high Andes mountains of San Juan, La Rioja, Catamarca, Salta, and Jujuy Provinces, and the complete lack of information about the location and characteristics of upcoming investments only further fueled speculation. Neither the private nor the public mining sectors of Argentina publish much information about mining operations under way, particularly in regards to exploratory activity. The Mining Ministry of La Rioja Province, for example, where many of Argentina's new mining projects are slated to advance, doesn't even have a website nor does it go about communicating information about mining activity in any other accessible or visible way.

The Mining Ministry of San Juan Province,[3] which is probably the most active mining province in the country, only publishes very limited information about merely six of its almost 200 projects in progress. No information is published via the Internet summarizing where mining concessions are located. Even the Federal Mining Secretariat, which is supposed to be carrying out nationwide mining strategy promotion, publishes each year the exact same mining map that was produced back in the 1990s.[4] The website is often out of date or dysfunctional.

The question of where mining projects are in relation to glacier and periglacial environments, although seemingly straightforward, was not so clear, not only to the public in general but even to the public agencies and authorities that now had to implement the new glacier law. The government would be roped into a vicious cycle that would largely impede the implementation of the law. That is, if the agencies responsible for mining do not inform the agencies responsible for carrying out glacier inventories about where mining is taking place, it is very unlikely that the priority inventories of glaciers near or at mining projects called for in 180 days by Article 15 of the new glacier law would be carried out.

The media, however, particularly the mining media and especially media focused on mining investments, was already announcing a number of new projects under exploration in San Juan Province, as well as in La Rioja, Catamarca, Mendoza, Jujuy, and Salta Provinces. These announcements would be key for civil society groups like CEDHA to gather information about where mining was occurring. And, as society became increasingly aware of the likely presence of glaciers in some of these provinces, questions naturally started to arise about the potential impacts of mining operations on these glacier and periglacial areas.

Some of the more prominent projects that began appearing in public discussion as possible locations where glacier impacts from mining might occur included:

- The Vicuña and Las Flechas projects (NGX Resources, in San Juan Province)
- The Amos Andres Project (AngloGold Ashanti, in San Juan Province)
- The Agua Rica Project (Yamana/Xstrata Copper in Catamarca Province)
- The Famatina Project (Barrick Gold and later Osisko, in La Rioja Province)

Four initial processes needed to be set in motion for the glacier law to move effectively to implementation:

1. The National Glacier Institute (IANIGLA) would have to initiate the national glacier and periglacial environment inventory—they had five years to carry out the full inventory, and they'd have to work with the provinces to do it (deadline: October 30, 2015—a few months after the publication of this book)—at the time of publication the inventory was not yet completed.
2. Risk areas had to be identified immediately (places where industrial activity is taking place at or near glaciers and/or periglacial areas),

particularly mining, but also road works, such as the projected Agua Negra Pass from Argentina to Chile and other industrial or commercial activity, such as, for example, the ski resort Las Leñas in Mendoza Province, which operates on an active rock glacier. Provinces would have to require companies in those glacier and periglacial areas to carry out specialized glacier impact assessments within 180 days (deadline for both priority inventories and specialized glacier impact studies: April 30, 2011).

3. Monitoring and measures or sanctions by the implementing authorities would have to be carried out where necessary; this meant that if projects were found to be impacting glaciers or periglacial environments, they would have to face sanctions, even suspension or termination (deadline: ongoing).

4. Federal and provincial agencies and jurisdictions would have to collaborate in carrying out inventories, monitoring, and engaging with companies (deadline: ongoing).

The problem in achieving these four conditions was that four more very significant circumstances worked strongly against all of these processes:

1. The provinces most in conflict with the law, namely San Juan, had no intention of handing over their glacier inventory work to the IANIGLA, and the IANIGLA would have a hard time carrying out their inventory without the collaboration of the provinces.

2. The law was legally incompatible with several important mining projects and investments, while mining companies would not stand idly by waiting, but would work to undermine the law.

3. Neither the federal nor the provincial governments wanted this law, nor did they want to implement it.

4. Neither the federal nor the provincial governments were willing to offer transparent information, particularly about mining activity location, needed to implement the law.

Both the political opposition and environmental groups realized that the scenario for real implementation of the glacier law was extremely poor and that, most likely, the law would not be adhered to unless the public agencies responsible for its implementation were forced to act. If government agencies did not follow through with their responsibilities, glacier vulnerability would remain, and glacier impacts by ongoing mining activity would be sustained.

The miners were set in formation ready to attack the glacier law if it would come into force. On November 1, 2010, the second business day after the publication of the new glacier law in the official bulletin, the AOMA (a workers union conglomerating mining personnel working for Barrick Gold) presented a legal complaint against the glacier law in a federal circuit court of San Juan Province accompanied by Barrick Exploraciones Argentina, SA (BEASA), and Exploraciones Mineras SA (EMASA)—basically Barrick Gold's local alias. They requested:

[That the glacier law] be declared unconstitutional and null ... [and that] the law be suspended for [Barrick Gold] ... the actors consider that the continuity of their projects currently underway are affected, ... as well as new projects ... in particular architectural and infrastructure works ... [the complaint goes on to cite conflicts of the law with provincial discretion over mining projects] ... creating a state of uncertainty over the faculties and competence of the Province of San Juan[5] ... [and finally that they request that] ... articles 2, 3, 4, 5, 6, 7, and 15 be declared unconstitutional as they violate the rights of [Barrick Gold] relative to mining exploration and exploitation known as Pascua Lama, [that the company] be excluded from the environmental impact study. And that as per article 15, the requirement of a project re-evaluation affects their acquired rights.[6] [unofficial translation]

Barrick Gold was legally attacking the glacier law, requesting specifically that the law be declared unconstitutional. The complaint raised eyebrows for several reasons. The company had stood firm, as had the province of San Juan, on the idea that it was not affecting glaciers or periglacial environments, but with this lawsuit, suggesting that the glacier law that protects glacier and periglacial environments was affecting their already acquired rights, the company was admitting that problems existed concerning operations at Pascua Lama and possibly Veladero in regards to glacier and periglacial impacts. Furthermore, the complaint was incongruous with what the company and the province had said on numerous occasions, which was that there were no glaciers in their mining activity areas. If there were no glaciers in their area, then why would they have to attack the glacier protection law in the courts? Clearly, despite the rhetoric maintained by both the company and provincial authorities, the company envisioned serious conflicts between the glacier law and their two flagship projects in Argentina, and the best solution was to attempt a legal derailment of the law.

Barrick Gold was also letting the court system, as well as the public, know that there were other conflicts in the pipeline with other mining projects and other works. These would later surface as other mining projects along the Central Andes and in San Juan Province (El Pachón, by Xstrata Copper; Los Azules, by McEwen Mining; Altar, by Stillwater; Las Flechas, Vicuña, Filo del Sol, El Potro, and Josemaria, by NGX Resources; Del Carmen, by Malbex (later Barrick Gold); Amiches and San Francisco (by AML), and several others, all in San Juan Province),[7] as well as a planned tunnel and international highway to be constructed through periglacial areas on the Argentine and Chilean border, in San Juan Province, at the Agua Negra Pass.[8]

The evidence linking mining operations to glaciers was popping up in many locations. As the resolution of Google Earth images improved, there was no hiding this reality. Anyone could find places that once were mere concepts in peoples' minds and see the glaciers and the impacts for themselves. CEDHA did extensive work to disseminate to Argentine society information similar to the images and information presented in Chapter 4 of this book. This dissemination helped educate other environmentalists to take note and act to protect glacier resources that they otherwise would not have known existed.

A RECORD COURT VERDICT GRANTS BARRICK SUSPENSION OF THE GLACIER LAW

A legal action of the type presented by Barrick Gold against the glacier law can take months, even years, to work through the court system. Incredibly, however, Barrick's case to derail the new glacier protection law was resolved in less than twenty-four hours, in favor of Barrick Gold and specifically in favor of the Pascua Lama project—and against the glacier protection law.

Federal Circuit Court Judge Miguel Angel Galvez of San Juan suspended Articles 2, 3, 5, 6, 7, and 15 of the glacier law at the request of the mining company. Accusations of corruption and favoritism, nepotism, and executive interference flew in all directions following the verdict, but, for the moment, the judge's decision stood.[9] Three days after Argentina had a glacier protection law, the law no longer applied for Barrick Gold. But the final merits of the case had not yet been defined.

Soon after the verdict, the province of San Juan asked to be party to the complaint on the side of Barrick Gold. This was a strategic move, purposefully done after the local verdict because by having the province enter the case at that point, it was automatically catapulted to Argentina's National Supreme Court. Had the province entered the case from the beginning, the local judge would not have been the one to decide on the injunction request, but rather the Supreme Court would have had to rule, forcing Barrick Gold to wait out an interminable process, creating delays, and casting doubt in the minds of investors at a critical pre-launch phase of the project.

THE SUPREME COURT OVERTURNS THE INJUNCTION: THE LAW STANDS

There was a period of limbo during which no one knew whether the law would be definitely buried by the Supreme Court or whether there might be a ruling in favor of glacier protection. Ultimately, however, about a year later, the National Supreme Court overruled the injunction granted to Barrick Gold, reprimanding the federal circuit court judge in San Juan for poorly reasoning his verdict and suggesting that there was no evidence to date that Barrick Gold's Pascua Lama project was suffering impacts from the glacier law. The reality was that the law had not been implemented yet for any project, and, as such, there was no imminent danger or immediate impact to Barrick Gold on which to justify granting an injunction order to suspend the law. The law would continue to apply, and we'd have to wait for the Supreme Court to address the merits of the case, which, as of this writing, has not occurred.

CIVIL SOCIETY FREEZE/THAW EFFORTS
FOR IMPLEMENTATION

Very few environmental groups jumped on the glacier law implementation bandwagon after the law passed. Only a handful of environmental groups were focused on glacier protection, and this was mostly lobby work to get the law through Congress after the veto. But even that work did not dig very deeply into glaciology. Few groups, in fact, were even focused on mining issues, much less on the specific mining impacts to glaciers.

Once the fanfare over the approval of the glacier protection law quieted, for the most part, the nongovernmental organization (NGO) community went into glacier retreat, periodically popping its head above the ice to complain about delays in the national glacier inventory or in the priority inventory areas that should have been reviewed in the immediate months after the law was passed.

A few isolated exceptions existed. Greenpeace, which was one of the more active environmental groups on the issue, maintained its team working on glacier advocacy and put together a glacier campaign mostly aimed at pushing the province of San Juan to implement the glacier law. The Fundación de Ciudadanos Independientes (FuCI) of San Juan Province maintained its legal case against Barrick Gold at the National Supreme Court. FOCO, a Buenos Aires-based group, filed an OECD Guidelines Complaint against Barrick Gold for alleged impacts to glaciers at Veladero and Pascua Lama. FARN, another Buenos Aires-based group, carried out periodic presswork focused on government delays in the implementation of the law.

CEDHA was another environmental group that specifically launched a glacier protection advocacy program called "democratizing glaciers," aimed not only at pushing for the implementation of the glacier law, but also on educating society and bringing information about glaciers to the general public. This book is a product of that initiative. Romina Picolotti, former Environment Secretary of Argentina, had returned to preside as president of CEDHA, the nonprofit environmental group she founded with yours truly in 1999.[10]

SETTING ADVOCACY BEARINGS

The largest challenge we faced in 2010 in promoting implementation of the glacier law was that, like all other environmental groups in Argentina, CEDHA had absolutely no experience working on glaciers. No one at CEDHA had ever even heard of a rock glacier or of the periglacial environment in 2008 when we jumped into the congressional sphere in an effort to get the glacier law back. That would change.

Sometime in late 2009, already out of government, Romina Picolotti met briefly (on other matters) with Richard Mott of the Wallace Global Fund (WGF), a Washington, DC-based philanthropic organization founded "to promote an informed and engaged citizenry, to fight injustice, and to protect the diversity of nature and the natural systems upon which all life depends."[11] This was exactly the aim of CEDHA's cryoactivism work. The WGF had given several grants to environmental organizations around the world to focus on environmental and social impacts from the extractive sector. They had at the time provided grants to several groups focused on Barrick Gold projects around the world, and the glacier issue in Argentina and Chile caught their attention. A small pot of money left over from WGF's 2009 budget became a start-up grant[12] for CEDHA to develop a glacier advocacy program focused specifically on the impacts of mining operations on glacier environments.

CEDHA used those funds in 2009 and 2010 to research, build internal capacity, learn, and publish information and to engage with Congress on glacier impacts from mining, and specifically to analyze and contribute to the debate on the various versions of the glacier law that were being discussed in Congress following the initial veto. Some of the very few analytical pieces that examined the substance and nuances of the glacier law versions before Congress and what each definition meant in terms of protection were coming out of CEDHA in 2010, leading up to the definitive glacier law 2.0 vote.[13]

THE DEMOCRATIZING GLACIERS INITIATIVE

I had been CEDHA's director practically since its founding in 1999, until 2006, when Romina Picolotti became Environment Secretary. I left that role in 2006 to assist Romina as her honorary Chief Strategic Advisor at the Environment Secretariat, and I served in that capacity until her departure in late 2008. In 2009, CEDHA's executive director at the time, Angeles Pereira, approached me and suggested that I return to CEDHA to lead its new Mining Environment and Human Rights Program, which I had helped design and launch.[14] It was a fantastic opportunity to return to CEDHA, not as an executive director (a very bureaucratic role) but rather as an environmental and human rights advocate, a job that's closer to the underlying reason why most people get into this line of work in the first place. It was a challenge I could not pass up. Mining was the emerging issue for environmental advocacy groups in the region, and practically none of the established environmental groups in Argentina had ventured into the sector. Although the financing at the time (the WGF grant) was not large, it was enough to start up work and put an advocacy plan in motion.

The big challenge for CEDHA in the launching of a new mining advocacy program was that mining was a very large sector, had few organizations working on the topic, and was so distributed around the country (with practically none of it in CEDHA's home province of Cordoba) that we really didn't know how best to begin. I decided to limit the mining advocacy work to a few tangible issues at the time, namely the industry's impacts on water and glaciers. Although ultimately not the same thing (as we have learned from this book), these issues were intricately related.

It was a way to narrow down a very broad sector to a focus area that could be managed and worked creatively using a small team and limited resources. It just so happened at the time that we had the very fortunate visit of an Italian scholar and skilled advocate with an expertise in mining issues, Flaviano Bianchini. He wanted to work for us and was specifically interested in Barrick Gold. Flaviano delved into Barrick's legal declarations on water contamination and found instances where the company had publicly published reports of excessive heavy-metal contaminants in streams coming from Pascua Lama and Veladero, including lead, mercury, aluminum, and arsenic.[15]

Flaviano's work on Barrick Gold's water contamination and parallel work that I started working on, examining the environmental impacts of mining projects in Catamarca Province would be the beginning of our mining advocacy work, work that would quickly turn to glaciers as our research got under way.

The first and most important phase of our work would be to learn as quickly as possible about the science of glaciers, which involved lots of reading and establishing contact with glacier experts (academics) who could advise our work and help inform our guidance material to congressional representatives. We also had to figure out where the crossroads between mining and glaciers existed. We knew about Pascua Lama and that the Barrick Gold project had problems with glaciers. But we really didn't know what the problems were or how to go about understanding the underlying environmental dynamics of these problems, since they were very technical and we had no experience on glacier science. You simply couldn't find much about mining impacts on glaciers in 2008–2009. So we started poking around. For most of 2010, we read, then read some more, and then some more. This phase concluded on the night of September 29, 2010, when I stayed up all night listening to the live feed from the National Senate. I held back the screams so as not to wake the kids when the final vote tally came in about 4:30 a.m.

We had a glacier law. And this meant a change of paths for CEDHA's work on glacier protection.

Despite my excitement that we now had a glacier law, I knew what the dynamics of implementation would look like. The national government, which was the implementing authority of the law and that would have to follow-up to ensure that the mining companies and provinces took measure to implement the law, had vetoed the law the first time around and really did not want this second version. We presumed:

- That the glacier inventory by the IANIGLA (Argentina's national glacier institute) might take longer than the five-year timeline set forth in the law and that the full inventory would be dragged out indefinitely
- That the 180-day timeline for "priority glacier inventories" in mining areas would not be met
- That the 180-day timeline established for specialized glacier impact reports of mining activity would not be met
- That we could not obtain necessary access to glacier and periglacial areas in order to determine for ourselves where the risk and impact areas were

In retrospect, it turned out that we were right on at least three of the four presumptions. We're still waiting on the first because the five-year deadline is still several months away. If someone didn't carry out the glacier inventories and make information about impacts to glaciers publicly visible, we would never be able to implement the law. It's very difficult to protect what does not formally exist! And until someone had a glacier and periglacial environment inventory that identified where the glaciers and periglacial environment were, it would be very difficult to take measures to protect the most vulnerable areas. Pascua Lama, where we knew there were problems, was impossible to get to, not only because it is in a very remote corner of the Andes, well above 4,000 meters (13,000 ft), but also because Barrick Gold has full control and discretion over everyone entering their access road up to Veladero, and they simply don't let you in. Even the Environment Secretary, back in 2008 with a police force present at her side, could not get into Barrick's project area up in the Andes. We simply couldn't get to the glaciers at Pascua Lama.

THE VILLANUEVAS

That didn't stop us from trying, however. In 2011, I made a personal, failed, but revealing attempt to drive up to Veladero. I stopped off in Tudcum, San Juan, the town at the base of Barrick's access road, where communities had carried out a roadblock to draw attention to their protests against the company. There, I met three brothers, the Villanuevas, who were probably my age (forty-seven), but looked about twenty years older, worn from life in the mountain environments. The Villanuevas are subsistence farmers, herding cattle and other livestock in the high wetlands in the area now practically all destined to mining operations, the first of which was Barrick's Veladero Project. They have lived in Tudcum for many generations and were/are the legitimate descendants of the landowner of all of the region stretching from Tudcum to the Chilean border, nearly 60 km (37 mi) away. Their story of how their father lost their lands to Barrick Gold is remarkable.

According to the brothers, when news broke of Barrick's imminent arrival in San Juan to exploit Veladero, Mr. Villanueva (their father) was ill and being cared for by his wife. They had been approached by an interested buyer for their land but had no intention of selling. One day, however, they were taking their animals up the mountain as usual and found a gate impeding their passage. They asked what the gate was doing there, on their lands, and were told that the land that had been in the family for generations was no longer theirs because they had legally sold the property to a third party, who was now renting the land to Barrick Gold for the access road currently in use.

Their father by then was too old and ailing to attend to the affair, so they hired a San Juan lawyer to bring a complaint before the provincial justice system. She discovered that there was indeed a sale registered by a public notary (called an *escribano*) in San Juan indicating that Don Villanueva had indeed sold the land, and that his Argentine wife, Mrs. Villanueva, had also signed the bill of sale. Curiously, they pointed out, their mother is not an Argentine citizen, but is instead Chilean. The bill of sale had been falsified, and the deal closed. The lawyer was threatened with her life if she did not drop the case, and that was the end of that. The Villanuevas now have to ask permission of Barrick Gold to allow their cattle to pasture on their own lands.

I arrived in San Juan that day in 2011 to visit the Villanuevas in their modest and very poor mud home in Tudcum. The wealth and riches promised by the government in exchange for producing gold seems not to have arrived to this little mountain town at the base of one of the region's largest and richest gold mines. We had a long chat, during which they told me their story. I suggested that we try to go up the mountain and that they should bring their deeds to the property, which they had in their possession. So the three brothers got into my car and we drove to the first entry gate. It was a Sunday afternoon, at siesta time, and no one was on the road. A security guard spotted my vehicle. As he approached, I asked Juan, a CEDHA staff member sitting in the passenger seat, to start his camera rolling. The security guard greeted us; I told him that I was with the owners of the road, that we wanted to go up to Veladero, and I asked him to lift the barrier of the gate so we could pass. He looked at me in disbelief, then looked at the three weathered men in simple clothes in the back seat. He nervously asked me to wait a moment and returned to his mobile container-type office nearby.

About five minutes passed, then another man came out. He was larger, sturdier, and more resolute in his demeanor. As he approached, I asked Juan to turn the camera back on and to make sure the approaching man noticed we were filming. Then I repeated the introduction and the request that they lift the barrier so we could drive up to Veladero. To our surprise he said, "Of course you can go on, we don't impede anyone's entry to the project. Give me a moment so that we can do the proper paperwork and you can go in." I was very taken aback by his answer. For a moment I naively figured he would lift the barrier and that we would simply drive on. But then he began explaining the importance of safety procedures and that he would be back in a moment with a series of safety regulations. Before returning to his office, he repeated, "we don't impede anyone's entry into the project." This sounded odd to me since he repeated it in the exact same words and tone as before, almost as if he had practiced the line or it was an automatic response given to everyone.

He took another ten minutes or so and then came back again with a list in his hand of all the requirements with which my vehicle and myself, as well as my passengers, needed to comply if we wanted to proceed. There were items ranging from special fog lights, to specifications of spare tires, to snow chains (it was summer), to fire extinguishers, to wooden wedges of a specific size, to steel levers, to oxygen tanks, to first aid kits—the list went on—and of course there were a series of medical exams that we all had to have approved by a competent authority before we could even consider passing the gate. And when we had complied with all of the items on the list, "of course," he repeated, "we could go to Veladero" because, as he said for the third time in the exact same words and tone, "we don't impede anyone's entry to the project." He said this as he impeded our entrance to the mine site. We thanked him and promptly left.

The Villanuevas, however, knew of an alternate entrance, and we simply drove back a few kilometers and took a parallel dirt road that led up the valley and toward Veladero. It seems the gate people had a visual on the road, because soon through the rear view mirror I could see the headlights of the security guard's pickup truck racing past the speed limit of the road, to catch up to us. I maintained a good distance at the same speed as we raced up the mountain toward the mine site. We could see the first glaciers appearing on the top of the mountains as we approached the entry to the canyon that would lead to the Conconta Pass.

About 36 km (22 mi) later, still chased by the security guard, we came up to the second road block, where several of Barrick's company vehicles had been set up across the road to ensure that the original landowners could not visit their own property. That was as far as we got that day. On the way back, the security guard asked me to travel with him, which I gladly accepted, and we had a long chat during which he told me all of the ways the company protects the natural environment in the region.

II. Inventories

MAPPING MINING

One of the first technical tasks we embarked on was to find mining projects, which seems pretty straightforward, but which turned out to be much harder than expected. Mining concessions are supposed to be public information, but public information is not always accessible to the public. In a province like La Rioja, for example, the Mining Ministry doesn't provide any public information over the Internet, it will not send information by e-mail, nor does it make it easy to get information even over the counter. You need to trek out to La Rioja (a seven-hour drive from CEDHA's offices) and visit the Mining Ministry's office, which is a tiny building in the quaint provincial capital of the same name. You can only visit the office for a few hours each day, you must pay for photocopies, and you must know exactly what photocopies you're looking for—if not, you spend most of your valuable time, the clock ticking, asking an uncooperative bureaucrat questions that lead nowhere while sifting through irrelevant documents pulled off the shelf to keep you occupied with something other than your objective. (The staff will not offer their knowledge about projects to you—you have to know what to ask for beforehand.) It's a catch-22 that is essentially designed to keep information out of the hands of those whom the Mining Ministry does not want to have it.

The mining bulletin board posting public information is a corkboard on the wall that has dozens, maybe hundreds, of single-page announcements literally hung by old tacks, announcing mining concessions typed onto paper. You're welcome to leaf through the pages, some of which have been on the board for months or even years, some of which are expired, some of which are abandoned, and maybe some of which are current. If you can locate an active concession, and if it has properly pin-pointed the coordinates, and if you can decipher and transpose the Gauss Kruger coordinate system used in Argentina to the more commonly used GIS system (there are courses you can take to learn to do this), you might be able to pin-point the real coordinates of the hundreds of concessions granted over the past several decades.

Or, you can present an administrative request to the mining authority to ask for a project location, as we did for the El Altar project in San Juan Province. In that case, a year passed with no reply from the province. At that point, we sent a new inquiry asking about the old inquiry we had made and the whereabouts of the El Altar project and whether the province knew if there were glaciers near or at the project. (We finally discovered on our own that indeed there were—and, not surprisingly, there were already signs of serious mining impacts to periglacial areas.) Again, several months passed with no answer. Finally, we called them about the inquiry a few weeks later, and they replied, audibly amused, that there was no project by the name of "El Altar" on record for San Juan, but that maybe we were referring to "Altar"—that is, without the article "el" before it. Yes, of course, that was the project, we answered, only to get a reply from the bureaucrat on the other end of the line that we should resubmit an information request to the ministry about "Altar" and not "El Altar." The error, said the bureaucrat, was ours, not theirs.

It's all set up so that getting information is costly, cumbersome, and complicated and more than likely to run into endless circles of red tape that lead nowhere. NGOs don't have endless resources to follow up on all possible fronts for our investigation and advocacy; so, ultimately, in cases like these, difficult choices must be made as to where to invest the little human and financial resources available to advocacy groups like CEDHA. Traveling is usually not an option, nor is spending time pursuing public officials so that they will respond to your inquiries. Most of the time, it's simply more time- and cost-effective to look for information through other channels.

We also approached several mining companies via the Internet, inquiring about project locations, glacier presence, and periglacial environments near projects, and whether the companies operating at the highest altitudes had carried out glacier studies. The results were a bit better, but not by much. We contacted local companies, headquarters of local companies abroad, foreign companies in home towns, and more. Generally (but not always), we would get an initial perfunctory reply saying, "thank you very much for your contact. Someone will get back to you." Sometimes they would, sometimes they wouldn't. The international bunch was not that much more efficient than the local bureaucracy although generally someone responded.

We wanted to know what they knew about glaciers in their projects' vicinity, what studies they had carried out, and if they had mapped glaciers and periglacial areas. If so, we wanted to know about risks and impacts. We contacted a few of the companies that appeared in financial news magazines trying to attract investors: Xstrata Copper, Yamana Gold, Anglo American, Barrick Gold, Minera Andes, NGX Resources, and a few others that we could identify, most of which were in the exploratory phases of their work in Argentina. Responses were meager, and every single one that did respond indicated that there were no glaciers at their mining project sites.

Although we couldn't obtain information from government sources about the locations of mining projects, there were other more effective ways to find them. We began relating and referencing stock investment information about mining projects with geospatial bits of information that the companies willingly revealed in the public domain to media and shareholders and, more importantly, to potential shareholders.

For example, a certain company wants to attract investors to XYZ project, which, they announce in a popular mining investor magazine is 50 km (31 mi) east of the capital of San Juan Province. Sometimes the company would provide a Google Earth snapshot of the project. With basic distance info and a Google Earth image, we became experts at quickly finding the project site. It's like building a puzzle. You have a piece, you know more or less where the piece goes; by moving the piece close to the location, with a little work, you can eventually find the site. Most of the time, the telltale signs are the mining exploratory roads and drilling platforms left by the mining companies, many of which are perfectly visible in satellite imagery. Look for conspicuous crisscrossing lines in the soil and you're likely at a mining site.

By this time, 2009–2010, Google Earth had improved immensely the quality of the images it was uploading. But not all sites on Google Earth had good images. Some dated back to the 1970s or 1980s, and even images from the early 2000s were not of very good quality. In full zoom mode, you could not distinguish a salt field from an ice field, or, for that matter, greenery from rocks. We were trying to find very specific features in the images, such as mining pits, rock piles, infrastructure, leach pads, roads, trucks, drilling equipment, and other mining infrastructure, and we needed to distinguish fresh, smooth powder snow from harder and harsher solid ice. All this requires a level of detail that was not available publicly until the late 2000s.

WHAT THE IRAQ WAR HAS IN COMMON WITH
GLACIER PROTECTION IN ARGENTINA

In Argentina, the National Space Agency (la CONAE) that handles satellite imagery is fortuitously located in Cordoba, not too far from my home (about a thirty-minute drive from CEDHA headquarters). The scientific lab at CONAE in Falda del Carmen (Cordoba Province), became globally famous when it was discovered that SCUD missiles in the Iraq War under US President G. H. Bush's administration had been designed in the 1980s by Argentine scientists in Cordoba working on the covert Condor project.[16] President Menem, who became one of Bush's golfing buddies, later agreed to scrap the missile project, and went along with the US request to pour thick cement on the installations that were hidden away from sight, buried under the natural seemingly virgin greenery of the Cordoba Mountains. What did remain at the site were the large satellite dish antennae that were functioning 24/7 capturing satellite images, mostly sold to the private sector, much of it to mining companies.

CONAE offers an excellent partnership program to nonprofit groups, which allows for NGOs like CEDHA to obtain satellite images at no cost, and CONAE will train interested parties on how to interpret the images as well. All you have to do is submit an application to the agency explaining what you want to use the images for, and if it is a legitimate request, it is usually approved. Buying the images from the CONAE is also an option, but for our small NGO budget, this would have been impossible.

Convinced that we had to learn how to read satellite imagery and that we had to obtain satellite pictures of all major mining sites, I personally visited CONAE with CEDHA's mining program team (there were eight of us at the visit) in late 2010, shortly after the entry into force of the glacier protection law. I was very amicably received by CONAE staff and shown all of the machinery and technology available. The staff was very friendly and eager to help. They were even excited about our specific request to provide us with images of glaciers near mining operations—these sorts of requests were not frequent, and they seemed earnestly eager to provide satellite pictures for an organization dedicated to protect the environment and the public interest. They also had a personal soft spot for CEDHA. Romina Picolotti, the former Environment Secretary and CEDHA's founder, had visited CONAE when she was Environment Secretary and had been on a tour of the site, where she asked many questions about the work and activity of the scientists. It was the first time such a high-level public official had taken personal interest in their work and visited the lab, and now they were returning the favor. They gave me the necessary paperwork and indicated that I should send the partnership request to their headquarters in Buenos Aires—a national office, whose boss is an appointee of the executive branch of government. Strike one.

I filled out the paperwork, sent in the information, and waited. Several months passed before we began to grow restless. We called the CONAE secretariat in Buenos Aires and even bothered the office in Cordoba several times before we started to get the sense that they did not want to answer our application. Strike two.

Finally, I called the CONAE director personally in Buenos Aires and confronted him with our application, which had gone unanswered now for more than six months. He answered that there was now a new procedure mandating that all requests from NGOs for partnerships had to pass through the Environment Secretariat for approval. We had to show that we had a legitimate request. CEDHA was the only NGO asking for satellite images to find glaciers in mining areas. We were also the only NGO actually studying glacier inventorying with the use of satellite images because we wanted to monitor and check the implementation of the glacier law. The new Environment Secretary's legal team answered our application and indicated that we had not satisfactorily shown that we had a legitimate request; they were rejecting our request for partnership with the CONAE. Strike three.

LET THE MINERS PAY FOR THE SATELLITE IMAGES

Ironically, it was thanks to the mining companies and that mining exploration was booming in Argentina that, in the mid-2000s, mining concession sites on Google Earth started to appear with very high-resolution images. This reason was quite logical. X company would approach a satellite image company, like the CONAE. They would provide project coordinates and ask for high-resolution images. The satellite company would adjust their satellites to take several photos of the location each time the satellite swung by the site in its orbit around the Earth. The pictures would be processed in different high-resolution packages. The company would review the images, discarding some maybe due to cloud cover and others because they were simply replicas of the shot they wanted. The company would buy only those images that it needed. The rest ended up in an archive on the satellite imagery repository. Eventually, images that lost their commercial value were sold off at a huge discount, or simply given away to . . . yes . . . Google Earth!

So, as mining exploration exploded, we started getting high-resolution satellite images of most of the areas where mining was taking place in Argentina. We started mapping mining projects one by one. Invariably, the first visible signs of mining activity would be intensely concentrated lines, randomly placed in the middle of nowhere (Figure 9.1). These lines are

FIGURE 9.1 *Site on Google Earth at 5,000 m (16,400 ft) showing impacts of mining operations on a mountain slope. This is the Del Carmen project in San Juan Argentina (by Malbex and Barrick Gold). GIS: 30°00'38.60" S 69°54'38.04" W.*

exploratory roads used by mining companies to search for minerals. They seem thin from high up, but in fact they are usually anywhere from 3 to 5 meters (10–15 ft) wide, and they typically display small circular joints (which are in fact flattened out drilling platforms). These roads crisscross the Andes in numerous places and are generally indicative of past mining exploration. In many instances, we could see the lines running through white blotches of what looked like snow or ice. We presumed each time they were glaciers, but we couldn't be sure.

GLACIOLOGY 101: CRASH TEST COURSE IN GLACIER STUDIES

I realized that we needed to learn about glaciers. Curiously, I soon discovered that a degree in glaciology does not exist. That is, there is no "faculty" of glaciology where one goes to become a glaciologist. Instead, geologists or geographers usually take an interest in the world of ice and spend a focused amount of time studying particular dimensions of glaciology, such as the physical dynamics of the interaction of rocks, water, and ice (as in geocryology) or the physical geographic characteristics of rock glaciers.

Thus, glaciology, because of how the scientific world has approached the issue, has been mostly reserved to the hard sciences, centered on those people who study the Earth's surface and climatology. Now, as the climate is changing, as society has begun to look more closely into the deep consequences of climate change, and as glacier protection begins to work its way into the law, we can expect other academic fields to become interested in glaciers, from areas as diverse as the social sciences, public environmental policy, environmental management, and more.

I had not received any training whatsoever on glacier dynamics, and I needed specialized training in order to launch many of the activities we wanted to carry out, including glacier inventories. We couldn't look at a Google Earth image and distinguish ice from snow, and without the collaboration of key government agencies that could give us the information, we were going to have to produce it ourselves.

Just as glacier law negotiations hit their final stride in August–September 2010, the United Nations Environmental Program (UNEP) announced a course for glaciologists in Chile scheduled for October 2010. This course, which would be taught by some of the region's most prestigious glacier experts, was meant for young professionals focusing on glacier studies. The idea was in part derived from the evolution in Chile and Argentina to develop policy and laws around glacier management and protection. Suddenly, there was interest in having trained personnel available to help study glaciers. Even the mining companies were requesting glacier experts, and there just weren't that many to go around at the time. It was a collaborative UN effort with the government of Chile, with CODIA (a European-based water agency), and UNEP. The trainers were Chilean, Argentine, Peruvian, and French.

We needed to be at that course. I was the person assigned to this task at CEDHA, so I would have to apply. My professional background was not in the hard sciences, although it was very much focused on environmental public policy over the past decade; the UNEP course was geared toward geologists, geographers, and hydrologists. I sent in my application to the UNEP Panama regional office anyway, and, not surprisingly, my application was rejected. I didn't fit the bill. Not affected by this rejection, as soon as I received the notice, I picked up the phone and called Isabel Martinez, the director of the UN water program in Panama who was coordinating the logistics of the training. I laid out the reasons why we had to be in this course: we were the only civil society organization seeking expert training on glacier recognition via satellite imagery, and we were the only group trying to inventory glaciers and monitoring, at a technical level, the implementation of the glacier protection law. I even sent her a file with 300 glaciers that I had already mapped in San Juan on Google Earth. Two weeks later, UNEP reversed its decision and even offered to pay for our travel. Home run!

The following month, I was swimming in glacier academia. I met thirty young and dynamic glacier specialists from Mexico, Venezuela, Colombia, Ecuador, Peru, Bolivia, Chile, and Argentina (I didn't even know at the time that Mexico, Venezuela, and Colombia had glaciers!). I got to spend a week tapping and drilling the minds of three of the most well-respected glaciologists on the continent and probably of the world: Juan Carlos Leiva (of the IANIGLA in Argentina), Cedomir Marangunic (of Geoestudios in Chile), and Benjamin Morales Arnao (of the Patronato del Museo, Huaraz, Peru). Among the three of them were more than one hundred years of field research and a wealth of unimaginable knowledge about glaciers and periglacial environments.

I racked my brain to focus on areas of glaciology I had never even heard of, definitions and classifications, structural movements of ice, erosion and transport, geology and the quaternary period, mass balance, the hydrology of glaciers, glacier management, glacier lake outburst floods (GLOFs), the periglacial environment, and several other cryology dimensions that were covered during that intense week of glacier studies. Throughout the week, I focused my attention on glacier recognition, categorization, and remote sensing, which was obviously what I could most use in our effort to find, inventory, and remotely monitor glacier impacts and risks. I also asked the experts many questions about the periglacial environment, particularly Marangunic who has worked on periglacial areas. One of the modules on the last day of the course was on "glaciers and the law," and, seeing that I was one the person in the room who had most engaged on glacier policy issues, more even than the instructors, the professors asked me to coordinate the module, give a presentation about what had transpired with the glacier law in Argentina, and lead the discussion among the students. Since that week, the course has been given on three separate occasions. I periodically return to teach the module on glaciers and legal aspects (law, policy, and society).[17]

The next year, when the course was given in Ecuador, I had the enormous pleasure of meeting Bernard Francou, a top-notch glaciologist from France working with the Institut de Recherche pour le Développement (IRD), who participated as trainer in the course. He was working on a film about glaciers and the Inca culture, called *The Retreat of the Gods*, focusing on the cultural links between local communities and the predicament of melting glaciers in the tropical Andes, which is where he has focused much of this academic work. It was Francou who, in reference to our work to protect glaciers and promote public policy and legal frameworks for glacier protection, coined the term "cryoactivism" in reference to the work CEDHA and I were doing to protect glaciers.

Two events occurred while I was away at the first glacier course that altered the Argentine political arena. Nestor Kirchner, the former president of Argentina passed away, changing the local political arena entirely, and the glacier law entered into force (the day after Kirchner's death), automatically and by procedure because it had not been acted upon by President Cristina Fernandez de Kirchner.

While in Chile, I had come across a study from 2008 by Alexander Brenning and Guillermo Azocar[18] of mining impacts to glaciers at Pelambres,[19] a copper project run by the Luksik mining company in Chile. Brenning and Azocar looked at the intervention and removal of mass from rock glaciers and from the periglacial environment at this border project in the Central Andes, west of the city of San Juan. This was one of the first academic reports available in modern times about mining impacts to rock glaciers and periglacial areas, one that approached the subject from an environmental impact perspective.

Brenning and Azocar not only completed a glacier inventory for the project area, showing where the company had intervened in rock glaciers, removed mass, and impacted the glaciers, but they also laid out the basic "typologies" of impacts that could result from this sort of activity at glacier sites and in the periglacial environment:

- Geo-chemical impacts, such as acid drainage, from the deposit of sterile rock on waste piles that could impact the water discharge of rock glaciers and periglacial areas
- Thermal and structural impacts to rock glaciers when debris is placed on them
- The acceleration and potential collapse of rock glaciers if sterile rock is deposited on them
- The loss of water stored in rock glaciers

In another of their works, published in 2010,[20] Brenning and Azocar noted that they saw impacts to rock glaciers in nearby projects in Argentine territory, namely, El Pachón, Los Azules, and Altar. This information would prove invaluable to map out much of our immediate future work on glacier monitoring and protection.

Azocar and Brenning's work on analyzing mining activity impacts to periglacial areas at the Pelambres project was very useful to our work because it showed us how to go about approaching a mining project in glacier terrain, how to identify key issues, and how to analyze the sorts of images that we could obtain from Google Earth.

After the glaciology course and upon my return to Argentina, I felt far more able to embark on glacier recognition using satellite imagery, particularly using Google Earth, which was the only real means we had at the time of collecting information, given that CONAE would not provide us with images.

By this time, volunteers at CEDHA had plotted some 500 mining projects on a Google Maps application, with information collected about each of them and, most importantly, GIS geographic coordinates that we could use to find the projects on Google Earth and start looking around the immediate vicinity for ice.[21] All we had to do was to cross-reference mining projects above a certain altitude, where visible white glaciers or rock glaciers could be identified on satellite images. Based on information readily available in private sector mining magazines of projects that were in the pipeline, we chose a preliminary list to examine more closely:

- Agua Rica (Yamana/Xstrata Copper) in Catamarca Province[22]
- El Pachón (Xstrata Copper) in San Juan Province[23]
- Los Azules (Minera Andes and later McEwen Mining) in San Juan Province[24]
- Famatina (Barrick Gold and later Osisko Mining) in La Rioja Province[25]

These projects were at the forefront of discussion in 2010 because they were in advanced stages of exploration or preparation. We knew that each of these were located at latitudinal heights where glaciers could be present. We were not sure which of these projects were truly in glacier or periglacial areas, but we were definitely ready to start looking at satellite images to find out.

CEDHA'S FIRST GLACIER AND MINING INVENTORY

Before leaving for Chile to participate in the glacier course, I had been in contact with and visited local communities in Andalgalá, Catamarca, concerned over the arrival of the Agua Rica gold and copper mining project (by Yamana Gold), and this was inspiration to begin work on a glacier inventory for the

Aconquija Mountains, where Agua Rica was located. Sergio Martinez, one of the community leaders, indicated that there might be glaciers up in the project area. There were the testimonies of bulldozer operators who had been hired by Xstrata Copper to carve out roads at high altitudes. These bulldozers had allegedly unearthed ice-rich grounds when they carved out the access roads to the Filo Colorado mine deposit. The local community did not yet relate this story to rock glacier or periglacial environment impact, but soon they would.

We did some searching and found a study from 2002 that included a partial glacier inventory of the Aconquija Mountains by a geologist named Ana Lia Ahumada.[26] Ahumada worked and lived in Tucuman Province in Argentina (just East of Catamarca Province), and, upon returning from Chile, I contacted her and introduced her to our organization and to our objective. We heard that she was working on an updated glacier inventory for the Aconquija Mountains (which she confirmed on our call). This is the border area between Catamarca and Tucuman Provinces towering up above 5,000 meters (16,400 ft), and it was precisely the area we were looking at for Agua Rica and the Filo Colorado projects. I spoke to her personally, introducing myself and CEDHA, and I mentioned our interest in helping monitor the implementation of the new glacier law. She was congenial toward our effort, but a bit hesitant to share her work. She also expressed her concern that activism around glacier protection might conflict with the mining sector, and she wanted to avoid philosophical and ideological extremes. In this regard, she didn't want her work to be used to prop up anti-mining advocacy.

I indicated to Ahumada that what we most needed was to gain access to geological studies that mapped out glaciers and periglacial forms. We were not anti-mining, but we did want the glacier protection law fully observed and glaciers protected. We indicated that we had contacted both Yamana Gold (leader on the Agua Rica project) and Xstrata Copper (leader on Filo Colorado) and that both indicated there were no glaciers in the Aconquija, but that since we knew of her past research on rock glaciers in the area, we realized that what the mining companies were saying was not true. Ahumada responded that she was updating her studies from the early 2000s and she would soon have an inventory completed, but that it was not ready for publication. I remember feeling very frustrated that no one seemed to be willing to share information. Ahumada's rejection forced us, and me specifically, to mobilize to do the inventory ourselves. They say that necessity is the mother of invention, and so, desperate to learn where Argentina's glaciers were, I started our first comprehensive glacier inventory of the Aconquija Mountains.

When I set out to do the first glacier inventory, I only had the short experience of the glacier course I had just completed in Chile and the few papers by Alexander Brenning that I was able to download online, particularly the study looking at the Pelambres project in Chile. Brenning stood out as perhaps the only glacier expert at that time who had dared to publish a strong critique against mining activity in glacier areas. Perhaps he could help.

I remember the day that I tracked him down at the University of Waterloo. I was driving home from a meeting with Ricardo Astini, Director of Graduate Studies of the University of Cordoba's Geology Faculty, where I was also seeking assistance from the department on glacier identification. Cordoba doesn't have glaciers (at least not in this geological era), but there is a strong interest in geology. Astini ran a group called CICTERRA (the Earth Sciences Research Center),[27] a research institute focused on geological studies. He indicated that a graduate student, Mateo Martini, member of CICTERRA and also part of the CONICET (Argentina's premier scientific association) was focusing his thesis dissertation on rock glaciers in Jujuy and Salta Provinces and that he could possibly help us out. Martini was not reachable at the time because he was off on a glacier field visit in Jujuy Province, but I noted his e-mail and telephone number.

I stopped on the side of the road to dial Brenning in Canada. He answered right off and listened quietly to my introduction and to my description of our work. We wanted to carry out inventories, as he had done at Pelambres, and show the public what was happening in the Andes. I asked him if he was willing to review our inventory work, which basically was plotting glaciers (or what we thought to be glaciers) on Google Earth in the Aconquija Mountains. Brenning noted that a geologist named Ana Lia Ahumada was working on an inventory and that she would probably be a great source of information and that we should contact her. I recounted my conversation with her and said that I didn't think she would share her inventory. As we closed the conversation, Brenning agreed to review our work. He asked me to send him the Google Earth file once I plotted the glaciers and drew the outlined polygons where I thought they were located.

That same week, I got a call back from Mateo Martini of CICTERRA who had just returned from Jujuy. I went through the same explanation as I had with Ahumada and Brenning, and in the short five or ten minutes that our call lasted, he was already on board.

I made one further contact that week to try to build an advisory team to guide our work, namely with Juan Pablo Milana. Milana was the geologist and glacier expert from San Juan who had been present at the meeting at the Environment Secretariat, when community members traveled to Buenos Aires to show the Environment Secretary Barrick Gold's impacts on glaciers at the Pascua Lama and Veladero projects.

To the surprise of many, Milana had come out strongly against the Maffei version of the glacier law and also against the Bonasso-Filmus agreement version. This was taken by the environmentalist community as a sell-out to Gioja and to the mining companies. Milana was accused of taking contracts with the mining industry and from Gioja, and his opposition to the more conservationist glacier laws was seen by many as a diametrical reversal of his earlier position, which was against Barrick Gold and against mining impacts to glaciers.

But Milana's problem with the mining law was not so much about mining impacts and definitions or conservationism. As a Sanjuanino, he is a strong federalist in nature, upholding the preeminence of provincial autonomy and control over natural resources. He saw the glacier law as an affront of the national government over the provincial authorities. He was protecting provincial autonomy. What is certain is that if it weren't for Milana, the communities of San Juan would never have had the evidence against Barrick Gold to start this whole glaciers and mining advocacy issue in the first place, and this is something that even many of Milana's detractors in the NGO world recognize.

Milana agreed to review our work as it advanced, and, over the past years, he has been one of our most trusted expert advisors, contributing regularly to our analytical work and to several of our reports.

THE ACONQUIJA MOUNTAINS

I also attempted to obtain information about mining operations from the Catamarca Province Mining Authority, with absolutely no luck. Unless we trekked out to Catamarca (a ten-hour drive from Cordoba) and badgered the Mining Ministry—with no guarantees that this would work—or initiated legal action based on their refusal to provide public information, it was clear we weren't going to get anywhere through that route.

We also made contact with the companies—specifically Yamana Gold. which was behind the Agua Rica project,[28] and Xstrata Copper that had explored and abandoned the Filo Colorado project[29] just a few kilometers from the Agua Rica site—requesting environmental impact studies. In both cases, they pointed us to the provincial authorities, indicating that this was public information and that we should approach the provincial government to request it. Obviously, this route also led to a dead end.

I remember personally opening up Google Earth and looking for Agua Rica. I started my Google search at Andalgalá,[30] just south of the towering Aconquija Mountains. I could see, immediately north, a snowy white peak,[31] but nothing that looked like a glacier. The elevation was a whopping 5,500 meters (18,000 ft). I didn't know there were mountains that high in this part of Argentina, which is nearly 200 km (125 mi) from the Andes Mountains. Snow was definitely visible a bit farther up the range, and these might be ice bodies,[32] but I wasn't sure. I consulted with a few local activists who also could not tell me if there were glaciers up in those parts, but they all referred to the "eternal snows" of the highest peaks, not sure if there might be glaciers near those peaks. As we've already learned, and as I knew by then, there's no such thing as eternal snow. These were definitely glaciers!

I could not get environmental impact studies from either the company or from the provincial authorities, and I could not even find information to determine precise and official project location. I'd have to find the project on my own. Local community members referred publicly in several interviews to media that Agua Rica was about 20 km (12.5 mi) as a crow flies away from the city and that the rivers the project would be polluting were immediately upstream. So, it should be easy: start at the city, work your way upstream for about 20 km, and we should find a mining project. It would be a mining project not yet in operation, which meant that there might not be much at all to see except for maybe the exploratory roads that I had already become accustomed to finding in San Juan Province on Google Earth.

Sure enough, as I wandered up one of the streams advancing from Andalgalá toward the snow-capped areas I had seen on Google Earth, I came across an area with numerous squiggly lines, clearly not a natural phenomenon. They were exploratory roads introduced into the mountain by Yamana Gold during their exploration of Agua Rica.[33] One notable characteristic was how the greenery abruptly ended right at the site with the concentration of exploratory roads. Most of the roads were in an area ranging from 3,000 to 3,700 m (9,800 to 12,100 ft), with a peak at 4,100 m[34] (13,500 ft) where one of the roads leading to the mine site crossed over a ridge. The heart of the project area was about 9 km (5 mi) southwest of the high mountain peak that seemed to have snow but no visible glaciers. From the comfort and safety of Google Earth, I explored the immediate area, particularly between the project site and the snow-covered mountain. I could see another road running alongside two lagoons, behind the ridge, seemingly unconnected to the road leading from Andalgalá to Agua Rica, that also had evidence of use as it wiggled through the high terrain. It led to a dead end area that looked also like a mining exploration site. I would later learn that this was Xstrata's abandoned Filo Colorado project.[35]

I used the Google Earth measuring tool to gauge the distance between the mining projects and the visible white peaks farther away. There were about 10 km (6 mi) to the first snowy area and almost 20 km (12.5 mi) to the second. Was this too far for impacts? Ana Lia Ahumada's article spoke of rock glaciers, not of uncovered white glaciers. If there were rock glaciers, then we were obviously in a periglacial environment (since rock glaciers are one element of the periglacial environment), but perhaps not in a *glaciated* environment. The rock glaciers would be underneath the surface of the earth, hidden from sight unless you knew what you were looking for. My hunch was that they had to be there, near the mine sites; that would coincide with what the bulldozer operators had indicated: the bulldozers had unearthed ice when they were carving out the road. I was certain that they had plowed right into frozen ice-saturated grounds of the periglacial environment.

The glacier training I had received had included a review of rock glacier and debris-covered glacier characteristics. I still wasn't sure exactly of the difference between the two, but I had seen dozens, even hundreds, of images of rock glaciers. They looked much like lava flow in the shape of a tongue, moving downslope, with a clear demarcated edge at the terminus, cut at 30–40 degrees. The surface could be very flat or have ridges, and the rock debris on the surface was generally abundant and evenly distributed (see Chapter 4 for images and descriptions of rock glaciers).

A quick browse through the highest peaks of the Aconquija Range on Google Earth turned up several such forms. A rather large area with many of these forms was located in the cirque hollow of a nearby peak not far from the Agua Rica location.[36] The adjacent mountain cirque with hollows to each side (to the southeast and to the west) also had several of these forms, although a bit less evident[37] (Figure 9.2). These areas were nearer to the Filo Colorado project and to the mining road that dissected several of these rock glaciers leading to Filo Colorado. The cirque slopes nearer to Agua Rica also had some of these forms, but they were less clearly formed.

I mapped about twenty-five of these bodies in the immediate vicinity and used Google Earth's polygon drawing tool to trace the edges of what I thought

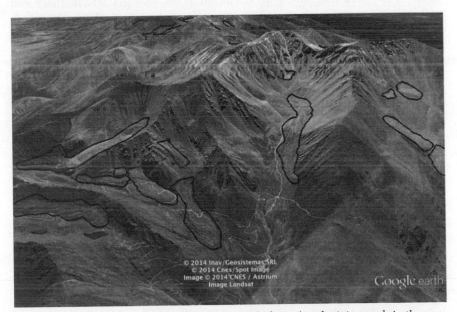

FIGURE 9.2 *Mountain cirques with rock glaciers (polygons) and mining roads in the Aconquija Mountains in Catamarca Province, Argentina, near Filo Colorado and Agua Rica projects. GIS: 27°20'18.15" S 66°12'34.64" W .*

were rock glaciers. I sent Brenning the .kmz file (the Google Earth file that contained all of the polygons I had drawn—anyone opening it would see on their desktop version of Google Earth the same polygons that I had drawn) and waited for his response. It was my first time identifying rock glaciers via satellite images and the first time I had ever created a .kmz file. I expected to be completely off and that my efforts would come back fruitless. Surely, I was wrong. Brenning answered within a few days, ruling out maybe two of the twenty-five bodies I had identified as rock glaciers, not because they were completely off, but rather because they looked like relict rock glaciers or possibly inactive ones (i.e., rock glaciers that have degenerated and that no longer have activity or ice in them). He added another ten or twelve bodies of ice in the immediate area that I had overlooked.

I had successfully identified rock glaciers in the vicinity of mining projects! For someone with a political science and economics degree, I was ecstatic. A month earlier, I didn't know where to begin with the identification of glaciers and mining project conflicts, and now I was actually mapping rock glaciers at mining projects on Google Earth.

But I needed more hands-on assistance to know if what I was doing was correct, and particularly more detailed training than Brenning could provide in our two-dimensional Skype conversations, which we held only a few times during that period. So, I followed up with my contact at the National University of Cordoba and scheduled a meeting with Mateo Martini, the young PhD candidate doing his dissertation on rock glaciers in Jujuy and Salta Provinces. We agreed to meet on campus, in his lab, which was a long narrow passageway in the basement of the university that had been converted into office space for graduate students. His desk was the last one in the corridor, by the window.

We sat down and I took out my computer to show him my findings. I needed him to train me on how to identify rock glaciers and distinguish them from normal debris talus (rock landslides) and also to distinguish an active rock glacier from an inactive one (a rock glacier with ice but with no movement) or from a relict rock glacier (a former rock glacier now with no ice). I decided to use the Aconquija Mountains as my classroom, and so we got started traveling through Google Earth through Catamarca and Tucuman's Aconquija Range, where I already felt very much at home because I was virtually spending many long evening hours there in search of perennial ice.

He was an eager teacher and patiently went through form by form. He hadn't ever explored the Aconquija Range for rock glaciers, so he was just as fascinated as I was about what we were discovering in these mountains. I made several visits to the university basement to review my evolving inventory. Each time I got more adept at recognizing rock glaciers, and each time my questions became more and more specific and more sophisticated about variations of the sorts of geocryological forms we were seeing. My most common mistake was identifying an inactive rock glacier for an active one, or a fossil glacier for an inactive rock glacier. These were minor details for me at the time because earlier I only saw a pile of rocks where there was ice, and now I was discovering a world of incalculable frozen hydrological wealth.

Although we started with twenty-five glaciers near the Filo Colorado and Agua Rica projects, once we completed our full glacier inventory for the Aconquija Range, we had plotted more than 200 glaciers. The reader can download that inventory by simply downloading the .kmz files that holds the polygons we drew for the inventory, in this case divided up into two separate files, one for each province (Catamarca and Tucuman). The files are readable on Google Earth and will plot the very same polygons that I drew at that time, on the screen when opened. Simply download them and then open them with the Google Earth program running.[38]

GOOGLE EARTH GLACIER INVENTORIES

As the rock glaciers started accumulating in my inventory, I sensed the need to give each of them a unique name. Most glaciers don't have a name like "Thwaites Glacier" or "Lyell Glacier," or "the Perito Moreno." When a glacier gets a proper name like that, it's because it has some major relevance to society, to history, or to local lore. For instance, maybe an explorer crossed a certain mountain pass and named the glacier when he passed the site. Or maybe it's in honor of a scientist who studied the area. Or maybe it's a glacier that everyone can see from town.

Glacier inventories are not usually about counting famous glaciers, but instead about registering the presence of perennial ice, and there are simply too many small perennial ice bodies out there to be giving each one a proper name. Furthermore, some of the smaller perennial ice features may come and go, and so proper naming wouldn't be the most efficient way of registering these bodies. Furthermore, you can imagine the complexity of giving proper names to the more than 10,000 glaciers or rock glaciers in a province like San Juan!

Instead, when inventorying glaciers, scientists name glaciers utilizing codes that tag the ice body with letters and numbers that represent different geographical characteristics, including continent, order of primary water basins, secondary basins, and the like. For instance, one of the glaciers in San Juan Province registered in the provincial glacier inventory has a reference name of JA1212001_3D. As a non-glacier expert, such a code is a bit daunting and really doesn't say much, and it's certainly not user-friendly. I challenge the reader to go find that glacier! In fact, even if you are a glacier expert, I challenge you to go find that glacier—you have a pretty good head start since I told you it's in the province of San Juan. But San Juan has thousands of small glaciers, so, I am sure it will take you a considerable amount of research to find it, if you can find it at all.

I found this scientific format not only very alien to me, but also extremely inappropriate if what we wanted to promote was better public access to information about glaciers. We were not carrying out a glacier inventory for academics and scientists to use, but rather an inventory for social education and awareness and for public policy purposes. What we needed was a naming system that would allow everyone and anyone to quickly understand something about each perennial ice body, its location, its value, and its relevance to them and to their ecosystem. We needed to bring glaciers, even the small ones, into peoples' living rooms and onto their computer screens. We were talking about the impacts of mining operations to the Toro 1 Glacier, to the Guanaco Glacier, to the Esperanza Glacier, and to many others that already had some degree of social recognition—at least by name. We said the Canito Glacier was soiled due to suspended particles from mining and that the Almirante Brown Glacier was dissected by a mining road. We made reference to hundreds of glaciers in mining territory, and we talked about rock glaciers that looked like lava flow. We said that satellite images revealed exploratory mining roads carving through glaciers.

But all of these references were merely words conjured up in peoples' minds written in black and white text in documents that didn't offer much in the way of visual education. In some cases, we had offered pictures that people could see in reports we could send around. But, for the most part, the glaciers remained in the abstract. Few had actually seen them or actively engaged themselves in seeing them. I remember the first time I used Google Earth to look at an area of the planet. It was my own home, probably just as millions of others have done. I tried to identify which of the cars on the street was my own. I tried to identify things in my yard that were distinguishable. I followed the streets around my house through my neighborhood, excited to see the world and my familiar surroundings from space. Seeing the world from a different vantage point, and being the one to actively seek out this vantage point, impacts us; it interests us, because we are seeing something familiar in a very different light, empowered by the new tool we are using to do so. This is interesting to us simply because it changes up the order of things.

I recall, back in 2005–2006, the first time I used Google Earth to examine an area in which we were engaged on the protection of a river, on the border between Argentina and Uruguay, due to the proposed installation of two mega-sized paper pulp mills at a small town in Uruguay called Fray Bentos. The Argentine community across the border river, at Gualeguaychú, was outraged, and began a long protest against the companies, against the Uruguayan government, which had unilaterally authorized the project, and against multilateral agencies (the World Bank), the private banks, and the countries (Finland and Spain) that were all set to finance the investment.

Seeing the area from above, understanding the emplacement of the mills, realizing how close communities and natural areas were to the proposed industry, following the river winding down through pristine environmental terrain, seeing nature in its vast expanse and the relationship of people, homes, schools, hospitals, and other places that are characteristic of the cultures affected by the projects, changed the way we understood the risks. Seeing the Earth in its vast but fragile extension and understanding the geophysical aspects of the risks we were placing on her changed the way we looked at the problem we were facing. At CEDHA, Google Earth changed forever the way we do our advocacy by allowing us to understand that this personal engagement with our Earth, made possible through modern technological tools, could inspire people to act or to at least become much more aware of the things that were occurring around them.

And so we looked more carefully at Google Earth and figured we could use this fantastic technical tool to bring glaciers closer to the people we were trying to reach. We wanted them to see where their glaciers were. We wanted them to go up to the mountains and fly like birds up and over mountainsides, through the ravines, and over their glaciered terrain, just as they would over their house and neighborhood. We wanted glaciers to become part of their neighborhoods, a part they had never been able to see because the terrain was far and largely inaccessible. Google Earth could change that!

We could show these magnificent ice bodies, their emplacement and relationship to river streams and lakes. We could show their vulnerability to climate and to man-made structures like roads and industry. We could show them the impacts of drilling for gold, of exploratory mining roads carved indiscriminately into ice crisscrossing glacier terrain and rock glaciers. We could show the public the dissection of glaciers and the abandonment of severely impacted glacier sites once miners had discarded the site because it held no minerals.

And, the best thing was, we could show all of these things almost effortlessly. And so, with those objectives in mind, we set out to develop our own glacier inventory nomenclature system, one that would be easier to use and catered to how everyday people would be engaging the inventories, online.

HARNESSING GOOGLE EARTH

This section is best read with Google Earth open on your computer. Google Earth[39] uses a GIS coordinate system, either offered in decimal format or in hour/minute/second format. It allows you to see the entire planet and any portion of it in fairly high resolution, at the whim of the user, and at the touch of a few keys. One moment, you can be looking at the Eiffel Tower, and seconds later you can be at the Great Wall of China, in the Grand Canyon, or standing in front of your home looking up at your own house or car in the driveway.

So why not use this versatility to look at glaciers in the same way? You could be looking at a glacier breaking up in the Antarctic, and, moments later, be flying through the glaciers of the Himalayas or climbing Mt. Everest. The problem, of course, is that you could very easily find the Eiffel Tower simply by typing "Eiffel Tower" in Google Earth's search box (try it, it works!)—but try to find Thwaites Glacier that way. Only a handful of glaciers in the world have searchable names in Google Earth (the Perito Moreno is one of them). Certainly, you could not type in the glacier's name in the previous example from San Juan (JA1212001_3D) to find it; wouldn't that have been easy!

Most people use Google Earth in two dimensions, which makes it basically the same as Google Maps, giving you the option of seeing the Earth in map format or in satellite view so you actually see the real Earth. But the true power and interest in Google Earth lies in its very user-friendly capacity to see the terrain in 3-D format, tilting the screen (by tapping on the arrows in the circular direction indicator in the upper right corner of the screen) and making the experience akin to flying or hovering over the site visited. You can then use the up/down/left/right arrows to fly around the screen. Google Earth provides an enormous amount of information to the user, including location, elevation, political boundaries, natural resource location, the possibility of measuring distance, the option to mark places that you visit, to draw polygons or other shapes on the screen, to insert overlays, and save them to your computer. It allows you to see a history of images of the site you're visiting, over several years, as well as a fantastic opportunity to engage with other users by sharing pictures and other information about a location. You can even very easily send to other users your stored information and place markers so that they can see the same markers and polygons that you've saved on your computer. The program even has a built-in filming tool that allows you to record your movements on the screen, so that if you've started your recording at Tudcum, San Juan, where Barrick Gold has its mining road leading through glacier territory, you can take the viewer on a tour all the way to Veladero, 180 km (112 mi) away, stopping at each glacier, hovering over the glacier, marking it, measuring it, indicating a point of impact, and then moving on to the next glacier. And you can even send that video to your friend, so she can also go through the tour at her own convenience, all the while never having left the comfort of her home. At CEDHA, we've prepared a five-minute glacier tour of the Veladero and Pascua Lama access road from Tudcum to the project areas that the reader can experiment with simply by accessing the Google Earth links available at our glacier inventory website.[40] It's quite eye-opening if you've never seen a Google Earth video!

CEDHA'S GLACIER NAMING SYSTEM

So as I sat down to figure out a way to get glaciers into people's living rooms, I jotted down the GIS location of each of the glaciers of the Aconquija Mountains that I had been registering. They looked like this:

- 27°20'06.36" S 66°12'14.05" W
- 27°18'39.41" S 66°11'02.38" W
- 27°15'37.32" S 66°07'27.80" W
- 27°13'27.70" S 66°06'02.94" W
- 27°13'14.50" S 66°04'13.68" W

These are the GIS reference numbers appearing at the bottom of the Google Earth screen. As you move the cursor over the screen, the numbers adjust, as does the elevation. You can also copy the exact number of the center position of the screen from the Edit drop-down menu.

When I compared the list of the eventual 200 glaciers we inventoried, to the list I had already come up with in San Juan, and as I thought of a simpler way to name the glaciers so that the system would not be so daunting for non-glacier expert users, something very simple jumped out at me that was very obvious from the GIS addresses. Here are three sets of samples, from San Juan, Catamarca, and La Rioja Provinces:

Five glaciers in San Juan:
29°22'10.25" S 70°00'15.08" W
29°20'37.12" S 70°00'34.49" W
29°21'33.75" S 70°01'10.83" W
29°21'54.58" S 70°00'47.51" W
29°17'41.98" S 70°00'01.12" W

Five glaciers in Catamarca:
27°20'06.36" S 66°12'14.05" W
27°18'39.41" S 66°11'02.38" W
27°15'37.32" S 66°07'27.80" W
27°13'27.70" S 66°06'02.94" W
27°13'14.50" S 66°04'13.68" W

Five glaciers in La Rioja:
29°01'45.63" S 67°49'48.59" W
29°01'20.32" S 67°49'04.27" W
29°01'05.63" S 67°50'03.78" W
29°00'51.81" S 67°49'39.65" W
29°00'07.28" S 67°50'16.45" W

I could intuitively tell you what province the glacier was in simply by looking at the first cardinal coordinate. After registering many glaciers, it was
intuitively evident to me that the first few digits of the coordinate already
identified what part of the planet you were in. For example, for the San Juan
list, all glacier locations begin with 29 for the S coordinate and 70 for the W
coordinate. For Catamarca, they all begin with 27 for the S coordinate and 66
for the W coordinate, and for La Rioja, they all begin with 29 for the S coordinate, and 67 for the W coordinate. Simply by looking at a GIS string, I could
quickly identify where the glacier was—more or less. This relationship could
be very helpful for the naming system.

Could we name the glaciers using these two sets of double digits? If so, the
name could be associated to these first digits and could help the user locate
the glacier with little effort. I tried entering the GIS address into Google
Earth simply by using the first coordinates, taking the first example in the
San Juan list,

29°22'10.25" S 70°00'15.08" W, this became
29 S, 70 W

That placed me 40 km (25 mi) away from the place the glacier was actually
located. Whereas on a planetary scale that wasn't bad, it was still too far
off—but even this was better than the scientific naming method, which is
simply useless to the lay user to figure out what part of the globe you're in.
So then I decided to take the first four digits of each S/W coordinate, that is:

29°22'10.25" S 70°00'15.08" W, which became
29 22 S, 70 00 W

Bingo! I was less than 300 meters (900 ft) away from the Canito Glacier!
I checked the other addresses. Using the first four digits of the address,
I could sometimes actually get on the glacier; other times, I ended up several
hundred meters off, but never at a distance so large that I couldn't see the glacier on the screen in an up-close image. Sometimes I only missed the glacier
by a few meters. We would name the glaciers using this methodology. The
user would be able to find any glacier we named on Google Earth, using only
its name! Once mapped, we'd register the full GIS address of the glacier in our
registry, but the glacier would be named using only the first four digits of each
coordinate to keep it simple.

So our inventoried glaciers were named (using the San Juan list as an example):

29°17'41.98" S 70°00'01.12" W: Glacier 2917-7000
29°20'37.12" S 70°00'34.49" W: Glacier 2920-7000
29°21'33.75" S 70°01'10.83" W: Glacier 2921-7001
29°21'54.58" S 70°00'47.51" W: Glacier 2921-7000
29°22'10.25" S 70°00'15.08" W: Glacier 2922-7000

If two glaciers had very similar GIS coordinates, so that the first four digits were repeated, then we could distinguish them by adding (a), (b), (c), and so forth after the name; For example:

29°21′04.92″ S **70°01′02.05″** W: Glacier 2921-7001
29°21′32.88″ S **70°01′09.90″** W: Glacier 2921-7001 (b)

I went back and tested the system with a basic sampling using the hundreds of glaciers we had already registered. It was flawless. The system worked. Anyone could visit any glacier quickly and easily using Google Earth simply by having its name! Now let's go back and look at how the scientists do it. Remember the example, the glacier from San Juan, identified by continent, basins, sub-basins, and the like? Here's the name: JA1212001_3D. Now go find it!

The four-digit naming is not exact. It may throw you several hundred meters from the glacier we've named, but it'll get you pretty darn close, close enough so that you can look for the glacier in the screen that pops up. Our idea is to bring glaciers closer to the people who want to see them; it's what I call "democratizing glaciers" (which was originally going to be the name of this book, but the editors didn't like it). Maybe, like most people, they simply didn't know what I meant by the title, which simply focuses on the importance that everyday people learn about glaciers and glacier environments and the role they play in our everyday lives and that we learn about them in simple participatory and engaging ways.

SMARTPHONE GLACIERS

We've been looking at glaciers in Google Earth. Google Earth is a fascinating program, but let's be honest: most of us get a bit lazy and for the most part our navigational program of choice is the two-dimensional map program that we use to figure out how to find a store that is just blocks away. In the past few years, I've noticed that the interns coming to CEDHA from around the world can't make it beyond the corner without their smartphones. GIS mapping systems now accompany most cars, and everyone is plugged into the planet through Google Maps or some related program.

So, can we see our glaciers on our phone with CEDHA's glacier naming system? Try it out!

Go to your map program (e.g., Google Maps). Put it in satellite mode—this is important to actually see the real ice—and let's see if you can find one of the glaciers in our inventory (its address is 29 18 00.24 S, 70 01 00.26 W: we call it Glacier 2918-7001). Take the first four digits for each coordinate and enter them in the following format in your phone's map program. Make sure you type the comma, the "S," and the "W."

29 18 S, 70 01 W

Zoom in to see the image more closely and give the program a moment to load the definition. You're looking at the Estrecho Glacier, one of the most controversial glaciers at Barrick Gold's Pascua Lama project on the border between Argentina and Chile. You can now slide around the image; zoom in further to see the mining roads carved into the mountainside. You can see roads going into glacier ice! For example, you can see exactly where Barrick Gold cut through the Toro 1 Glacier at:

> 29 19 56.53 S, 70 01 14.44 W

Let's look at some more places of concern because of glacier impacts, right on our phones (we've used three sets of digits for a more precise location):

- Here's the place on the Conconta Pass in San Juan, Argentina, where Barrick Gold ran a road through the Almirante Brown Glacier: 29 58 37 S, 69 38 01 W
- Here's the place where Xstrata Copper ran roads through rock glaciers in the Aconquija Mountains of Catamarca to get to mineral deposits at Filo Colorado: 27 19 51 S, 66 13 27 W
- Here is the place where the Chilean Codelco mining company is intervening in dozens of glaciers at the Andina project in Chile, near the capital of Santiago: 33 09 47 S, 70 15 15 W
- Here is a place in Peru where climate change is causing large ice-boulders to fall into glacier lakes and create massive waves that rush into valleys below, causing glacier tsunamis that have killed thousands: 9 23 20 S, 77 22 28 W
- Here is the place in Kyrgyzstan, where the company Centerra dumped millions of tons of sterile rocks on a moving uncovered glacier, diverting its path (note the S/W is changed for N/E, as we are in the Northern and Eastern Hemispheres, and remember to give your phone a little extra load time to acquire definitions as it loads the new satellite maps): 41 52 40 N, 78 12 29 E
- Here is an active rock glacier and glacier lake in Armenia, in the vicinity of the Kajaran mine. Local communities asked us to do a glacier inventory of the rock glaciers in the mine's vicinity to ensure that the activities of the mine were not intervening ice: 39 05 16 N, 46 03 26 E
- Here is the place in Indonesia where Freeport operates on the fringes of the country's last remaining glacial environments: 4 03 20 S, 137 07 15 E
- Here's a place in France where a ski resort operates on top of a glacier (the Argentiere Glacier): 45 56 57 N, 6 59 26 E
- Here is a place where another ski resort (Las Leñas) operates in Argentina, on top of a rock glacier: 35 08 31 S, 70 08 25 W

- Here is a place in Bolivia where a ski resort operated on the Chacaltaya Glacier but had to close because climate change killed the glacier: 16 21 12 S, 68 07 53 W
- Here is Thwaites Glacier in Antarctica, which recently began an irreversible collapse and where you can see the thousands of glacier pieces that are beginning to float off into the ocean, melting and causing eventual sea level rise: 74 50 26 S, 107 16 35 W

You can see all of these anthropogenic impacts to glaciers right on your cell phone!

III. Mining Impacts

NASCENT CRYOACTIVISM

Once we had completed our first glacier inventory in mining territory, the inventory for the Aconquija Mountains, we then turned to examining mining projects near the glaciers to look at impacts. Our first focus area was for the Agua Rica (Yamana) and the Filo Colorado (Xstrata Copper) mining projects and the potential impacts these projects might have on glacier and periglacial environments.

With analytical help from glacier experts such as Alexander Brenning, Juan Pablo Milana, and Mateo Martini, we prepared a report called *Impacts to Rock Glaciers and Periglacial Environments by the Filo Colorado and Agua Rica Mining Projects.*[41] This report, published in February 2011, was the first of a series of reports by CEDHA's team on mining impacts to glaciers and periglacial environments. It was well in advance of the date established by the glacier law that mandated companies with operations in glacier areas to produce environmental impacts studies to determine glacier impacts and risks of their operations, which was April 2011. If a small NGO on a minimal budget working with volunteers could produce such a study in such a short time, there was no reason that the public authorities couldn't oblige mining companies to produce their own reports.

CEDHA's report basically concluded that:

- There were rock glaciers and periglacial environments in the vicinity of both the Agua Rica and Filo Colorado projects.
- There was evidence that, at least in the Filo Colorado project, the roads introduced by and utilized by Xstrata Copper traversed and dissected rock glaciers and periglacial environments.
- Agua Rica (by Yamana Gold) was adjacent to periglacial environments, and we could not ascertain if there was a direct relationship between exploratory activity at Agua Rica with periglacial environments, particularly at an area visible at: 27°20′43.38″ S 66°16′29.69″ W.

- This situation called for a specialized glacier environmental impact study, as mandated by the new argentine glacier law (Article 15).
- Neither company had identified glaciers, rock glaciers, or periglacial risks in their environmental impact studies—except as a risk to the working environment due to possible landslides if the terrain were intervened.[42]

We then approached the provincial authorities with our findings as well as both companies (to this day, they have not answered us). Xstrata Copper's Filo Colorado project was by far the most compromised by rock glacier impacts. They had run their access road straight through several rock glaciers and through permafrost grounds. From the account of the bulldozer operators, we had confirmation that they had indeed removed ice from these glaciers. We could not know the extent of damage, however.

In December 2010, two months prior to releasing our report, we contacted Xstrata Copper and Yamana Gold to give them a chance to contribute their views on the material and evidence that we were revealing and analyzing in our report. We also offered to publish these views in our report. Their reactions were both worthy of mention.

GLACIERS AND PERIGLACIAL AREAS
AT FILO COLORADO: XSTRATA COPPER

Xstrata Copper went back and forth with us about whether or not they would give us their environmental impact assessment (EIA) for Filo Colorado. They indicated that we should ask the provincial authorities of Catamarca for the documents. Seeing that the government of Catamarca was not answering our inquiries, we ended up threatening the company in Australia with an information access lawsuit if they did not provide the information, after which they sent us their EIA immediately.[43] The e-mail received on December 17, 2010, from Xstrata's communications manager at the time, read:

> Thank you for getting in touch with me with in regards to the Filo Colorado project and the glacier inventory that CEDHA is conducting. The EIA for exploration at Filo Colorado was approved in October 2006 (please see copy of the EIA attached). After conducting eight diamond drill holes, the company ended the exploration program at the project in 2007. The exploration team found no evidence of glaciers at the project site.

What was especially noteworthy was the information contained in the EIA. Annex 9 of the document Xstrata Copper sent to CEDHA[44] had several maps of the project's geological risks. Ironically (considering the e-mail message received), three of these maps clearly indicated both "glacier areas," as well as "risks" from glacier areas!

But what was even more ironic was the photo on the cover of the EIA. The cover showed the Filo Colorado mineral area with a visible road leading to the mine site. Inside the EIA, on page 86 were two additional photographs of this same site, revealing more of the surrounding landscape. This was the road that bulldozers hired by Xstrata Copper had carved out to reach the mine, since no road existed in this area prior to Xstrata's intervention.

There were two small frozen lagoons visible in the image. The photographer was standing on rocks off to one side of the valley, looking down at Filo Colorado in the background. I had spent hours, days even, on Google Earth, scouring the Aconquija Mountains, trying to locate glaciers and more specifically rock glaciers, and I was sure I had seen those lagoons before. I double-checked with Google Earth references and with the polygons that I had mapped and found the exact place from which the photograph had been taken.[45] I was extremely surprised with what I was discovering, so I picked up the material and headed to the University of Cordoba to meet with Mateo Martini, our rock glacier advisor.

An hour later I was walking down the hallway in the basement of the geology faculty at the National University of Cordoba, and found Mateo working diligently on his own glacier inventory of Jujuy. I walked straight up to his desk and laid Xstrata Copper's Filo Colorado EIA report in front of him. "What do you see in this picture on the cover?" I asked Mateo.

"A road that goes by two lagoons, and several rock glaciers at the foot of the slope of this mountain," he said pointing to some dark rocks and the access road leading to Filo Colorado.

I turned to page 86 of the EIA to show Martini the amplified images, "here's the same area but with more terrain visible as well as the rock glaciers and the road."

"Yes, the road is cutting right through the rock glaciers ... in fact, the photographer is probably standing on the rock glacier to take this photo," he answered matter-of-factly.

The evidence was irrefutable. The access road to Filo Colorado sliced through a rock glacier, and the evidence was on the cover of Xstrata Copper's EIA report!

On December 27, 2010, I sent another e-mail to Xstrata Copper, which read:

> Thank you for sending us the EIA and annex info, which was very helpful. We had some preliminary sense that the Filo Colorado project (or some complementary parts of the project such as the access road) were projected in glacier territory, and this has been confirmed by our satellite imagery analysis, which was consulted at length with several geologists and glacier specialists.
>
> ... we are convinced beyond doubt that there are glaciers, and many of them, in the immediate project area and relative to complementary

components of the project (such as the access road). This is not only our view, but the Project's EIAs also confirms it. You'll see in Annex 9 of your EIA, references to glacier sites and risks that coincide with the access road. They are explicit.

I'd like to have a conversation with you, to share some aspects of our findings regarding the extent to which exploratory work on Filo Colorado has occurred in the vicinity of glaciers. We are likely to release our findings next month, and as such, it would be helpful to talk to some of your technical experts that worked on the EIAs to answer some of the questions that we have, so that we can be as precise as possible on our own commentary.

As information started to come out about Xstrata Copper's impacts to glaciers at Filo Colorado, and later at another project we analyzed and at which we found over 200 rock glaciers in the project influence area (El Pachón in San Juan Province),[46] the company started to retrench and cut off communications entirely. CEDHA eventually filed an international complaint in Australia against Xstrata Copper for their impacts to rock glaciers and periglacial environments at both El Pachón and Filo Colorado.[47]

Since our contact, neither Xstrata Copper nor the authorities of Catamarca have taken any measures whatsoever to evaluate or repair the damages to rock glaciers left by the Filo Colorado access road. In December 2013, a team of experts went up to the glacier area near Filo Colorado, including glaciologists of the IANIGLA as well as personnel from the Municipality of Andalgalá, to review glaciers in the area. While they noted the abandoned roads dissecting rock glaciers and periglacial environments, they did not carry out any impact studies related to Xstrata's or Yamana's activity in the region.[48]

GLACIERS AND PERIGLACIAL AREAS AT AGUA RICA: YAMANA GOLD

The experience with Yamana Gold at the Agua Rica project was different but no less startling, for different reasons, and it goes to show just how difficult it is for environmental groups with environmental concerns to address corporate actors on sensitive environmental impact issues.

The Agua Rica project,[49] then owned and operated by Yamana Gold (now owned by Xstrata-Glencore), is adjacent to periglacial areas. CEDHA's findings could not determine if exploratory work at the project had impacted or was impacting rock glaciers or periglacial areas. One mountain cirque hollow area was close enough to operations to merit study.[50] The others seemed too far off for there to be any likely impacts. We did not know, however, if the company had also ventured into areas beyond the visible exploratory roads, which seemed all to be located below the likely permafrost zones because there was visible greenery in all of the area, indicating a low likelihood of permanently frozen grounds.

We had not received any response to e-mails we had sent in late 2010 to Yamana Gold authorities in Canada. So, when we decided to travel to Andalgalá to present our findings to the community on March 4, 2011, we contacted the company, which had a local office in the town of Andalgalá. Community relations with the company in Andalgalá were extremely tense. Each weekend, hundreds of people marched around the town plaza in opposition to Agua Rica. The company hired local security forces, which would come to the marches and stand alongside the plaza to face off the peaceful protestors or would drive their vehicles around the plaza during the march in repudiation to the protestors; they saw and defined themselves as *counter-marchers.*

I personally contacted Roberto Vallejos, the public relations officer for Yamana Gold in Andalgalá. We set up a meeting for the very day I was arriving to town, at which I would publicly present our report and hopefully get a chance to speak to his team, principally his geologist, who would know of any periglacial studies carried out by the company (which I assumed she would have carried out herself). I indicated to Vallejos that our report was more favorable to Yamana and Agua Rica than it was to Filo Colorado, largely because it seemed from our findings that Agua Rica was next to (but not in) and below the periglacial environments in the vicinity, but that, nonetheless, there was one cirque hollow up the mountain from the mine site that we had some doubts about. We also did not know if Yamana had explored other areas that we may not have covered in our analysis. Vallejo was very amenable and very open to our meeting. I assumed the meeting would be amicable, with no confrontational circumstances, and that I'd probably be referred to their technical staff for answers to my questions. But this is not how things played out.

We met that day by chance at a café on the central plaza. This sort of occurrence is common in small towns where everyone knows everyone. We chatted briefly, and, since I was on foot, he offered to give me a ride back to the Yamana offices where he wanted me to meet his principal geologist. I accepted the ride and off we went. We chatted briefly about his work in Andalgalá as we arrived at Yamana Gold's modest building just four or five blocks away from the café. "We could have walked," I thought as we entered the building.

I sat in a small office with walls made of movable temporary panels, as if they could simply pick up everything and leave on a whim (which sometimes occurs when financing for mining exploration suddenly vanishes). Vallejo made a call via the internal office phone announcing to someone at the other end that I had arrived, and moments later a young woman appeared in the doorway. He introduced me to Selva Ahumada, Supervisor of the Department of Geology and Environment of the Agua Rica Mining Company. She barely looked at me as she sat down at the table with us. I remember thinking that she looked nervous and rather agitated, but she remained silent for the moment.

I explained who I was and gave some basic information about the organization I represented, indicating that we were monitoring the implementation of Argentina's new glacier law. And that, in this process, we had chosen both the Agua Rica and Filo Colorado projects to analyze first. I mentioned that we had read studies by the geologist Ana Lia Ahumada (a different Ahumada and apparently not related) about there being rock glaciers in the Aconquija. I summarized our findings indicating that we had found extensive conflicts between the Filo Colorado project and periglacial areas and that, in the case of Agua Rica, the findings were inconclusive. I said that we hoped that Yamana Gold could provide us with technical information about their own studies. I also pointed to the mountain cirque hollow closest to the project where we thought there might be a rock glacier and, very possibly, frozen grounds.

What followed was completely unexpected. Ms. Ahumada had been getting visibly and progressively more nervous in her chair; I could see that she was very anxious to respond, and, as I spoke, that anxiety only built up. When I finally turned the conversation over to her she pounced at me, yelling: "Who do you think you are to produce a glacier inventory?! How dare you?!" I was taken completely off guard by this aggressive reaction. She continued, "You have no legitimacy or knowledge to produce such a report! I will not respond to your report or even read it! You will have to answer legally to this. I will file a complaint to the Chamber of Geologists! How dare you cite an academic publication about glaciers?! Did you get permission to cite Dr. Ahumada?! You will have to answer to this! You have no idea what you are talking about! Why don't you go back to your province [Cordoba] and worry about the problems you have there instead of poking your nose around here?!"[51]

I remember thinking to myself that I had thought this would be the easy company to engage!

These were just some of the highlights of her diatribe against CEDHA and against me personally. She went on for what must have been ten minutes of insults and aggressions. Vallejo, the head of community communications, hired to build constructive relationships with the community and with environmental groups (like ours), sat bewildered, not knowing what to say to either of us. He tried to calm her down, but there was no stopping her. She needed to get all of her anger out about our report and about me not being a geologist and not having (in her view) any authority to comment on glaciers (albeit considering that, as we have seen, glacier experts are not necessarily geologists nor is there a degree that offers such "authoritative" grounds from which to comment). I let her go on, and, as she finished, I turned to Vallejo and said how sorry I was to see that this was the response I was getting from the company's technical team. I turned back to Ahumada, welcomed her feedback on our report, then got up and left the building.

I later communicated the incident to Yamana's corporate executive office in Canada, sorry to see that this company employee had aggressively insulted and denied our right as stakeholders to request information and that their employee had so rudely insulted us and me personally. I received a half apology seventy-two hours later from Ahumada that read:

> By express request of the company to which I proudly belong, I express my apologies for the way in which I treated you during your visit to our company. As a good Christian I recognize that beyond any differences we must conserve our good treatment of others. As regards the technical questions, I maintain my position, but I would like to clarify that on a personal level I have no intent of offending your person. I clarify that I apologize for the form but not for the content of our discussion. I trust that in the context of good faith, you accept my apology, in the understanding that as persons of good, we end this inconvenience. Attentively, Selva P. Ahumada.[52]

We never again heard from Yamana Gold regarding periglacial impacts (or lack thereof) from the Agua Rica mining project.

GLACIER CINEMA

I had arrived in Andalgalá on the very same day I had the very unfortunate meeting with Yamana Gold. My first meeting that day was with the mayor's office. Being a very small town, the mayor was aware of everything that is going on. I was scheduled to speak that evening before a public audience at the municipal theater. The subject was our report on rock glacier and periglacial environment impacts by Agua Rica and Filo Colorado. It would be the first time that residents of Andalgalá and nearby towns would see the glaciers that existed up in the mountains above their homes—and for most, if not all, it would be the first time they would hear of something called "rock glaciers" and of "periglacial environments." Local residents referred to the snow-capped peaks of the Aconquija Mountains visible from their town as "eternal snow." They did not realize that the snow they were seeing up in the mountains were glaciers, and they had absolutely no idea that there were invisible glaciers up there called rock glaciers, as well as freezing creeping soils that also contained water and ice. But they would soon find out.

My bus came into town that day early in the morning, and I was at the mayor's office by 8:00 a.m. The municipal office is in a small building downtown, with its entrance at the corner and the mayor's office off to one side. There were television crews interviewing someone near his office, probably the mayor himself. I entered through the public entrance and asked the woman at the help desk to see the mayor. I explained who I was and why I was there, at which point she sent someone, who ran off nervously, presumably to inform him of my request to see him. The mayor was known to be very much in favor of Agua Rica and was not on the good side of the environmental movement in the locality, so I figured he would not be happy about my presence or about the reasons that had brought me to his town, but I nonetheless wanted to share our findings with him as a matter of political courtesy. Moments later, the messenger who had disappeared returned to tell me the mayor was not in (which I later found out was not true), but that I should leave my number and he would contact me later. I acquiesced, and gave the woman at the help desk my cellphone number and stepped out. He never called.

I noticed the television crews were wrapping up their session (probably interviewing the mayor) and that they were about to get into their van, so I walked over and introduced myself, indicating that I had just met with the mayor's office to inform him that I would be speaking that evening at the municipal theater and showing images of dozens of glaciers in the Aconquija. The journalists were impressed because they had never heard of glaciers "up in those mountains." "Do you mean the eternal snows?" they asked, to which I responded, "No, I'm talking about invisible glaciers beneath the surface of the earth called 'rock glaciers.' There are hundreds of them up there that provide this whole town with drinking water, and the mining roads they're introducing cut right through them," I added. They were fascinated, and offered to do a short interview for a local television news program about my visit. I agreed, and they set up the camera right there on the sidewalk just outside of the mayor's window—so that, when the interview ran, the mayor's office was our set! I sent three messages to the community that were broadcast minutes later to most of Andalgalá. One: that we had carried out a study that found hundreds of glaciers up in the mountains above the city. Two: that there was extensive evidence of mining impacts to glaciers in the area. And three: that everyone was invited to the municipal theater that evening to see these glaciers.

The airing of my interview must have been soon afterward because about one hour later I started receiving calls from local radio stations to get more information and to go on the air about our findings. I went to at least two radio interviews before my scheduled meeting with Yamana Gold, which was set for 3 p.m., after lunch. One of those interviews was re-edited later as a documentary used to educate local communities about the Aconquija Mountain glaciers.[53]

That evening, I met up with one of the Algarrobo Assembly environmental movement leaders, Sergio Martinez. Martinez was a local public administration employee turned environmental advocate fighting against mining investments near Andalgalá. The Algarrobo movement, named after a tree with extremely hard wood typical of Argentina, had set up a roadblock on the main access road to Agua Rica, the one I had followed on Google Earth up to the mining project and to the rock glaciers. Sergio had provided us with much information about the Agua Rica project. He had seen our report, which I provided to him before it went public. He knew there were glaciers in the Aconquija, but he had no idea how many, nor had he ever heard of these enigmatic rock glaciers beneath the surface of the earth. He was astounded. He convinced me to come to Andalgalá to present our report, and he had organized the theater event, which was to start at 9 p.m., giving residents ample time to stroll downtown after work to come see for the first time in their lives the glaciers that were literally just up the mountain, the closest one just 20 km (12.5 mi) away. But when he picked me up at my hotel around 8 p.m., he was nervous.

When he went to the municipality to pick up the keys to open the local theater, he was told that the mayor had ordered the theater fumigated that evening. It wouldn't be open to the public for at least three days. It was the first time in years that he could remember that the theater was fumigated! The closure was obviously planned to derail the presentation of the report. They didn't want the people of Andalgalá to see the glaciers. Martinez deliberated with his fellow Algarrobo Assembly members and came up with an alternative. They would blockade the street just outside the theater, set up a projector on the street, and project the images onto the wall of a building. The show must go on!

With the commotion about the theater closure and people arriving downtown to its closed doors, we moved the time of the presentation to 10 p.m. and set up shop on the street. Fortunately, it was a nice end-of-summer month in Andalgalá, and the temperature mild. There were some verbal altercations with the police who showed up to dismantle the roadblock, but, in the end, they acquiesced and let the show go on; several of them actually wanted to see the images and stayed around for the show. We finally got started around 11 p.m., by which time hundreds of people were gathered on the street, sitting at curb-side and in the middle of the street as I started up the presentation. The looks on their faces when the first images of rock glaciers went live were memorable. Eyes fixed, pondering, in awe.

As I spoke, I noticed a young woman sitting on the curb and jotting information down in a notebook. I thought she was taking notes about my presentation. Maybe she was a journalist or even a spy from the mining company—of which there were a few present, I was told later. As it turned out, it was Vanesa Martinez (no relation to Sergio), the singer for a local rock band, and she was writing a poem about the enigmatic rock glaciers she was just discovering:

> Cómo no esconder mi alma blanca
> Del vacío que se asemeja
> Al terror de verme florecer para dar lujo
> Y no para dar vida
> No poder salir de la piel
> Para volver a recrearte
> Con el frío con mi fuerza natural
> Desigual a las voces que toman las enredaderas

del peligro destruido y hecho de mal
Darle a ese mal abrigo
Mas imponente es el cielo
Que refleja mi condición
Tan inútiles son, que sus armas me han descubierto.[54]

The presentation went on for about an hour and a half. We viewed more than 100 images of different types. Many were from Google Earth, showing the hundreds of glaciers we had found in the Aconquija Mountains. Others were of other rock glaciers (some of the same images in this book), showing the ice within a rock glacier, its dimensions, its relevance, and the source of water that comes from such ice and rock bodies. This was all very new information for the people of Andalgalá, just as it had been new for us only a few months earlier. The viewing and the discussion went on until nearly 1 a.m., taking questions, discussing our findings, answering inquiries about the new glacier law and what it meant for Catamarca. We also discussed the glacier protection bill that had been passed by the Catamarca Provincial Senate, on August 5.[55]

This bill was very important for glacier protection in the province, but it left out much of the periglacial area that we were showing because it covered only rock glaciers and omitted extensive areas of frozen ground with probable high ice content, the type of ground that Xstrata had destroyed with bulldozers near Filo Colorado.

The next day, I was invited by several radio stations to interviews about the images we had projected the night before. One woman called the radio program while I was on the air, live, and insisted that I go to her home immediately after the program and have some of her homemade *empanadas*, a token gift to me for having produced our report revealing their glaciers. Being the small town that it is, I could not turn her down; in fact, I didn't have lunch plans, so I gladly went to her home straight after the program and sat down along with several other neighbors that had tagged along to hear about glaciers and to a delicious feast with local grapes and wine.

The Algarrobo Assembly also invited me to go to the roadblock and show the presentation for those who had spent that night at the roadblock and did not see the images. The roadblock against the Agua Rica project was (and still is) a 24/7 affair, and there are always two volunteer assembly members manning the blockade to ensure the mining company trucks don't have access to Agua Rica.

I went to the tent that they had set up at the roadblock, we made a makeshift screen, and I repeated the presentation. Vanessa Martinez, the rock singer and writer, was there, and together we proposed naming one of the rock glaciers, to give it identity among the local folk. None of the rock glaciers in the Aconquija had names. There was one glacier that was very large, nearly 900 meters (2,950 ft) long and 300 meters (984 ft) wide. The rooting area of the glacier, where it was fed snow and ice, looked like the branches and leaves of a tree. The choice of name was easy, with everyone overwhelmingly in agreement "El Algarrobo,"[56] the name of the Argentine hardwood tree, a symbol of strength and perseverance, and also the name of the environmental movement fighting for environmental protection in Catamarca—and now also for glacier protection in Andalgalá.

Under a newly elected mayor, who promised the electorate to fight for the protection of glaciers, CEDHA's report on mining impacts to rock glaciers and periglacial environment at Agua Rica and Filo Colorado was eventually declared of public interest[57] by the city council of the municipality of Andalgalá. The community later filed legal action with the Supreme Court of Argentina against the province of Catamarca for not complying with its due diligence under the national glacier protection law to protect the province's glaciers. CEDHA's report was submitted as evidence, and we were asked to draft an amicus curiae (a friend of the court) statement outlining our findings and revealing the risks we saw to glaciers from mining operations in the vicinity, which CEDHA presented to the tribunal on Earth Day, April 22, 2013.[58] Cryoactivism was in full swing!

After the Agua Rica and Filo Colorado report, we were anxious to move on to our next glacier inventory. We published the report in February 2011, then continued with our glacier inventory, mostly centered on San Juan Province, but also along areas farther north, including La Rioja, Catamarca, Salta, and Jujuy Provinces. In addition to the glaciers that I had inventoried in Catamarca, Tucumán, San Juan, and La Rioja, CEDHA's glacier inventory had already registered more than 1,000 glaciers.

CEDHA went on to produce a series of reports on mining activity and glacier impacts:

- *Glacier Impacts at El Pachón* (Xstrata Copper), published May 23, 2011. A report detailing the extensive rock glacier and periglacial presence surrounding the copper project by Xstrata Copper in Southern San Juan Province.[59]
- *Glaciers and Mining in the Province of La Rioja*, published April 2012. A report revealing more than 400 glaciers and perennial ice patches in La Rioja Province, despite the governor's claim[60] that there were no glaciers in his province.[61]
- *Glacier Impacts at Los Azules* (McEwen Mining), published in May 2012. A report showing drilling impacts to periglacial environments at another copper project just north of El Pachón. [62]
- *The Periglacial Environment and Mining in Argentina*, published November 9, 2012. A report to help the public understand what the

periglacial environment is and what the risks of mining are to the periglacial environment.[63]

- *Glaciers and Periglacial Environments in Diaguita-Huascoaltino Territory, Chile*, published December 2012. A report revealing more than 400 glaciers in the autonomous territories of the Diaguita-Huascoaltino Indigenous Community, in Chile, as well as risks and impacts from the Pascua Lama (Barrick Gold) and El Morro (Goldcorp) mining projects.[64]
- *Barrick's Glaciers*, published May 20, 2013. A report showing extensive glacier presence, impact, and risk by Barrick Gold's two projects, Pascua Lama and Veladero.[65]
- At the time this book went to press we were working on yet another report focusing on the Cerro Amerillo project in Mendoza Province (by Meryllion), studying risks and potential impacts of that project on rock glaciers and periglacial environments.

Each of these reports entailed a lengthy engagement with interested stakeholders, with the governments of each jurisdiction, and with the mining companies in question. In each case, these engagements were complex, with a number of actors interceding.

In some cases, we had the collaboration of some of the actors; in others, they were mostly unwilling to enter into dialogue and provide neither the information we requested nor access to their mining sites to review our findings and address our concerns. I briefly mention some of the more salient highlights in each case.

GLACIERS AND PERIGLACIAL ENVIRONMENTS
AT EL PACHÓN (SAN JUAN): XSTRATA COPPER

El Pachón was a project mentioned in Brenning and Azocar's report on the Pelambres project in Chile. The authors had indicated visible evidence of rock glacier impacts in the El Pachón project, as well as at Los Azules and Altar,[66] so we already had the project on our radar screen for likely mining impacts to glaciers and periglacial environments. We decided to focus on projects in the pipeline, and El Pachón was definitely one of them, set to begin operations sometime between 2012 and 2014.

We had already engaged Xstrata Copper on the Filo Colorado project, with meager results, perhaps because they had already closed up shop and abandoned the project. Discussions continued with El Pachón. In March 2011, we asked for the environmental impact studies for the project. The company answered, as usual, that we should inquire with the province of San Juan, which we had already done to no avail. They simply would not answer, and there was no information about the project on the provincial website. After a few back-and-forth e-mails, we decided to press the company with the threat of legal action, which, as usual, got us better results. Within forty-eight hours, I received a message from San Juan's Mining Minister himself, Felipe Saavedra, suggesting that the province published all information about projects on its website (to this day, this is not true). When I visited the Mining Ministry's website, the El Pachón EIA had been uploaded! An Xstrata Copper employee later confided that they had called the Minister after our inquiry, requesting that the report be uploaded.

By that time, we had already conducted our glacier inventory for the area around El Pachón and identified nearly 250 rock glaciers.[67] There were no uncovered white glaciers in the area. In the project EIA, there was a geomorphological map produced by a consulting firm called URS that provided just the evidence we were looking for. The company indeed had information about the rock glaciers present in the project vicinity. They had mapped more than 200 rock glaciers in the project area.[68] One rock glacier was in the project pit area—that glacier would be completely destroyed by the advancing project.[69] Another three rock glaciers were in the area destined for the project's sterile rock pile.

In our report we recommended that:

First and foremost, all activity of El Pachón, including exploratory work, project preparation, or any other activity should immediately cease until past, present and future impacts to rock glaciers and periglacial environments of any mining activity at the El Pachón site could be determined.

Second, Xstrata must immediately produce a Glacier Impact Assessment, of past, present and future activity, as mandated by Argentine federal law as well as San Juan's provincial law.

Third, all past impacts to rock glaciers and to periglacial environments caused by El Pachón mining project should be repaired (including existing roads affecting rock glaciers or periglacial environments), as best possible to their original state, prior to any mining or other anthropogenic intervention.

Fourth, Xstrata should establish and clarify what procedures it would use in any future mining activity at El Pachón or other sites where glaciers, rock glaciers, or periglacial environments exist near operations and where any of its mining activities might potentially impact any type of glaciers, white uncovered glaciers, rock glaciers, periglacial environment, etc. This would include reconsideration of pit location excavation, relocation of mineral tailings waste deposit sites, infrastructure or other elements related to the project.

Fifth, Xstrata should be absolutely transparent and share all information about its mining operations taking place in glacier territory, and any past, present or future studies on glacier impacts.

Soon after our report, Xstrata Copper's CEO in Argentina, Javier Ochoa, began appearing in local media arguing that there were no "ice glaciers" at

El Pachón or at any other mining project sites.[70] This was a clear distortion of the facts. His reference to "ice glaciers" was erroneously skewed to suggest that there were no uncovered glaciers at El Pachón, which would look white to anyone visiting the site. Instead, the hundreds of rock glaciers at El Pachón were underground and had colossal ice content, visible only due to the active characteristics of rock glaciers, something Ochoa was trying to hide. The company criticized CEDHA's work as "desktop research" and insisted that there were no glaciers at El Pachón, claiming that what there was (they didn't ever refer to rock glaciers) wasn't even related to glaciology. He later changed those claims recognizing the presence of rock glaciers and permafrost, indicating that the company was amending its studies to comply with the new Argentine glacier law.[71]

Curiously, Xstrata Copper included a section in its 2010 Sustainability Report for El Pachón that never mentions the word glacier but makes indirect reference to and criticizes our impact report and our cryoactivism:

> The cryological evaluation of El Pachón was initiated spontaneously in 2008, and has the objective of describing the presence and state of geoforms and periglacial phenomenon present in the area where the El Pachón is located defining the limits of the cryosphere and the area of influence of the project, and monitor its geo-dynamics through different methodologies. It is important to note that despite the fact that much has been spoken [they are referring specifically to CEDHA's work] that this work can be done with satellite images, unless one uses high-resolution images and specialized work protocols in the field, any scientific conclusions are mere conjecturing. (El Pachón Sustainability Report 2010, p. 42) [unofficial translation]

I later entered into dialogue with Jorge Sausset, the environmental director of the El Pachón project. He confided that our findings were valid and that there were indeed rock glaciers at El Pachón. He wanted to engage with us and collaborate on a solution, which, in fact, they were already working on. I suggested to him that together we could come up with a Protocol for Mining in Glacier Areas, which could serve not only as a policy for Xstrata to follow in all of its high-altitude operations, but which would also provide some critical guidance to other companies in Argentina and around the world for such operations. CEDHA could bring its technical advisors to the table, and Xstrata Copper could bring theirs, and we could come up with a very important guidance document to help the industry, protect glaciers, and help mining companies comply with the law. It would also be a great tool for public officials to have to set up their own controls and monitoring systems for mining in glacier areas.

Sausset was interested in the proposal and even got his team to start working with CEDHA on the idea. We drafted an outline of what the Protocol would entail, and Xstrata went as far as to translate the outline. It was divided into eleven chapters:

- Definitions
- Description of Glacier and Periglacial Environments
- Risks from Mining Activity to Glaciers and Periglacial Areas
- Corporate and Public Policy for Glacier and Periglacial Impacts
- Glacier and Periglacial Impact Assessment
- The Exploration Phase
- The Extraction Phase
- Mass Balancing and Monitoring
- Impact Mitigation
- Glaciers as Basin Regulators
- Glaciers, Communities, Public Participation, and Human Rights

We identified experts who could help develop the protocol, and Sausset promised that, when finished, Xstrata would adopt a glacier protection policy derived from the protocol. But when we were about to start drafting the guidelines, the company's lawyers stepped in and terminated our collaboration. They were going to file a lawsuit like Barrick Gold's to try to derail the glacier law.

CEDHA eventually filed a complaint against Xstrata Copper in Australia for the company's reiterative impacts and failed due diligence on glacier protection at both the Filo Colorado and El Pachón projects.[72] Just a few months before this book went to press, Glencore-Xstrata decided to unilaterally walk away from mediation in the case.[73]

We do know that despite Xstrata Copper's public statements ignoring glacier impacts, they have proceeded with their glacier impact study and have glaciologists currently working to address the damage to glaciers that CEDHA revealed in its reports. We also know that they have taken specific measures to repair damaged glaciers. However, none of this information is in the public domain and hence we could not confirm it at this stage.

GLACIERS AND MINING IN LA RIOJA PROVINCE

When Beder Herrera said to a reporter from *La Voz del Interior*, a regional paper published in Cordoba, that there were no glaciers in La Rioja Province,[74] we were furious. We had been carrying out a glacier inventory that included numerous glaciers near the provincial border between La Rioja and San Juan[75] and on the international border between Argentina and Chile, in La Rioja Province. We had also been looking closely at the Famatina Mountains area (part of La Rioja) because it was the site of much social conflict over another mining project (Famatina) originally planned by Barrick Gold and later passed on to Osisko Mining.[76] I promised myself when I heard Beder Herrera's comments that at least one of the reports we would do would focus on La Rioja, if merely to publish a glacier inventory to refute Beder Herrera's incomprehensible comment.

Gathering information about La Rioja's mining projects turned out to be the hardest task for our report. The province's Mining Ministry didn't even have a website. There was no official information about mining under way in the province.

We ended up relying on old reports we were able to dig up about mining in La Rioja, a few national reports that were also outdated but which published some concession names, and we drew most of our information from foreign mining publications that periodically announced discoveries, exploration, and other bits of information about mining operations taking place in the province. In this way, we identified some forty potential projects in the province. In time, we were able to plot most of them on Google Earth.

For our glacier inventory, we centered on the region along the border with Chile as well as on the Famatina Range. We counted more than 400 glaciers and smaller perennial ice patches in all, some quite spectacular, like the El Potro Glacier[77] at the tripartite union of La Rioja, San Juan, and Chile. El Potro is La Rioja's largest glacier.

At about that same time, several news programs had identified our evolving work on glaciers and mining and would periodically touch base with us to see if we had anything new to share. We were just about to go public with the La Rioja report and mentioned it to a national environmental news program produced by a well known environmental reporter named Sergio Elguezabal and Maximiliano Heidersheid. They asked to see the report before we published it. It turns out that the community at Famatina was creating quite a stir against the advance of the Barrick Gold Famatina project, which had been sold to Osisko Mining.

The news station was doing a story on the protest and found that CEDHA's report was just in line with what they were looking for. The community was claiming concerns over glacier impacts, but, as with the case of Andalgalá Catamarca, no one had ever revealed the glaciers in the Famatina Mountains. With CEDHA's report in hand, which revealed more than 400 glaciers or perennial ice bodies in the province, several dozen of which were in the Famatina Mountains, the journalist traveled to La Rioja and carried out a series of interviews, one with the province's Environment Secretary.

This incredible interview, reproduced in its entirety below and viewable still on YouTube, revealed the absolute ignorance (or deliberate denial—which would be worse) of the provincial authorities as to glacier presence in their province. Remarkably, the public official speaking is the highest environmental authority of the province. The section of interview reproduced here begins at minute 3:45 of Part I:[78]

SERGIO ELGUEZABAL (REPORTER): "Are there glaciers in La Rioja?"
NITO BRIZUELA (ENVIRONMENT SECRETARY OF LA RIOJA PROVINCE): "So far the study shows that there would be **one** glacier. In the area . . . uhhh, . . . at

the border area with Catamarca Province." [this would not be the El Potro Glacier, which is the province's largest glacier, but another glacier further north]

REPORTER: "And the province is sure that there is **only one** glacier in La Rioja?"

ENVIRONMENT SECRETARY: "Until now, yes. In any event, we are waiting to work on the Glacier Inventory that the Federal Government has to do, and which will be done with the province. With the province, we will determine the types of glaciers there are."

In relation to a question as to whether there are glaciers in the controversial Famatina Mountains, where communities have protested against a mining project, at minute 7:03 of the video, the Environment Secretary makes no sense whatsoever in his reply:

ENVIRONMENT SECRETARY: "What happens in Famatina is that the area is very removed from any possibility of there being snow on the mountain. This is the idea, everyone believes this is close to the Snowed Mountain of Famatina, and this is quite far." [he suggests that Famatina is far from itself]

REPORTER [referring to the Famatina Mountains, where there are some thirty visible white glaciers]: "There is no snow at that place?"

ENVIRONMENT SECRETARY: "No, no, no, no, no . . ."

REPORTER [after showing the audience an image on Google Earth that shows the mining project 5 km (3 mi) from a glacier]: "How far is the mining project from the glaciers?"

ENVIRONMENT SECRETARY: "About 25 km (15.5 mi)." [the Secretary not only misses the distance by five times, but contradicts his earlier statement indicating that there were no glaciers in the province]

REPORTER [insisting on the snow question]: "And there is **no** snow there at any time of the year?"

ENVIRONMENT SECRETARY: "No, no, no. Not in that area."

REPORTER: "There are **no** bodies of ice?."

ENVIRONMENT SECRETARY: "There are no bodies of ice. That's the most important thing, there are no bodies of ice."

When asked whether there were mining operations near glaciers, the Secretary replied (at minute 5 of Part II):

ENVIRONMENT SECRETARY: "For the moment there are no projects in natural reserve areas, in the case of forests, that are tangible, so for the moment, no." [his words don't make sense in Spanish either]

REPORTER: "Have you sent the information?" [referring to the obligation under the national glacier law that the province send information to the

IANIGLA regarding the presence of mining projects in glacier areas;
information that, in fact, was never sent]

ENVIRONMENT SECRETARY: "Yes. We have." [this was a lie]

REPORTER: [making reference to La Rioja's largest glacier, the El Potro
Glacier, which is more than 100 meters (328 ft) thick] "El Potro, that's
another glacier, right?"

ENVIRONMENT SECRETARY: "Yes, . . . what we don't know is if it is a glacier
or if it is eternal snow. There is a difference, there is a conceptual difference" [actually, that's not true: there is no conceptual difference—there
is no such thing as eternal snow, since that simply would be glacier ice]

REPORTER: "The province is not sure if El Potro is a glacier or not?"

ENVIRONMENT SECRETARY: "Yes, that's right. According to their definition
[he's referring to the IANIGLA], it would be eternal snow."

REPORTER: "And what activity is there up there?"

The Environment Secretary fumbled with this question, contradicting
himself by first recognizing that there is mining going on around glaciers and
then denying it, all in one sentence:

ENVIRONMENT SECRETARY: "Principally mining prospecting activity. For
mines . . . that are . . . that is, uhh, . . . there are no perforations for gold,
that is . . . there are no projects, but rather other activity."

This interview with the Environment Secretary of La Rioja Province, Nito
Brizuela, is an incredible revelation for many reasons. The interview was
filmed in April 2012, a year and a half after the glacier law was passed and
in force, nearly two years since La Rioja adopted its own glacier protection
law, and nearly four years after public officials starting debating glaciers in
Argentina. This is a province with glaciers, many of them, and it is a province that even adopted a glacier protection law at the provincial level—why
would it adopt a glacier protection law if the province has no glaciers, as the
Secretary suggests?[79] This was a time when glacier debates were heated, and
they focused on the problems faced by provinces like La Rioja and San Juan.
The Congress had spent two years in a deadlocked debate about glaciers and
mining. With all of this having recently transpired, it's remarkable that the
province's highest environmental authority simply did not know what a glacier was and if his province had any or not, or that he would ignore the presence of the province's largest glacier, the El Potro.

Only one of two possibilities exists: either he was ignorant about glaciers,
in which case, his case is truly lamentable because there is no excuse at this
stage and after so much public debate for an Environment Secretary to not
know some basic facts about glaciers; or, he is hiding the presence of glaciers
to protect mining interests, which would be worse. Whatever the reason, the

situation is alarming but also goes to show just what local society and groups like CEDHA face in trying to protect glaciers.

The interview with the provincial Environment Secretary kicked off a series of public reactions not only in La Rioja but in many locations in Argentina. CEDHA released its report on glaciers and mining impacts the day following the airing of the interview on national television, and reactions to the report, especially in La Rioja, were explosive. The mayor of Famatina used the report to file a legal complaint to the courts asking to suspend Osisko's operational permit until a proper glacier impact study could be carried out in the Famatina Mountains. The court granted the suspension. Like in Catamarca Province, cryoactivism is at work in La Rioja.

Later, glaciers came up widely in local media, generating much local debate about the province's future, about mining impacts to hydrological resources, and about glacier impacts and need for glacier conservation. We were later invited to the provincial book fair in La Rioja's capital, where we were able to hold a workshop on glacier presence in the province.[80]

SAN JUAN PROVINCE AND BARRICK'S GLACIERS

Nowhere has the debate around glaciers and mining been so confrontational as with San Juan Province. I purposefully do not say "in San Juan Province" because the debate has been mostly outside the province, with the political arena and media in the province unwilling to engage openly or transparently. This is most unfortunate since it is the one area in South America—and maybe in the world—where mining is at most conflict with glacier resources. It is where such conflict spawned the first debates over the need for a glacier protection law (in Argentina), and it is the place that defined the content of Argentina's national glacier law.

This book would not exist if Barrick Gold had not proposed dynamiting glaciers at Pascua Lama. It was this idea that spawned the movement in Chile to protect Andean glaciers from mining activity, which in turn crossed the Andes to engender the world's first national glacier protection law in Argentina. San Juan had no policy or laws to protect glaciers. Even its detailed water law didn't mention glaciers. Glaciers became an issue in San Juan because Barrick Gold made it an issue, and the environmental movement followed.

The provincial authorities were not interested in glaciers or in glacier protection. In fact, before this discussion emerged, the provincial government didn't require any special attention to glaciers or periglacial areas by mining companies operating in these regions. Nor did they view glaciers or periglacial features as critical reserves that needed protection.

When the glacier law appeared, Barrick Gold was the first to react and the first to mobilize, not only leveraging a presidential veto the first time around (2008), but also with a legal attack on the law once it held firm after the second congressional vote and approval of the law in September 2010. Barrick repeatedly denies glacier presence in Pascua Lama's zone of influence and refutes all claims that it is impacting glaciers. We will not dwell on refuting this point because it is not the objective of this book, except to say that if there were no conflicts between Barrick's Pascua Lama project with glaciers, then the company would never have attacked the glacier law in the courts and the Chilean government would never have closed down Pascua Lama due to ... glacier impacts! If you are interested in a detailed report prepared by CEDHA on the damage and risks by Barrick Gold to glaciers and periglacial environment, consult our report entitled *Barrick's Glaciers*.[81]

That point cleared, we can move on to examine the dynamics among Barrick Gold, San Juan, the national government, society, and environmental groups that lent themselves to the nature of the evolution of glacier protection in Argentina.

Barrick Gold started two adjacent projects without any laws or regulations on glacier and periglacial environment protection on the books, Veladero and Pascua Lama. The original EIAs of these projects did not mention glacier impacts nor did the company refer to glaciers when they mentioned the "natural resources" that might be impacted by the projects. They saw ice and frozen grounds as problems to be resolved rather than as resources to be protected. This is an important distinction because we often see this view from industry, and particularly from mining companies, when confronting glacier impacts. As mentioned earlier, this was the case with Xstrata Copper's Filo Colorado project. Periglacial environments are hazardous for equipment and heavy transport, and this the mining companies understood perfectly.

The glacier law process threw several wrenches into the state of things in San Juan. First, it undermined provincial discretion over how mining was viewed and allowed (or not) to proceed. San Juan had chosen a development path grounded on mining. The glacier law told the province that it could not promote mining in glaciated areas or in periglacial environments, which for San Juan meant a practical ban above 3,500 meters (11,500 ft). This affects probably about 80% of the 180 (or more) projects that are currently in either the exploration (the majority) or extraction phase.

For Barrick Gold, the problem was immediate, not only because of Pascua Lama, which was one of the two most important projects it had in the pipeline worldwide, but also because its existing project, Veladero, could also run into retroactive problems because Article 15 of the law could conceivably be used to stop activity there. It was expected that Barrick Gold, as well as the province of San Juan, would react against the law. And it was expected that the province of San Juan would want discretion over the glacier inventory, over the definition of glaciers and periglacial environments, and over the decision of what is to be protected; finally, it would want control over the decision of which projects could go forward if they were located in periglacial or glacial areas. Several multibillion-dollar projects in glacier and periglacial areas were already advanced beyond the exploratory phase, including El Pachón, Altar, and Los Azules, and these were all in doubt if the glacier law stood. At least a dozen other projects, also in glacier and periglacial areas, were also in the pecking order for further advancement.

It was also logical to assume that San Juan would not want a conservationist glacier protection law because it would mean losing more than 80% of its projects. If it were to maintain its "pro-mining" mindset for development, San Juan would have to opt for an "environmental services" approach to the law.

Second, a traditional mistrust exists between provincial politics and Buenos Aires-focused federal politics. Argentina has a long history of conflict between the *unitarios* or "unitarists" preferring a strong centralized government and the *federales* or "federalists," which were led by provincial strongmen who put provincial independence far above federal unity. This dynamic plays out in almost every phase of life and generates a constant political tug-of-war between provinces and the national government. For the provinces, the glacier law was yet another instance in which the federal government was meddling in provincial affairs. For San Juan, it went straight to the heart of their newly found development path, mining.

Third, the federal glacier protection law caused quite a stir in academic circles and in terms of glacier academia independence. The opposition to national government meddling in provincial glacier issues, stemmed principally from San Juan, and was largely personified in geologist Juan Pablo Milana, then the province's leading expert on glaciers and periglacial environments. He stood very much alone and independent from the IANIGLA and didn't want to see the national glacier inventory process turned over to the province's competition.

SAN JUAN'S GLACIER INVENTORY

These three dynamics governed politics around the glacier law nationally and in San Juan. The province's mining authorities, which also oversaw environmental matters involving mining operations (the Environment Ministry of San Juan does not have mining operations under its jurisdiction and so environmental permitting, auditing, fines, and promotion are all run by the same political figure, the Mining Minster—at the time and currently Felipe Saavedra).

When the glacier law first passed, Milana was charged with carrying out San Juan's glacier inventory. He was ideal for this role. Under the auspices of the National University of San Juan and equipped with several young and upcoming geologist-glaciologists, he started counting ice.

San Juan kept to much of the timing set forth by the National Glacier Law; in fact, on the inventory, it got ahead of schedule, publishing the first draft of the San Juan Glacier Inventory in December 2010.[82] But by then Milana had run into trouble with the governor and with authorities of the university.

Despite Milana's many critics (ranging from civil society, government, and the private sector), he stood his ground in pointing out problems with mining projects and glaciers, and he was quite displeased with what he saw at Pascua Lama. He is also someone who readily speaks what he believes to the media, and, in the world of politics, this doesn't pay off well in the long term. He's also willing to defend mining, which for environmentalists is generally blasphemy. Forces within the university and within government started working against him. A short time before the publication of San Juan's first draft inventory of glaciers, upset about how the politics of the university were lining up regarding funds channeled for glacier work and seeing a political undercutting of his discretion over the glacier inventory happening before his eyes, Milana resigned as director of the San Juan Glacier Inventory. He made several public accusations of corruption against university representatives, including against the dean, and against the government, and went out with a bang.[83]

The person chosen to replace him was Silvio Peralta. As soon as he took his position at the helm of the INGEO,[84] the agency carrying out the provincial glacier inventory, he made some startling comments that showed that the province was set on affirming that there was no problem between mining and glaciers. The comments were specifically targeted to the various systematic critiques heard at the national level about mining impact to glaciers:

> We have not seen any glacier affected by mining activity, nor by any other industrial activity, by tourism or by public works. We have seen that the glaciers are over there [sic], near, but that mining activities don't reach or affect the glaciers.
>
> Behind all of this [says Peralta to the reporter] is a series of false damaging concepts, such as that mining contaminates, that mining activity destroys glaciers, something that we have absolutely not seen anywhere. They say that mining projects are advancing more into glacier areas but until now, this is not true. I don't see any mining company pulverizing a glacier to expand its area of extraction.[85]

But in terms of the draft inventory itself (never updated publicly), several points are worthy of note. Milana had done a thorough job of both designing and advancing the inventory. He hadn't finished it, but it already listed more than 3,500 glaciers without counting rock glaciers (far fewer than what the province would reveal in December 2010: 2,500 glaciers). But what was especially troublesome was that the province did not provide any means of actually seeing where the glaciers were. The inventory did not provide any

GIS coordinates, such as CEDHA does for each of the glaciers it inventories. San Juan released no data whatsoever that could be used to actually locate a single one of its glaciers; convenient, of course, if you don't want to show mining conflicts with glaciers.

The San Juan Inventory published information about the types of glaciers, the number of glaciers, the surface area they cover, and the general basin areas they are in, but no information that would allow any user, even a versed professional in glacier academics, to actually find them. Finally, although the provincial inventory included some maps, the images were taken from too high an altitude to distinguish any useful information (with glaciated areas showing in red and blue) and left off the last several kilometers of the province's most northern extreme, where most of the mining in glacier areas is located. I asked Milana personally if location information was easily publishable for the inventory, knowing well what his answer would be. He responded, "You simply have to add a column to the Table that has the GIS coordinate. Obviously they have that information! Publishing it however, is another story."

In a word, the glacier inventory of San Juan was a sham because it didn't show the glaciers.[86]

The province was able to maintain a lid on the glacier and mining situation for several reasons. One, it has full control of the media. The governor has been known to buy up all magazines reaching the province (they generally come in by bus to the central terminal) when articles appear in popular magazines criticizing his administration. Two, media such as television and printed matter is generally under the financial wing of the governor's party, which guarantees that the press will not portray the debate openly or ratio nally. Finally, the mining projects themselves are inaccessible to the public, which keeps glacier information from interested stakeholders. No one can go up to Pascua Lama without express authority to do so, and only Barrick Gold decides who goes up. Furthermore, other mining projects either use Barrick Gold's access road to Veladero, or the roads are controlled by provincial authorities, which also limits public access to the mines.

The glaciers, at least until they melt away, are under lock and key.

Barrick Gold's Pascua Lama project undoubtedly created the most fallout in San Juan. This was the project of all projects, a mega-billion dollar investment first valued at US$3 billion, then US$5 billion, and then at more than US$8 billion. Here, the rubber hit the road: there was already extensive, visible, quantifiable damage to glaciers, particularly of glaciers at the pit site and along the access road to the mine site on the Argentine side of the project. This was the Kumtor[87] of Argentina (Kumtor is the mining project in Kyrgyzstan, undertaken by Canadian mining company Centerra, that has had enormous impacts directly onto glaciers). It was the case where a mining company wrote in black and white that it needed to bulldoze and dynamite glaciers because they were a threat to the environment.[88] It was the project that leveraged a presidential veto. It was the reason Argentina had gotten into this mining and glaciers issue in the first place.

In a nutshell, the glacier issues that shaped the discussion at Pascua Lama (and, for that matter, for other mining areas as well) could be summarized as follows:

- Exploratory roads and exploratory drilling can impact glaciers due to ice penetration, soiling, fuel emissions, and more.
- Mining activity in the pit area will invariably destroy any ice that is in or immediately adjacent to the area.
- Sterile rock deposited on glaciers or periglacial areas creates acid drainage and instability problems for those areas and can cause severe impacts to the natural evolution of the glacier or periglacial area (such occurred in Kyrgyzstan).
- Blasts from operations emit dust particles and debris into the air that are deposited on glaciers, soiling and darkening them, and leading to accelerated melt due to albedo changes.
- Snow clearing from roads can strangle glaciers by eliminating their accumulation.
- Contaminants from mining activity compromise glacier water quality.
- No one had carried out a baseline assessment to know where glaciers are located or their health at the onset of operations.

Barrick Gold obviously set up its defense, as have other mining companies, around several of the following arguments:

- Climate change is melting glaciers; therefore, glacier size decreases since it started operations was due to exogenous forces and not due to the company's activity.
- Its roads or their maintenance don't impact glaciers in any way because the company waters the roads to reduce dust to avoid dust deposits on glaciers.
- The glaciers immediately adjacent to the pit area, Toro 1, Toro 2, and Esperanza, are fully covered in dust and debris (a phenomenon that has occurred since the company began preparatory operations) due to natural causes and not to Barrick Gold's activities.
- No acid drainage will occur because the company has introduced the necessary works to avoid it.
- Only seven glaciers exist in the project influence area and the company is duly monitoring them to assure they are not impacted.

Whichever side you end up on regarding the debate, some inescapable issues jump out at us:

First, mining activity of all types obviously has significant environmental impacts. The question is not whether mining impacts the environment, but rather how much we are willing to tolerate the environmental legacy of mining operations in exchange for the wealth and benefits that it may bring. Road presence in glacier areas and roads running through glaciers are undeniable. The mining blasts and resulting impacts in an area with hurricane-force winds are also undeniable and have led to several project closures on the Chilean side of Pascua Lama due to dangerous amounts of suspended particles in the air, which also meant that glaciers were obviously being impacted by such air contamination (Figure 9.3). And, finally, climate change makes glaciers vulnerable, which would also imply that mining impacts to glaciers would aggravate already vulnerable natural resources.

Barrick was just as new to the glacier debate as local actors, and it attempted several strategies to minimize its responsibility. The company focused its initial arguments on nomenclature, indicating, for example, that a certain body of ice was not actually a glacier but rather some other type of ice, such as a glacieret or a perennial snow patch. That strategy was short-lived and ultimately

FIGURE 9.3 *Dust and debris from mining activity soils glacier surface, altering albedo and accelerating melt.*

Source: Jhon Melendez.

nixed in Argentina by the glacier law, which recognized glaciers of any size or form for what they are and not what they are called: namely, these are water reserves important not only because they store water but because they regulate natural water provision to downstream environments during warm seasons or especially dry years.

Barrick also tried to limit the discussion to only those glaciers that were in the immediate pit area. The company only identified a handful of glaciers, but, in fact and as Milana had indicated earlier, there were at least fifty glaciers on Argentine soil near the mineral deposits. And that didn't even take into account the glaciers along the access road, which Barrick had already accepted were in the influence area when it carried out studies at the Conconta Pass glaciers.

When it comes down to it, the question we should answer is really how much the company impacts these glaciers, and how much of this impact the government (and society) is willing to tolerate. In Chile, just across the border, the discussion starts with a presumption that glaciers must be protected for their environmental value but also for the use of society and industry. The Argentine glacier law took a different approach and established an outright protectionist and conservationist policy, imposing a ban on all activity that impacts glaciers in any way. This, in the end, was the bottom line for the company, and it did not provide wiggle room for negotiating impact. In Argentina, the impacts by mining activity that groups like CEDHA were already registering were illegal under the glacier law. Ironically, perhaps, in the end, it was the Chilean government (that didn't have a glacier law) and not the Argentine government (that did) that would close Barrick Gold's Pascua Lama project due to glacier impacts.

WHAT TO DO WITH MINING IMPACTS TO GLACIERS?

As per the discussion on the glacier law, if you're on the side of the environmental services approach (where much of the industry situates itself), in which a balance is sought between environmental protection and social and industrial use of glaciers, then you probably would design a law that sets more flexible boundaries. You may decide to allow activity in the periglacial areas, or even right up to glacier ice. You probably think that there are ways to contain blasts from harming nearby glaciers, and you probably trust industry and its comptrollers to do the right thing: to contain contaminated water and provide open-ended guarantees that nothing will ever go wrong with the cyanide leach pads left behind in sensitive periglacial environments. You might rationalize that you can remove glaciers and transplant them to other sites (this was proposed) or that you would be willing to sacrifice a few glaciers for the millions of dollars of revenue that the project will generate over a lifetime. You might argue that debris cover from blasts actually protects glaciers from climate change. Who knows, you may even think you can drill into glaciers and not harm them, as many mining companies have done all up and down the Central Andes Mountains. In Chile, government agencies are now wrestling with many of these questions. They have temporarily suspended Barrick Gold's Pascua Lama project, precisely because the company is not meeting the necessary due diligence to completely address these questions.

If you're on the side of the conservationists, as I am, and you see glaciers as a vulnerable and fundamental natural resource at the top of the hydrological pyramid and you value their hydrological environmental and ecosystemic importance above and beyond services they might provide to particular stakeholders, then you probably will lean more toward a full ban of all activity in glacier and periglacial environments because your gut and your logic, as well as the evidence available, tells you that it is the right thing to do. If you are in this camp, then moving glaciers is a natural travesty, dynamiting them is an affront to nature, blasting mountains too close to glaciers can do nothing but pollute the air and soil them—even ones that are faraway—drilling into ice is a complete no-go, and you fully reject the idea that there is any rationale in economic progress through environmental destruction for the banal consumption of jewelry or for wealthy individuals and nations to hoard gold bars for financial security. Although this opinion may sound extremist to some, I draw attention to recent comments made by the president and founder of Barrick Gold, Peter Munk, to his shareholders, which reveal precisely this justification for gold mining:

> Gold in contrast to all other commodities is driven entirely by the comfort level of people with money globally. India, Abu Dhabi, New York, Venezuela, that's what gold is all about. What takes up 2,000 tons of gold is the uncertainty of people who control money trying to protect their wealth. And let me just say to you, . . . We print more money to make us feel better. . . . There have been only three reserve currencies, the German Mark (before the Euro), the US dollar, and the British Pound. . . . and not one of them have managed to get there without having a predominance of overwhelming gold reserves in their treasury. . . . China has less than 3% of its enormous accumulation of US dollars in gold. . . . If you think that most of the world's problems have been solved . . . then you should be pessimistic about gold. But if you believe in the long term protection of what you and your children own, whether you are a country or a person then do what the cautious people do, buy the commodity that cannot be reprinted.[89]

So, I ask, should we trade our glaciers for gold? So that wealthy people around the world "feel better" knowing that their savings are safe? I don't think so!

I like to quote one of our rivals on this debate, the governor of San Juan, Jose Luis Gioja, who says it brilliantly, "the Andes are enormous." He was suggesting that some minor mining activity would not impact the environment. To this

I say, if they're so enormous, the mining companies can go look for minerals elsewhere, but not where the glaciers are!

FINAL WORDS ON THE IMPLEMENTATION
OF THE ARGENTINE GLACIER LAW

The courts have ruled on two occasions to uphold the glacier law. All of Argentina's glacier protection legislation, both provincial and federal, are firmly in place. The National Supreme Court still has to rule on the merits of the Barrick Gold case, in which Barrick Gold and the province of San Juan request that the glacier law be ruled unconstitutional.

Barrick Gold, for the moment, has faced insurmountable hurdles with Pascua Lama, in Chile, not in Argentina. This is ironic, perhaps, because the Chileans never passed their glacier protection law, whereas Argentina now has both a provincial law and a federal law protecting glaciers from the sort of impacts Barrick Gold is causing. Pascua Lama is fully suspended in Chile, the company faces numerous fines, and it is currently carrying out new infrastructure works to show that the impacts to periglacial areas and to uncovered glaciers has been eliminated. Only then will the government lift suspensions. Meanwhile, the cost of operations has skyrocketed, and Barrick finds itself in a difficult financial position with the project and with macroeconomic circumstances in Argentina, which are not favorable for the repatriation of wealth that Peter Munk talks about in his address to shareholders. In Canada, Barrick Gold faces a US$6 billion lawsuit for allegedly lying to shareholders over environmental impacts (including impacts to glaciers) at Pascua Lama.[90]

The national glacier inventory is behind schedule, and although the IANIGLA inventory advances at a steady pace, there is little or no collaboration from provinces such as San Juan, which do not want to fully reveal the information they have collected about their glaciers.

The priority inventories in mining areas have not been conducted, or at least they have not been publicized, and the big question remains: if the Supreme Court sustains the glacier law, what will happen with projects like El Pachón, Los Azules, Altar, Pascua Lama, Famatina, Vicuña, Josemaría, Las Flechas, Del Carmen, and a long list of others that are squarely in glacier and/ or in periglacial environments?

Meanwhile, society is becoming more informed about glaciers. There have already been numerous complaints filed with the court system and extrajudicial forums, such as the OECD Guidelines for Multinational Enterprises or the Canadian CSR Ombudsman for the Extractive Sector, against projects for violating the glacier law and for their general failure to meet basic due diligence on glacier protection.[91] These cases will only increase as society becomes increasingly aware of and sensitized to the vulnerability of glaciers to local anthropogenic impacts.

In this regard, I am confident that the future will be one where we are more attuned to glacier vulnerability and where we work to reduce our impacts to glaciers and periglacial environments.

Notes

1. See http://www.youtube.com/watch?v=DN8X-HjaP4Q&list=PL5A353061CDEB3FE6.

2. See http://www.boletinoficial.gov.ar/Inicio/index.castle?s=1&fea=28/10/2010.

3. See http://mineria.sanjuan.gov.ar.

4. See http://www.mineria.gov.ar.

5. See Judge Galvez's verdict, which cites the original complaint text presented by Barrick Gold: http://wp.cedha.net/wp-content/uploads/2011/11/Fallo-Galvez-Glaciares-San-Juan.pdf.

6. The latter half of this quote was taken from Bonasso (2011, *El Mal*, p. 432), in which the original complaint was quoted.

7. Projects in San Juan that could be in serious conflict with the glacier law:

El Pachón (Xstrata-Glencore): 31°44'53.36" S 70°25'51.24" W
Altar (Stillwater): 31°28'41.99" S 70°28'58.56" W
Los Azules (McEwen): 31°03'47.42" S 70°13'57.21" W
Del Carmen (Malbex): 30°00'39.19" S 69°53'58.33" W
Pascua Lama (Barrick Gold): 29°19'08.80" S 70°00'37.31" W
Las Flechas (NGX Resources): 28°42'53.09" S 69°39'24.82" W
Vicuña (NGX Resources): 28°26'46.45" S 69°36'31.91" W
Josemaria (NGX Resources): 28°25'20.34" S 69°33'01.71" W
Filo del Sol (NGX Resources): 28°26'57.89" S 69°38'41.46" W
Amiches (AML): 30°16'07.05" S 69°43'12.44" W
San Francisco (AML): 30°32'05.16" S 69°50'43.77" W.

8. The Agua Negra Pass is at 30°11'19.57" S 69°49'26.82" W.

9. For press reactions to the verdict see:
(1) Critical of the verdict: http://www.8300.com.ar/2010/11/03/san-juan-suspenden-la-aplicacion-de-la-ley-que-protege-los-glaciares/ ; (2) in favor of the verdict: http://www.diariodecuyo.com.ar/home/new_noticia.php?noticia_id=430767.

10. See http://www.cedha.net.

11. See http://www.wgf.org.

12. The grant was for US$60,000 for one year to work on advocacy against the adverse impacts of mining activity.

13. A list of these analytical documents can be viewed and downloaded at http://wp.cedha.net/?page_id=1277.

14. See http://wp.cedha.net/?p=12684&lang=en.

15. Bianchini Report on Water Quality at Veladero and Pascua Lama: http://wp.cedha.net/wp-content/uploads/2011/09/IMPACTO-DE-LOS-EMPRENDIMIENTOS-VELADERO-Y-PASCUA-LAMA-SOBRE-LOS-RECURSOS-HIDRICOS-DE-LA-PROVINCIA-DE-SAN-JUAN-CEDHA-2011.pdf.

16. See http://www.nti.org/country-profiles/iraq/delivery-systems/; http://www.nti.org/country-profiles/argentina/delivery-systems/.

17. See

   (1) Chile 2010 training: http://www.pnuma.org/agua-miaac/CursoGlaciologia.php.

   (2) Ecuador 2011 training: http://www.pnuma.org/agua-miaac/CursoIberoamericanoGlaciares_Present.php.

   (3) Chile 2012 training: http://www.pnuma.org/documento/Glaciologia/Dossier_Glaciologia__2_.pdf.

18. See http://wp.cedha.net/wp-content/uploads/2012/10/Azocar-Brenning-2008-Pelambres.pdf.

19. Pelambres is at 31°41'41.22" S 70°30'13.17" W.

20. See Brenning and Azócar, 2010: http://wp.cedha.net/?attachment_id=8283.

21. The reader can access the Google Earth mining database from a link on the following page: http://wp.cedha.net/?page_id=6332&lang=en.

22. Agua Rica: 27°22'13.77" S 66°17'04.82" W.

23. El Pachón: 31°45'14.46" S 70°25'41.47" W.

24. Los Azules: 31°06'19.22" S 70°12'48.55" W.

25. Famatina: 29°01'20.04" S 67°46'36.34" W.

26. See http://erth.waikato.ac.nz/antpas/pdf/AHUMADA-PRINT.pdf.

27. See http://www.cicterra-conicet.gov.ar.

28. Agua Rica: 27°22'13.77" S 66°17'04.82" W.

29. Filo Colorado: 27°22'11.31" S 66°12'38.85" W.

30. Andalgalá city: 27°35'59.81" S 66°18'58.67" W.

31. Snowy area visible on Google Earth: 27°19'09.66" S 66°11'25.22" W.

32. See 27°15'18.57" S 66°08'09.95" W.

33. Heart of project area with squiggly lines (exploratory roads) at 27°22'24.25" S 66°16'41.03" W.

34. Ridge crossing to Agua Rica: 27°21'04.67" S 66°17'59.54" W.

35. Filo Colorado site: 27°21'51.35" S 66°12'25.66" W.

36. See 27°20'04.00" S 66°12'03.12" W.

37. See 27°20'43.73" S 66°11'49.16" W.

38. Download .kmz files, then open them in Google Earth.
Catamarca Province glaciers: http://wp.cedha.net/wp-content/uploads/2012/04/Glaciares-Catamarca.zip.
Tucumán Province glaciers: http://wp.cedha.net/wp-content/uploads/2012/04/Glaciares-Tucuman.zip.

39. Read about Google Earth's features at http://en.wikipedia.org/wiki/Google_Earth.

40. For glacier tour of the Veladero and Pascua Lama access road and project areas, visit: http://wp.cedha.net/?page_id=14255.

41. See http://wp.cedha.net/wp-content/uploads/2011/09/Informe-Glaciares-de-Aconquija-Impactos-de-Mineria-Agua-Rica-y-Xstrata-Final-feb-18-2011.pdf.

42. In the case of Xstrata Copper, the company had identified risks to the periglacial area but not in terms of environmental importance assigned to the water reserve. Instead, the risk was measured in terms of risks to Xstrata's operators, relating to the ground instability in the area (i.e., potential risks of landslides resulting from the instability of rock glaciers and frozen soils). This was the same approach Barrick Gold had taken at Pascua Lama, ignoring altogether that the mining operation was destroying hydrological reserves.

43. Filo Colorado's EIA: http://wp.cedha.net/?attachment_id=14260.

44. Annex 9 of the Filo Colorado EIA: http://wp.cedha.net/?attachment_id=14261.

45. Filo Colorado photo taken from 27°19′33.55″ S 66°14′40.65″ W, looking southeast.

46. El Pachón project site: 31°45′22.05″ S 70°25′54.56″ W.

47. For CEDHA complaint against Xstrata Copper, see http://wp.cedha.net/wp-content/uploads/2011/06/Specific-Instance-xstrata-copper-glaciers-June-1-2011-final-english.pdf.

48. For information on the visit, see http://www.comambiental.com.ar/2013/12/el-municipio-de-andalgala-releva.html.

49. Agua Rica project location: 27°22′10.56″ S 66°16′52.60″ W.

50. Questionable cirque hollow at Agua Rica: 27°20′45.84″ S 66°16′23.97″ W.

51. Words spoken by Selva P. Ahumada, Yamana Gold's principal geologist for the Agua Rica project, to Jorge Daniel Taillant, Director of CEDHA, at the Yamana Gold offices in Andalgalá Catamarca, on March 4, 2011.

52. E-mail received from Selva P. Ahumada on March 28, 2011.

53. See https://www.youtube.com/watch?v=-iiSj5HiDYY.

54. Translation of poem about rock glaciers by Vanesa Martinez:

> *Why wouldn't I hide my white soul,*
> *From the abyss that appears,*
> *In face of the terror of seeing me flourish to give luxury,*
> *And not to give life.*
> *Unable to leave my skin,*
> *To create you anew*
> *With the cold of my natural strength*
> *Indifferent to the voices taken by the entangled paths of destroyed danger and*
> *made of evil*
> *To give to this bad cover*
> *More intense is the sky*
> *That reflects my condition*
> *So useless they are, that their weapons have uncovered me.*

55. Glacier Protection Bill for Catamarca: http://wp.cedha.net/wp-content/uploads/2011/04/Ley-de-Glaciares-para-la-Provincia-de-Catamarca.pdf.

56. The Algarrobo Rock Glacier is at 27°18′47.10″ S 66°11′04.35″ W.

57. See http://prensaelalgarrobo.blogspot.com.ar/2011_05_01_archive.html.

58. See press release about the amicus brief: http://wp.cedha.net/?p=13174#.

59. For CEDHA's report on El Pachón's impacts to rock glaciers and periglacial areas, see http://wp.cedha.net/wp-content/uploads/2011/08/Glaciar-Impact-Report-el-pachon-xstrata-FINAL-english-version-may-23-2011.pdf.

60. For Governor Beder Herrera's denial of glacier presence in La Rioja, see http://www.lavoz.com.ar/noticias/politica/dejen-joder-con-criticas-mineria.

61. For CEDHA's report on glaciers in La Rioja Province, see http://wp.cedha.net/wp-content/uploads/2012/04/Glaciares-y-Miner%C3%ADa-en-la-Provincia-de-la-Rioja-English-Summary.pdf.

62. For CEDHA's report on Los Azules impacts to rock glaciers and periglacial areas, see http://wp.cedha.net/wp-content/uploads/2012/07/Glaciar-Impact-Report-Los-Azules.pdf.

63. For CEDHA's report on periglacial environments and mining in Argentina, see http://wp.cedha.net/wp-content/uploads/2012/11/El-Ambiente-Periglacial-y-la-Mineria-en-la-Argentina-English.pdf.

64. For CEDHA's report on glaciers in the Diaguita-Huascoaltino territories, see http://wp.cedha.net/wp-content/uploads/2013/02/Informe-de-Glaciares-y-del-Ambiente-Periglacial-en-Territorio-Ind%C3%ADgena-Diaguita-Huascoaltino-english.pdf.

65. For CEDHA's *Barrick's Glaciers* report, see http://wp.cedha.net/wp-content/uploads/2013/05/Los-Glaciares-de-Barrick-Gold-version-20-mayo-2013-ENGLISH-small.pdf.

66. See Brenning and Azócar, 2010, p. 154.

67. For El Pachón glacier inventory download: http://wp.cedha.net/wp-content/uploads/2012/04/El-Pachón-Valley.zip.

68. For Xstrata Copper's geomorphological map showing rock glaciers, see http://www.cedha.org.ar/contenidos/MAPA%202.6.1-AM-GEOMORFOLOGIA.jpg.

69. Rock glacier in pit area of El Pachón at 31°44'58.73" S 70°25'41.96" W.

70. The press articles where this statement was quoted are no longer available online.

71. For Javier Ochoa's amended statements on glacier presence, see http://www.mining-press.com.ar/nota/69593/ochoa-xstrata-pachon-una-minera-nacional-no-es-amenaza.

72. Complaint filed in Australia against Xstrata Copper for glacier impacts; see http://wp.cedha.net/wp-content/uploads/2011/06/Specific-Instance-xstrata-copper-glaciers-June-1-2011-final-english.pdf.

73. See: http://wp.cedha.net/?p=14455&lang=en.

74. For Governor Beder Herrera's denial of glacier presence in La Rioja, see http://www.lavoz.com.ar/noticias/politica/dejen-joder-con-criticas-mineria.

75. See tripartite border area of La Rioja, San Juan, and Chile at 28°23'07.12" S 69°36'36.29" W.

76. See Famatina area at 29°01'19.10" S 67°49'28.34" W.

77. The El Potro Glacier is at 28°23'05.92" S 69°36'38.01" W.

78. Interview with La Rioja's Environment Secretary, who incredibly denies that there are glaciers in La Rioja province:

> Part I: https://www.youtube.com/watch?v=itVm72-bHFA.
> Part II: https://www.youtube.com/watch?v=ro-095CdEFI.

79. For the La Rioja Provincial Glacier Protection Law, see http://wp.cedha.net/wp-content/uploads/2011/09/Ley-de-Glaciares-La-Rioja.pdf.

80. The presentation generated an interesting debate with one of San Juan's most notable opponents of the glacier law, director of a key mining association, and former Congressman Mario Capello. See http://www.datarioja.com.ar/imprimir.php?id=4858&PHPSESSID=54ada83a3a81e76f9af1dc70e973c7e0.

81. For press release on report, see http://wp.cedha.net/?p=13135&lang=en.

For *Barrick's Glaciers* report, see http://wp.cedha.net/wp-content/uploads/2013/05/Los-Glaciares-de-Barrick-Gold-version-20-mayo-2013-ENGLISH-small.pdf.

82. San Juan's Preliminary Glacier Inventory: http://wp.cedha.net/?attachment_id=10975

83. See http://www.tiempodesanjuan.com/notas/2012/6/10/calientes-glaciares-11872.asp.

84. The Institute of Geology and Exact Sciences of the University of San Juan.

85. Taken from *Diario de Cuyo*, December 30, 2010: http://www.diariodecuyo.com.ar/home/new_noticia.php?noticia_id=439842.

86. For a full commentary on San Juan's preliminary glacier inventory, see http://wp.cedha.net/?p=720.

87. The now infamous project in Kyrgyzstan impacting glaciers.

88. See sections 1, 2.2, and 2.4: http://wp.cedha.net/wp-content/uploads/2011/11/Plan-de-Manejo-de-Glaciares-Barrick-english.pdf.

89. See minute 1:18: http://www.gowebcasting.com/events/barrick/2013/04/24/2013-annual-meeting-of-shareholders/play/stream/7102.

90. Seehttp://www.mining.com/canadian-firms-sue-barrick-over-divisive-pascua-lama-project-88378/.

91. See complaint filed to the CSR ombudsman against McEwen Mining for glacier and periglacial impacts from the Los Azules project: http://wp.cedha.net/wp-content/uploads/2012/07/Request-for-Review-McEwen-CEDHA.pdf.

10

The Human Right . . . to Glaciers?

Yes. The right to glaciers! The very title of this presentation creates a sense of discomfort for some legal experts, much like the discussion about the "right to water" did several years ago. This discomfort is intentional. Hopefully, by the end of this chapter, you will agree that, at the very least, we do need to have a discussion on the role glaciers play in terms of human rights realization, one that may lead us to deepen this discussion on the human rights dimension of glaciers.

Glaciers are melting: we know that. And climate change, including natural ecosystemic millenary cycles of climate change, is causing glacier melt. But so is anthropogenic climate change, which is accelerating natural melt at alarming rates. Glacier melt will lead to both flooding in greatly populated areas—particularly downstream from rivers born in the Himalayas—and to the disappearance of massive water reserves in our glaciers and polar ice-caps, which will in turn cause sea levels to rise and flood many low-lying island states. Some entire populations in the South Pacific, such as islanders on Tuvalu, are in fact already looking for a new nation. They simply have to move or be submersed by the sea.

Glacier melt is also an enormous risk to the stability of massive ice bodies in high mountain altitudes. As these bodies deform (as often occurs due to melt), they can collapse and come rushing down-mountain with ice blocks as large as skyscrapers, sometimes pounding into glacier lakes formed by natural dams (formed in turn by moraines left by receding glaciers). These impacts can cause tsunami waves many meters high, taking out anything in their path. In the not-so-distant past, glacier tsunamis have taken thousands of lives in the mountainous areas of Peru, in parts of the Himalayas, and in certain parts of Europe.

This chapter is reprinted from *The Journal of Environmental Law and Litigation*, 2013, 28(1): 59–78; adapted from a presentation on September 28, 2012, at "New Directions for Human Rights and the Environment: A Symposium Inspired by Svitlana Kravchenko."

We read and hear about such predicaments in the media daily and are pretty much desensitized to this issue, although we know very little about the specifics or technical aspects of glacier melt. Most of us envision huge polar ice sheets breaking off into the oceans to become massive, floating icebergs, eventually melting off as they flow into warmer water. We imagine, with some stretch of the imagination, that this ice melt will somehow raise the sea level, although many of us have a hard time accepting just how this impact will play out and with what magnitude. Can the melting of a big iceberg really raise the ocean level? Yes, it can!

Regardless of just how this will occur, most of us will generally conclude that glaciers today are a vulnerable natural resource and that, because of what is happening to glaciers, we will likely be faced with catastrophic tragedies. We conclude in this context that our priority should be to avoid glacier melt, which naturally takes us to a discussion about global climate change. Industry is contaminating the environment and, more specifically, the air. Carbon dioxide (CO_2) emissions cause global warming, and its impact is that ice warms and melts. A lesser-known impact of growing CO_2 emissions is that the miniscule carbon particles emitted into the air are deposited on ice. This darkens the glaciers and, just as when we wear a black shirt on a hot day we immediately feel the heat, glaciers likewise melt faster when they are stained by black carbon.

So, in 2013, the year in which this chapter was originally written and published, we are concerned about climate change, and we would like to see CO_2 emissions reduced. We know that a warming climate is melting our glaciers and that this is impacting our renewable ice bodies. We generally think of glacier melt as an indicative variable proving climate change, but we actually talk little or nothing at all about glacier protection. In fact, even the few cases we have seen linked to climate change talk about communities or animals (like the polar bear) whose habitats are affected by a changing climate. No one seems to be talking about the need to protect glaciers or even linking glacier melt to the direct consequences it brings for affected communities. The closest the argument comes is the emerging link between polar ice cap melting, sea level rise, and endangered populations.

The Tuvaluans have surely already understood this linkage in terms of the human rights implications of glacier melt. Other more affluent communities, like California's coastal property owners will surely take up this agenda as soon as their multimillion dollar homes begin to collapse due to sea-level rise and unusually adverse weather.

But these circumstances seem not yet to have generated a human rights discussion about glaciers, or more specifically, "the right to glaciers." The rationale for this discussion nonetheless is as colossal as the ice we're talking about.

A staggering fact that we sometimes overlook is that nearly all of our water is in the oceans and is very salty. In other words, most of the planet's available water is not drinkable unless we invest lots of resources in desalination plants. Only 2–3% of our planet's water is actually freshwater. That's a miniscule—but very precious—amount. And in addition to this alarming and largely ignored statistic, 75% of this available drinking water is in the form of ice in our glaciers, mostly in the polar icecaps but also in mountain glaciers. All of this ice is presently melting due to climate change. That's alarming.

One would think that such a rare resource would be closely protected. One would think that most countries would have very strict water laws and that water protection would include the protection of ice, which we generally think of as water in one of its forms (liquid, ice, or gas). One would think that most countries would have long ago established a human right to water. Some have. Most have not. And one might even think that, somewhere in all of this legislation, there may surely be at least a mention of glaciers, and—why not?—even a law to protect glaciers. Wrong.

Until very recently (2010), there were absolutely no laws to protect glaciers anywhere in the world. It's actually quite remarkable that with so many other laws focused on natural resources—on flora, fauna, national parks, sensitive ecosystems, and the like—no laws existed anywhere to protect our most important natural resource, particularly when glaciers hold three-quarters of that most precious resource: water. In our research (before the passage of Argentina's glacier law in 2010), we were not able to find a single water law, or any other law, that mentions the word "glacier" or refers to glacier ice. (If the reader knows of any, please send the author such information because we are compiling resources to protect glaciers and such legislation would be exemplary.)[1]

When I first started researching glaciers in 2007, I presumed that the Swiss, French, Norwegians, Swedes, the people of Greenland, Russians, Icelanders, Kyrgyzstani, Chileans, Argentines, Peruvians, Bolivians, Canadians, Americans, Pakistani, Nepalese, Chinese, Mongolians, Japanese, or Tanzanians, would either have glacier laws or regulatory frameworks that included glacier protection because all of those countries (and several others) all have important ice reserves. They don't.

Not a single country in the world had a glacier law in 2007. The first glacier law was passed by the Argentine National Congress in 2008 and was vetoed just a few days later due to pressure from a large mining company operating on the border between Argentina and Chile, a company whose project is surrounded by hundreds of glaciers. In 2010, that law came back, stronger and more stringent than its earlier version, surviving a presidential veto on its second appearance and becoming the world's first national glacier law. A few local provincial governments had anticipated the national law, and, in fact, the province of Santa Cruz in Argentina's Patagonia region introduced the planet's first glacier legislation. Argentina's national glacier law protects glaciers as a public good, for their water storage and water basin regulation value.

So, how do we get to the point where we consider that we need to establish a right to glaciers? Is this just some arbitrary decision we make because one day someone realizes that glaciers are at risk and that they need legal protection? Let's consider a few related issues and take the right to water as a corollary, in part because some might say that the right to glaciers is implied within and by the right to water. Ice is, in the end, water in one of its forms, and the right to water, for many reading this article, may seem to have the necessary substantive doctrinal underpinnings to suffice for glacier protection.

As a global society, we are pretty convinced (although we were not just a few years ago) that collectively, individually, and in specific communities, we have a right to water. This collective and progressive recognition was not the product of an arbitrary decision or a spontaneous proposition. It came along with a global awakening that took several decades to mature over the awareness of our worsening anthropogenic impact on natural resources. We were and we are, in many instances, destroying our natural planet in an unsustainable way, and, along with it, we are contaminating our water.

We are 98% water by some estimates (a number that is curiously identical to the amount of saltwater available to us in the natural environment); hence, water is a fundamental ingredient to our existence—not to mention that, in liquid form, we need a significant daily intake of water to survive. Without it, in mere days, we would shrivel up and cease to exist.

When discussions on the right to water surfaced in the early 2000s, there was much resistance to the concept in the sphere of international law, mainly from corporate actors who commercialize water and from states that tend to resist any advancements on globally recognized rights, mostly over the concern of losing national sovereignty over the management of their natural resources.

But the discussion on the need to establish a right to water was both rational and logical, and, thanks to initiatives by many civil society groups, supported by institutions such as the United Nations Office of the High Commission on Human Rights. Specifically through the help of UN General Comment 15 on the right to water, today we are pretty much in global agreement that consolidating and establishing the right to water was a good thing. The understanding of this right and the necessary substantive characteristics of policy needed to realize that this right is still evolving, and while we haven't yet sorted out all of the concerns that the debate warrants, we are on the road to substantively defining and fully consolidating this fairly new right.

To begin our discussion on the right to glaciers, we should remember that law follows cultural or social custom. Societies exist, people interact, their actions have consequences, and when we collectively realize that those actions begin to have an undesired impact on the collective well-being (on the public good) or if they begin to infringe upon individual or community rights, we establish rules about how we want to coexist. Many of these rules evolve into formal norms, regulations, or laws to control the social behavior that is having the undesirable consequence.

It is important to stress the origin of laws and regulations because we tend to forget that laws do not appear arbitrarily; they exist because we are witnessing a behavior that we would like to change for the good of everyone. Even laws that appear spontaneously by the will and initiative of some legislative actor are generally a response to something that the legislative representative has seen in society that she or he would like to modify, presumably for some public benefit. Laws are acts of formal governmental power that wish to avoid or change a behavior that is damaging the public good. In the end, they are intended to protect the public good, the community, and the individual from harm.

This brings us to the discussion around glaciers and the "human right to glaciers." We thus propose three sets of questions to address in the discussion around the need (or not) for a right to glaciers addressed in the following sections:

1. Aren't glaciers water? And, as such, wouldn't a right to water already comprise an implied right to glaciers? This question also leads to another basic question: "What are glaciers?"
2. What is the relationship between glaciers and society? If melting glaciers are one of the most visible consequences of climate change, and if they are so important to society, and if there is so much freshwater in glaciers, why haven't we enacted a law to protect them? Or are water laws sufficient for this protection?
3. Are glaciers suffering some specific impact or risk that would make them especially vulnerable, so that we need a law to protect them? If that vulnerability exists, is it different from the sort of vulnerability faced by water resources? We mentioned general climate vulnerability, but are there other more specific vulnerabilities stemming from anthropogenic action resulting in impacts to glaciers that we can and should address?

What Is a Glacier, and Would the Right to Water Adequately Cover Glacier Protection?

First, we consider the definitional issue and the relationship of glaciers to water, which forms the beginning to our discussion. Glaciers are comprised of water in one of its forms, ice. We recall that water comes in three basic varieties: liquid, gas, and ice. In this sense, we could say simply that if glaciers are made of ice, and ice is one of the forms of water, then glacier protection would be covered by a right to water. This is not exactly correct, however.

Technically speaking, water in liquid form is not exactly the same as water in a solid ice state. Lots of self-evident characteristics that most of us know about differentiate ice from water; ice floats on water, ice takes up more room than its equivalent in liquid state (it is 8% less dense than water); ice is considered a mineral due to its crystalline structure; ice is colder than water (it only exists below 0°C/32°F); and, from a molecular standpoint, the relationship between the hydrogen and oxygen molecules in ice is different from their relationship in water. So, scientifically, there is some argument to suggest that ice is not actually a form of water, and, as such, it merits special consideration. From a legal standpoint, at the very least, we should consider if the technical differences between water and ice merit differentiated legal considerations.

When we consider the risks faced by water sources and the measures needed to protect them, we quickly realize that they are indeed different from the risks faced by ice and measures needed to protect it. When we talk of water protection, we're usually concerned over the quality of the water we drink in our home, the accessibility of that water in our home, the contamination of this water from industrial effluents, and the transport and fair pricing of this resource.

We can guess that most of these dimensions of discussion in relation to glaciers would be quite different. We don't usually bring glacier ice into our home. We are not generally transporting glaciers (although some mining companies have proposed this). We don't usually consider the price of ice (unless we are on vacation at the beach), and we generally do not think about ice as affected by industrial effluents (although it sometimes is). Generally, the only risk and dimension we usually hear about in regards to glaciers is risk due to climate change. Conversely, we don't usually talk about our right to water in relation to climate change.

As we delve into understanding glacier ice, the context for our discussion and treatment of glacier protection takes on a very different path. Glaciers exist in places different from those where we generally come into contact with water. These places are generally not very hospitable to human life (high altitudes or very extreme planetary latitudes, such as the poles). Our relationship with ice in its natural and permanent state (glaciers) is quite unique and, for most people, very rare. It is different from our relationship with water, which we can easily interact with in our home, in our garden, at our place of work, or practically anywhere we carry out our daily lives. Water is everywhere around us. Glaciers are not.

Some of these differences, including the alienated nature of natural perennial ice, have conditioned the way in which we have organized ourselves in terms of water conservation, which is very particular to the location of our water reserves. Conversely, this has also resulted in our disregard for the need to protect natural ice reserves, which are generally at different locations and in different form.

What Is a Glacier?

In very simple, very basic, and very vernacular terms, a glacier is ice that survives in the natural environment through the summer months; in fact, it survives for the entire year and generally for several years, even hundreds of years.

We can get much more specific about this definition, and scientists definitely do, because we need to distinguish and categorize this surviving ice that can exist in many forms, some of which are considered by scientists as glaciers and some of which are not. But, basically, a glacier is formed from snowfall when the snow accumulates, compacts, turns into ice, and, if the conditions are right (generally if the temperature of the outside environment remains below 0°C/32°F for most of the year), then this ice will survive the warmer summer months even though some portion of it may melt off during the warmest days. When winter comes back around, the ice body receives more snow, and the cycle begins all over again. Pretty cool, huh? Pun intended.

Over time, the glacier may grow (if more snow falls on it than melts away) or decrease (during especially dry years) in size, and it may actually be moving if it is on an incline (such as a mountainside). Parts of the glacier generally melt as it moves downhill and the front end reaches a warmer environment because part of the glacier may be in contact with water (as in icebergs) or because the ambient temperature surrounding the glacier warms, as is occurring with climate change. What makes the glacier ice different from other surrounding seasonal snowfall is that, for some ecological reason having to do with the glacier's immediate ecosystem, the snow that falls where the glacier is located has the necessary conditions to survive beyond the summer. At that location, the winter snowfall recharges the glacier so that it doesn't actually lose any mass.

Glaciers form over many seasons of snowfall and seek equilibrium with their surroundings. As long as an outside force (such as global warming or an especially long dry spell) doesn't affect the glacier, it will survive over time: for tens, hundreds, and even thousands of years.

One critical aspect of the process, which is very important to the natural environment, is the cyclical recharging and slow melting of part of the glacier. This is actually one of the most important aspects of glacier ice for many dry ecosystems. Glacier ice stores water, and the slow melting of the ice during dry and very warm summer months is a critical feature of the natural environment that ensures water provision during the driest parts of the year. If the year happens to be especially dry, or if there is an extended drought in the area, the glacier may be the only active source of water for much of the ecosystem. This is particularly the case in the Central Andes region of Latin America, well known for its especially dry climate, such as that found in the Atacama Desert of Chile, the driest place in the world.

We will see later when we discuss Argentina's glacier law that the legislation protects precisely this glacier feature; that is, glaciers as "regulators of water basins." This means that glaciers "regulate" the flow of water into the ecosystem.

We sometimes use the image of a water faucet to describe this function. We can imagine that the mountain has a very large water faucet in the glaciers, and, when it is especially dry and warm, such as in the summertime or during a drought, the faucet (the glacier) is slightly open. The ice begins to melt with the heat but doesn't melt right away as seasonal snowfall would melt. It is cold enough at the high altitude where the glacier is located so that the melt is slow. As such, the ice melt from the glacier provides a slow and steady flow of water into the streams and rivers. Without glaciers, in a dry year, once the winter snow melted, the mountain would be dry and water would cease to flow into the ecosystem.

There is one other dimension to this discussion that we will not get into too deeply, which has to do with permafrost or permanently frozen grounds. The Argentine national glacier law also protects what are called "periglacial environments" in which permafrost is located. Periglacial environments act like glaciers in terms of water provision. Permafrost is earth (which can also have a high water content) that freezes and, just like glaciers, conserves ice ready to be used as water when the temperature changes and the ice melts. Permafrost in high mountain altitudes can be very extensive. Entire mountains can be at temperatures well below zero. Any humidity contained in the mountain is converted to ice (and should be considered water in storage), ready to be used as a "regulator of water basins" when the environment needs the water feed. As such, along with glaciers, permafrost zones are very significant water reservoirs in our natural ecosystems and also need protection. A more extensive discussion is necessary to address permafrost, relevant to the discussion we are engaging in here, since permafrost areas are just as important (or even more important) as glacier cover for water basins. Such a discussion is necessary, and a "right to permafrost" can be considered a corollary to the "right to glaciers" discussion and should be the focus of future work.

Hence, we now understand the importance of the role glaciers play in conserving water. They are natural water reservoirs or dams that Mother Nature has created to conserve water and make it available when rain is short and the weather is hot and dry.

In some cases, snow and water make their way into the ground, through gaps, crevasses, and other openings. When the water is frozen into ice, the ice expands, breaking the ground and creating new gaps, which are in turn refilled with water and new snow. Rock debris from mountainsides can also cover the ice, and, in some very special cases, a very large ice glacier can form below the visible surface of the earth. In the dry and hot summer months, the rock cover protects the ice from warm temperatures, and the ice may in fact survive below the surface of the earth for many years, even hundreds or thousands of years. Parts of that ice, generally those parts that are closer to the surface of the earth or at lower altitudes, may have active layers that melt and freeze in cyclical fashion, providing regular water flow into the ecosystem. Other parts of the ice buried more deeply under the surface of the earth may be permanently conserved. These natural phenomena are commonly referred to as "rock glaciers" or "debris-covered glaciers." In the middle of summer, you might be standing on top of a rock glacier with billions of cubic feet of ice below you and not even realize it is there because there may be no visible snow or ice anywhere in the area.

Does the Right to Water Suffice to Protect Glaciers?

So, now we know what glaciers are: they are bodies of ice that survive throughout the year, storing and providing water to us all year long. Hence, does the "right to water" adequately protect glaciers? Or do we need a specific "right to glaciers"?

Society derives its laws from cultural practices that merit consideration of the protection of the public good. We collectively act in a certain way, and, when that way of acting places our individual or public safety at risk, we take measures to protect that good. The discussions around the "right to water" appeared only very recently, as we began to realize that human activity was degrading the quality and availability of our water.

With the intent to protect this inalienable right to a natural resource as important to us as air and food, we must begin to discuss the risks to our water resources and then begin to discuss ways to ensure their protection. Once we realize that we need to establish a right or a law, we must then proceed to think about the specific elements necessary to effectively ensure that protection.

At about the same time the right to water discussion was in full force, our global society awoke to our critical climate problem. We were pushing for the "right to water" in the mid-2000s at forums like the World Social Forum or the World Water Forum, when Al Gore, then-vice president of the United States, started his crusade on climate change.

Melting glaciers all over the world were identified by the Nobel Peace Prize winning Intergovernmental Panel on Climate Change (IPCC) as one of the most visible features indicating that our planet has a fever. Glaciers are melting fast, and much of the reason for this is our growing industrial expansion based on the burning of CO_2-emitting fossil fuels.

Surprisingly, however, no one ever mentioned the need to "protect" glaciers; we have simply remarked that it is a given fact that they are melting. We do not speak of protecting the ice, but rather we focus on the need to revert the practices that led to that ice melting. We talk of reducing CO_2 to stop global warming. There is nothing wrong with this conclusion. In fact, the objective of eliminating fossil fuels may be one of the most important challenges of our time. But let us focus for a moment on the implications of this trend on the natural resource victim, glaciers. We only speak of glacier melt as a visible fact of life and the hope that we can somehow stop glacier melt. But nobody speaks of glaciers as vulnerable resources that need direct protection, nor do we talk about glacier melt in terms of increased vulnerability to "other" potential glacier impacts. Finally, we do not talk about stopping glacier melt, regenerating glaciers, or repairing damage to glaciers, all of which can be done.

Perhaps this omission is simply because it seems impossible to detain glacier melt. That is not actually true, as we have already seen that in places like Switzerland, concerned ski resorts are covering mountainsides with reflective sheets to increase albedo and reflect sunlight (and heat), with the value that the glaciers melt much more slowly. And it works.

Others are experimenting with covering parts of glaciers with sawdust (something my grandparents used to do in Argentina to conserve big blocks of ice left by the "ice man" who was as popular (or even more popular on hot days) than the milkman—this was before refrigerators arrived in many homes). Still others build snow walls on mountainsides to divert wind flow and generate snowfall accumulation where otherwise the snow would simply have swept over the terrain, thus creating instant glaciers in a single winter season. Actively protecting and even generating glaciers is possible, although clearly at a global scale in the face of global warming the added glacial benefit might seem minuscule.

For the most part, however, we are not out there protecting glaciers. Growing scientific evidence, however, is showing us that many communities depend on glacier melt and periglacial environment areas for their water provision. Particularly in temperate high-mountain climates, the perennial ice in glaciers and in permafrost areas are a critical water reserve, and the same ice in lower elevations where ice forms and melts cyclically are a fundamental source of water. In some areas, it's the only source of water for local communities.

The science on just how much water a glacier provides to downstream ecosystems varies widely. The numbers generally increase for especially dry years. Hence, glaciers act as reserves for dry years. Some glacier experts have shown that in especially dry climates like Mendoza or San Juan Provinces in Argentina, in dry years, glaciers can provide up to 80% of river water. Other scientists have argued that glaciers actually provide very little water because much of the ice in uncovered glaciers sublimates (vaporizes) instead of melting.

What is undeniable is that even very small glaciers can store enormous amounts of water. A quick calculation taking a glacier as small as the size of an average football field and just a few meters thick, would provide a typical family of four with the total water consumption it needs for subsistence (as determined by the UN) for the entire lives of all of the family members! That's a lot of water. In a province such as San Juan, which is estimated to have more than 12,000 glaciers of all sizes and types, the amount of water stored in these mountain ice bodies is colossal. A single larger glacier such as the El Potro Glacier on the border between San Juan and La Rioja Provinces could provide the entire Argentine population with drinking water for more than a year.

The warming trend of our global environment and visibly accelerated glacier retreat suggests that glaciers are melting as the 0°C/32°F isotherm moves up mountainsides. In such a context, we can presume that melting glaciers are not in ecological balance and are actually acting as positive water providers. Hence, in a warming climate, glaciers that might otherwise be in equilibrium and not melting much are actually important sources of water to our ecosystem that will last for many years as significant water providers.

Whether we can agree or not on the theory that glaciers are not in equilibrium, they nonetheless act as water reservoirs and, as such, are important water resources.

Rock glaciers, protected by rock cover, are also significant water reserves. Not only do they hold massive ice content deep in their interior, but they actively contribute water provision to ecosystems through their active surface layers that move and have cyclical freezing and melting phases.

For communities farther away from the perennial ice, such as those in large urban areas, the direct provision of glaciers to the water of those communities may represent a smaller percentage. However, discontinuous permafrost zones that cyclically freeze and melt logically contribute much larger percentages to downstream communities since the area they cover is substantially, even exponentially, greater. From a scientific standpoint, we do not have the studies to determine the percentage or reach informed conclusions on volume, but the direct relationship between permafrost melt and ecosystems is clearly undeniable.

So what are the risks to glaciers posed by anthropogenic forces (aside from climate change) that we should consider in this discussion? There are several.

Glaciers are generally in equilibrium or are seeking to find equilibrium. In this sense, their volume, weight, position, water and ice content, and their rock or other debris content, all work together to create an ecological balance that is conducive to the glacier's survival. Because they are often on slopes, they creep downhill. The lower parts of the glacier then begin to enter warmer elevations above 0°C/32°F, and they begin to melt and, eventually, disappear. Newer snow enters the glacier from above, regenerating the portion that has moved downhill. This is a cyclical process that is repeated indefinitely with each season.

Furthermore, a glacier's immediate surroundings are very important to its very existence. The glacier formed where it did because of its surroundings. For some reason, either due to local snowfall, wind patterns that brought snow to that location, a nestled location in a rounded mountain ridge, at the base of an incline, along a high mountain valley, especially low temperatures in the specific spot where the glacier is located, or limited exposure of the glacier to the sun (such as on hillsides that face toward the Earth's poles) are all reasons that a glacier may form in one place and not in an immediately adjacent one.

Science has not yet established a specific term to define this specific "glacier ecosystem" necessary for the glacier to form. Because it is important to the implementation of specific public policy geared toward glacier protection, I have established and defined a term to distinguish this area. I call it the *glaciosystem*. In brief, the glaciosystem can be defined as:

> The glacier and its surrounding ecosystem that influences its constitution and composition, with respect to its water and ice accumulation and ablation, determining its biological process, its natural evolution during its periods of charge and discharge, and which, if affected, could impact or cause the alteration of the glacier and/or impact the ecosystem in which it exists.

If we impact the glaciosystem, or if we carry out activities that destroy part of the glacier, we can have significant impacts on the glacier or on its evolution, which could lead to its collapse or accelerated melt and disappearance.

Some critical glacier-threatening impacts include:

- Modifications to the glaciosystem that is necessary for the glacier's natural formation and sustainability. This could include altering the mountain hillside or ridges where the glacier is located or altering the wind patterns that favor the accumulation of snow at the glacier site.
- Ice mass removal, such as carving out a section of a glacier. If a section is carved out near the glacier's end (the lowest elevation end), the large mass behind it may accelerate forward, and the whole glacier may come crashing forward. Conversely, if we carve out a section above it, we may affect the entry of new snow and disrupt its natural regeneration.
- Soiling of glacier surfaces with contamination (such as black carbon/ soot); this could darken the glacier and result in surface warming and accelerated melt.

- Excessive weight placed on glaciers; this could alter structural balance and lead to collapse, as might occur from placing millions of tons of sterile rock on the ice from mining operations.
- Surface impacts to rock glaciers from roads introduced on top of them; these may alter the glacier's ecological balance.

We might also see impacts from certain activity that could alter the "water provision" function of the glacier, such as:

- Any of the above-mentioned impacts that lead to water flow alterations
- Acid drainage from industrial operations that falls on glaciers and subsequently contaminates the ice and eventual meltwater
- Melting away of water reserves

These cited potential impacts to glaciers are not theoretical. There is extensive evidence already available from certain areas, such as the Central Andes in Argentina and Chile and the mountains of the Kyrgyzstan Republic, where such impacts are already occurring.

Returning then to the "right to water" corollary and our question as to whether a "right to water" adequately protects glaciers, we should ask ourselves if the sort of regulatory frameworks to protect water sources that exist today are sufficient to protect glaciers and periglacial environments. The argument presented here is that they are not.

The right-to-water debate, and even laws or regulations attempting to protect water, has no focus on glaciers or periglacial zones. The right to water debate has not addressed the need to protect ice, or more specifically, glaciers. Nor do most people working on "the right to water," either from a policy, legal, or civil society angle, focus at all on glaciers or permafrost. None of the risks just listed comes up in debates about water protection. As mentioned earlier, science has not even come up with a term that is functional to the public policy debate to understand the glaciosystem, the natural area surrounding a glacier and all of its natural characteristics (mountain ridges, positioning, wind patterns, etc.) that are necessary for the glacier to form in the first place. Glaciers are simply not on the radar screen of the "right to water" debate, nor are they included in systemic measures, policy, or laws to protect natural resources.

The right-to-water world is mostly focused on urban water management, storage and transport, household access, use, quality, pricing, water contamination, water storage (in human areas), and sanitation. Nobody is talking, however, about the fact that much of this water derives from ice bodies way up in the mountains, where in many cases few people have ever been. The world's second-highest peak (K2 in Pakistan, or otherwise known as Savage Mountain and covered with glaciers) was not even discovered by humans until the past century. It was simply too far out of reach to humans. It is such a treacherous environment that one out of every four mountaineers who tried to reach the summit perished. In fact, knowledge of glaciers, periglacial environments, and rock glaciers is extremely scarce and limited to a handful of academics who, as a collective group, are quite removed from broader policy discussions about natural resource conservation and, even more so, water management.

Even the science around glaciers fails to address the relationship between glaciers and downstream ecosystems. Glaciologists cannot even come to terms and agree on how much water glaciers provide to rivers, much less permafrost (which we can safely guess will be a much greater number).

What does this all mean for water policy? Essentially, that public policy, regulations, laws, norms, and cultural practices related to water leave out glaciers as a point of discussion. Glaciers are simply ignored when it comes to policies and laws. It is for this reason that the world needs laws and policies to protect glaciers and periglacial environments: specifically, to protect them as vulnerable natural resources and to protect the function they serve as water reserves and regulators of water basins.

We mentioned earlier that the first glacier law ever passed was in 2008, in Argentina. It was vetoed just days later by the president due to strong opposition from the mining sector. It was not until 2010 that the first glacier laws were formally instated in Argentinean subnational government jurisdictions.

Argentina's national glacier law, promulgated officially in October 2010, establishes "a minimum standards regime for the preservation of glaciers and the periglacial environment," which is also the actual name of the law.

The objective of the law, as cited in Article 1, is to protect glaciers and periglacial environments "as strategic freshwater reserves for human consumption; for agriculture and as sources for watershed recharge; for the protection of biodiversity; as a source of scientific information and as a tourist attraction. Glaciers constitute goods of public character."

As mentioned earlier, glaciers are important because they store water for future use and regulate water flow. The law rightly captures these functions as the critical and underlying value of glacier ice in places like the Central Andes, where dry climates are extremely challenging for the sustainability of human life. Without glaciers and frozen grounds (periglacial environments), the subsistence of human life would be much more difficult, if not impossible.

This function of glacier ice and frozen grounds in temperate high-mountain climates is quite different from the function of the glacier ice in the polar icecaps, and the difference should be noted. In the former case, water from glacier melt is critical to human survival, and local ecosystems are dependent upon that melt. In the case of the polar ice caps, glacier melt could alter ocean ecosystems and the ecological systems of local fauna, but is not likely to affect immediate downstream communities because no people live at the polar icecaps.

In both cases, however, glacier melt could pose life-threatening risks to certain communities. In the case of glaciers in the high Andes of Peru, glacier melt and collapse has led to mountain tsunamis that have taken out entire villages and cities, killing thousands of people. The melting of the polar icecaps may not affect people at the poles, but it will cause sea level rise and affect the livelihoods and habitat of thousands, or even millions, of low-lying coastal homes in other parts of the world.

So, given the vulnerability and importance of glaciers, what is to be done, from a legal and administrative (policy) perspective, to protect glaciers and their function as a public good? The substantive elements identified in the above-mentioned law help us understand the necessary steps to guarantee effective protection.

The Argentine glacier law establishes the obligation of:

- Registering glaciers in an official inventory (Articles 3 and 4)
- Prohibiting certain activities (such as mining) that might impact glaciers through the deposit of contaminants or construction of works (Article 6)
- Conducting specialized glacier impact studies for activities taking place near glaciers (Article 7)
- Applying the law retroactively for activities currently taking place near glaciers (Article 15)

The first three elements cited (registering glaciers, prohibiting activities, and conducting studies) are three fundamental steps or tools to ensure glacier protection and to respond to the fact that, for the most part, the functions and value that we are trying to protect are largely unknown.

No glacier inventory existed in Argentina—and you cannot protect what you do not even know exists. Few people in Argentina even knew there were more than just a handful of glaciers. In fact, most Argentines prior to the debate on glaciers could probably mention only one glacier (the Perito Moreno Glacier) in Patagonia, because it is a very popular glacier and gets much media coverage when its front disintegrates upon touching land. Most Argentines were completely oblivious to the fact there were actually thousands of glaciers, particularly in provinces such as Salta, Jujuy, Catamarca, La Rioja, and San Juan. As mentioned earlier, in San Juan, there are more than 10,000 small glaciers, many of which are rock glaciers beneath the surface of the Earth, whereas in each of the other provinces, the official glacier inventory will reveal quantities in the several hundreds for each.

We also had very little or no information on anthropogenic activity, such as mining, affecting glaciers. Information about the impacts of activities such as mining and construction is beginning to rise, and this new area of information is pushing the regulatory framework.

Finally, glacier impact studies are a novelty for environmental impact assessment (EIA) exercises. As information about glacier impacts becomes public, official agencies are now pushing for glacier impact studies from public and private projects that could affect glacier well-being.

One question that is worth considering is why Argentina suddenly decided to embark on debating, negotiating, and eventually adopting a glacier protection law. Several reasons could be cited:

- Climate change has placed glaciers at risk, making them vulnerable resources.
- Argentina has many glaciers, upwards of 25,000, many of which are small but very significant water reservoirs, so that a law to protect glaciers makes sense.
- Anthropogenic activities are impacting glaciers.

Although we would like to imagine that society spontaneously generates acts geared to protecting the public good before disaster strikes, in this case, the last reason was the underlying one for the appearance of the legislation. In fact, it was not even the local community that first introduced the issue and concern, but a reaction to a situation in Chile that spawned the response across the border in Argentina.

In the early 2000s, preparatory studies for the binational mining project Pascua Lama, owned by Barrick Gold, got under way. Barrick Gold found gold deposits right at the border, underneath three glaciers. They were small glaciers that had never drawn anyone's attention, not even that of the region's principal glacier academics. These glaciers (the Toro I, Toro II, and Esperanza Glaciers) were nestled at nearly 5,000 meters (16,400 ft) above sea level in the high Andes, nearly 200 km (124 mi) from the nearest human settlements on the Argentine side of the border.

When the Diaguita indigenous group and local farms in the Huasco Valley in Chile learned that Barrick's gold deposits were sitting underneath glaciers that were at the top of the chain of their hydrological ecosystem and that Barrick's idea was to dynamite and haul off the glaciers in dump trucks to get at the gold, they launched a massive campaign against Pascua Lama and against Barrick.

Barrick argued that the relatively small glaciers were not actually glaciers but rather perennial ice patches. Losing this argument, Barrick shifted the terminology to "glacierets," which would imply "small glaciers"; as such, the company implied that their loss would be irrelevant. Later scientific studies debunked this theory and showed that these smaller glaciers actually provide more meltwater than nearby larger glaciers.

The debate over saving the ice went full force, largely led by Chilean stakeholders of Pascua Lama. Across the border in Argentina, concern was less manifest but slowly growing thanks to the attention raised across the border. A renowned local glacier expert, Juan Pablo Milana, approached Argentina's National Environment Secretary, Romina Picolotti, founder of the Center for Human Rights and Environment (CEDHA) and a strong advocate for the right to water, and brought to the attention of the national government thousands of existing glaciers in the Andes highlands, including the invisible "rock glaciers" underneath the Earth's surface. Argentina awoke to glaciers that year, and a process was put in place that resulted in the unanimous approval of the first National Glacier Law in 2008, vetoed shortly thereafter but reapproved in 2010.

Today, Argentina is set on inventorying its glaciers. There have been numerous reports by civil society groups revealing glacier impacts by mining projects and public road work in the high mountains. Provincial governments, which are at legal odds with the national government over who should manage glaciers (and more specifically who should approve mining projects near glaciers) are actually beginning to ask for glacier impact studies from companies carrying out mining operations, both exploratory as well as extractive work.

Societies are learning about glaciers. Societies are learning about the function of glaciers, about where they are and why they are important to local ecosystems. Most ignored these facts. With the creation of Argentina's National Glacier Law and the elevation of glacier resources to "public good" stature, glaciers today are taking on a new life.

Glaciers are important to local ecosystems, and they are at risk. Glaciers are important to local communities, and their water contribution to those communities is at risk. As such, we need glacier protection. We need glacier protection laws and policy and yes . . . why not? We need a "right to glaciers."

Note

1. Since the original publication of this paper, a few mentions of snow and ice have been found in relatively old water laws of some Latin American countries, simply indicating that the "snowed mountains" belong to the state. Furthermore, one subnational jurisdiction in Austria in the early 2000s mentions the need to protect glacier environments.

11

Final Words

In the preceding chapters of this book, we've traveled through a world of ice that was probably largely uncharted for most of us. Hopefully, we've learned a little bit about these fantastic frozen natural resources that play such a fundamental role in the sustainability and balance of our global ecosystem.

Glaciers are melting. They are in danger because we have placed them in danger and, as such, we need to take note of and responsibility for this vulnerability, not only to protect glaciers but also to protect the very essence of our global habitat.

Glaciers have been unprotected because they are obscure, removed, alien to our daily lives, located in far away places that are for the most part inhospitable to our way of life. And yet, they are a fundamental and integral part of our way of life. With modern tools like the Internet and programs like Google Earth, we can get closer to these fabulous vulnerable resources, to learn about them and work to protect them.

The world is challenged today to address global climate change. If we envision a sustainable and harmonious environment in our future, we must progressively move away from fossil fuels and introduce a more balanced and sustainable mix of energy sources grounded on renewable energy. We must find solutions to generating, harnessing, transporting, and managing renewable energies, and we must progressively phase out oil and gas from our daily lives. It *is* possible; it just takes personal and collective conviction to set ourselves in motion to achieve this goal.

Glaciers are a majestic resource, inspiring awe and wonder in a world of frozen beauty that awaits our discovery but that also alerts us to our excesses and indifference. We are losing our glaciers because we have ignored the extreme vulnerability of our planetary ecosystem, and we now must face difficult decisions about policy, consumption, and lifestyle changes that shake the foundations of our society.

Global climate change for many seems intangible. How do we assign responsibility to the deteriorating world, and who must make the adjustments to reverse our predicament? These are questions that are hard to answer, but, in each case, the answer clearly begins with a look to our inner selves. We need an attitude change, and only through individual commitment to make these changes can we hope to find collective solutions to our environmental emergency.

Glaciers will melt eventually. Perhaps more rapidly than we would like, or more rapidly than they need to melt due to natural causes. We should have glaciers around for many more centuries, and even millennia, to provide us with critical freshwater for our consumption and healthy existence; however, we are modifying nature, creating a human-defined era that we call the "Anthropocene," one in which geological, environmental, and climactic conditions are determined by the human race and not by the natural evolution of things.

To the extent that this human-defined era is at the root of many of our developmental problems and constraints, this should worry us and drive us to change our modus operandi. Politics need to be more transparent. Communities need to engage more in determining their development. Democracies need to be more democratic, and politicians need to be more responsible for working for the public good and not for self-interest and re-election, as many do. Companies need to better balance their objectives between the quest for economic wealth and the rational exploitation *and protection* of nature. Laws need to be more focused and better respected, while policy makers and implementers need to carry out their due diligence in good faith and good reason.

And we, the individuals who comprise our global society, going about our daily lives often indifferent and tolerant of the intolerable, need to look beyond our disposable coffee cups and see the ice, because it's melting.

This book has offered a small window into our cryosphere, one that helped me make this leap in hopes that it can awaken our engagement and our interest in our changing climate in ways that we may not have yet understood, so that each of us, through our appreciation of nature, and in small and larger ways, can build a new and constructive relationship with Mother Nature.

{ APPENDIX }

The Argentine National Glacier Act

Law 26.639
Minimum Standards Regime for the Preservation of Glaciers and the Periglacial Environment
Promulgated Officially: October 28, 2010
(unofficial translation)*

Article 1: Subject

The following law establishes the minimum standards for the protection of glaciers and the periglacial environment with the objective of protecting them as strategic freshwater reserves for human consumption, for agriculture and as sources for watershed recharge; for the protection of biodiversity; as a source of scientific information and as a tourist attraction.

Glaciers constitute goods of public character.

Article 2: Definition

As per the present law, we understand glaciers to be all perennial stable or slowly-flowing ice mass, with or without interstitial water, formed by the re-crystallization of snow, located in different ecosystems, whatever its form,

---

Translation by the Center for Human Rights and Environment (CEDHA), Argentina; edited for scientific accuracy by A. Brenning, University of Waterloo, Canada. More info: jdtaillant@gmail.com. Original text: http://www.infoleg.gov.ar/infolegInternet/anexos/170000-174999/174117/norma.htm.

dimension and state of conservation. Detritic rock material and internal and superficial water streams are all considered constituent parts of each glacier.

Likewise, we understand by the periglacial environment of high mountains the area with frozen ground acting as regulator of the freshwater resource. In middle and low mountain areas, it is the area that functions as regulator of freshwater resources with ice-saturated ground.

Article 3: Inventory

The National Glacier Inventory is hereby created, in which all glaciers and periglacial landforms that act as freshwater reserves on national territory shall be identified along with the pertaining information that is necessary for their adequate protection, control and monitoring.

Article 4: Registration of Information

The National Glacier Inventory shall contain information about glaciers and the periglacial environment by watershed, by location, by surface area and by morphologic classification of the glaciers and periglacial environment. This inventory shall be updated no more than every 5 years, verifying changes in the area of glaciers and the periglacial environment, their advance or retreat, and other factors that are relevant to their conservation.

Article 5: Implementation of the Inventory

The inventory and monitoring of the state of the glaciers and the periglacial environment shall be carried out by the Argentine Institute of Snow, Glaciology and Environmental Sciences (IANIGLA), in coordination with the national implementing authority of this law. The Foreign Ministry shall participate when border zones are concerned in which the international demarcation is still pending prior to inventory registration.

Article 6: Prohibited Activities

All activities that could affect the natural condition or the functions listed in Article 1, that could imply their destruction or dislocation or interfere with their advance, are prohibited on glaciers, in particular the following:

a) The release, dispersion or deposition of contaminating substances or elements, chemical products or residues of any nature or volume. Included in these restrictions are those that occur in the periglacial environment;
b) The construction of works or infrastructure with the exception of those necessary for scientific research and to prevent risks;
c) Mining and hydrocarbon exploration and exploitation. Included in this restriction are those that take place in the periglacial environment;
d) The installation of industries or the building of works or industrial activity.

Article 7: Environmental Impact Evaluations

All activities planned on glaciers and in the periglacial environment, that are not prohibited, shall be subject to environmental impact evaluations and environmental strategic evaluations, depending on the scale of intervention, in which public citizen participation must be guaranteed as per established in Articles 19, 20, 21 of the General Environment Law (Law 25.675), before authorization and implementation is granted and conforming to existing norms.

The following activities are excluded from these requisites:

a) Rescue activities, as a consequence of emergencies;
b) Scientific activities, taking place by foot or on skis, with eventual sample taking, that do not leave waste on glaciers or in the periglacial environment;
c) Sporting activities, including trekking, mountain climbing, and non-motorized sports that do not perturb the environment;

Article 8: Competent Authorities

As per the present law, the competent authority shall be that authority that each jurisdiction chooses. In the case of protected areas under law 22.351, the competent authority shall be the National Parks Administration;

Article 9: Implementing Authority

The implementing authority of the present law shall be the institution with the highest national environmental jurisdiction.

Article 10: Functions

The functions of the national implementing authority shall be:

a) Formulate actions conducive to the conservation and protection of glaciers and the periglacial environment in a coordinated manner with competent provincial authorities, within the Federal Environmental Council (COFEMA), and with the ministries of the National Executive Power in their respective areas of competence;

b) Contribute to the formulation of a climate change policy relative to the objectives of glacier protection, both at the national level as well as with international climate change agreements;

c) Coordinate the execution and updating of the National Glacier Inventory, through the Argentine Institute of Snow, Glaciology and Environmental Sciences (IANIGLA);

d) Prepare a periodic report on the state of existing glaciers on Argentine territory, as well as the projects or activities that are taking place on glaciers or in their zones of influence, which shall be submitted to the National Congress;

e) Advise and support local jurisdictions in monitoring programs, controls, and glacier protection;

f) Create programs to promote and create incentives for research;

g) Develop campaigns to educate and produce environmental information conforming to the objectives of the present law;

h) Include the principle results of the National Glacier Inventory and its updates in national information sent to the United Nations Framework Convention on Climate Change;

Article 11: Infractions and Sanctions

The sanctions for non-compliance of the present law and the regulations that shall be introduced, beyond other responsibilities that might apply, shall be those that are established by the jurisdiction according to its corresponding policing power and which shall not be lower than those established here. Jurisdictions that do not have a sanctions regime, shall apply the following sanctions which correspond to the national jurisdiction:

a) Warning;

b) Fine of 100 to 100,000 minimum incomes of the entry level national public administration wage;

c) Suspension or revocation of authorization. Suspension of activity could be from (30) days up to (1) year, as merits and according to the circumstances of the case;

d) Definitive ceasing of activities.

These sanctions shall be applicable following substantiated summary proceedings in the jurisdiction where the infraction took place, and shall be regulated by the corresponding procedural administrative norms, assuring a due legal process, and they shall be incremented according to the nature of the infraction;

Article 12: Re-incidence

In the case of re-incidence, the minimum and maximum sanctions stipulated by paragraphs b) and c) of the previous article could be tripled. It shall be considered re-incidence when within a period of (5) years following the commission of an infraction, the party has been sanctioned for another infraction of environmental cause;

Article 13: Solidarity in Responsibility

When the violator be a juridical person, those that are in positions of direction, administration, or management, shall be responsible in solidarity of the sanctions established by the present law;

Article 14: Destination of the Fines Collected

The sums collected by the competent authority, in the concept of fines, shall be directed in priority, to the protection and environmental restoration of glaciers affected in each of the jurisdictions;

Article 15: Transitory Disposition

In a maximum period of sixty (60) days beginning with the sanction of this law, the IANIGLA shall present to the national implementing authority a chronogram for carrying out the inventory, which shall commence immediately in such zones where due to the existence of contemplated activities in Article 6, are considered priority. In these zones, the inventory stipulated in Article 3 shall be carried out in a period of no more than 180 days.

With respect to the competent authorities, these shall provide all the necessary information pertinent that the cited institute requires.

The activities described in Article 6, in progress at the moment of the sanctioning of the present law, must, in a period of no more than 180 days from the promulgation of this law, submit to an environmental audit in which potential and actual environmental impacts to glaciers are identified and quantified. In the case of verification of negative impacts to glaciers or the periglacial environment, contemplated in Article 2, the authorities shall order the pertinent measures so that the present law is complied with, and could order the ceasing or relocation of the activity and protective measures, cleaning and restoration as appropriate.

Article 16: Argentine Antarctic Sector

In the Argentine Antarctic Sector, the application of this law shall be subject to obligations assumed by the Republic of Argentina in view of the Antarctic Treaty and the Protocol to the Antarctic Treaty on the Protection of the Environment.

Article 17: Dating of the Law

The present law shall be regulated in a period of 90 days from the publishing of the law in the Official Bulletin.

Article 18:

Inform the Executive Power

{ BIBLIOGRAPHY }

Aedo, Maria Paz, and Teresa Montesinos. *Glaciares Andinos: Recursos Hídricos y Cambio Climático: Desafíos para la Justicia Climática en el Cono Sur*. Chile Sustentable. 2011.

Ahumada, Ana Lia. Periglacial phenomena in the high mountains of northwestern Argentina. *South African Journal of Science* 98, March/April 2002. Pp. 166–170.

Ahumada, Ana Lia, S. V. Paez, and G. Ibañez Palacios. Los Glaciares de Escombros en la Alta Cuenca del Río Andalgalá, SE de la Sierra de Aconquija, Catamarca. VIII Congreso Geológico Argentino. May 2011.

Ahumada, Ana Lia et al. El Permafrost Andino, Reducto de la Criósfera en el Borde Oriental de la Puna, NO de Argentina. *Asociación Argentina de Geofísicos y Geodestas. Ciencias d ela Tierra*. 2009. Pp. 249–255.

Anderson, Don, and Carl Benson. The Densification and Diagenesis of Snow. In *Ice and Snow*, edited by W. D. Kingery. MIT Press. 1963.

Arenson, Lukas, Silvio Pastore, and Dario Trombotto Liaudat. Characteristics of two Rock Glaciers in the Dry Argentinean Andes Based on Initial Surface Investigations. *GEO*. 2010. Pp. 1501–1508.

Azócar, Guillermo, and Alexander Brenning. *Intervenciones en Glaciares Rocosos en Minera Los Pelambres, Región de Coquimbo, Chile*. University of Waterloo. 2008.

Bahr, D. B., and V. Radic. Significant Total Mass Contained in Small Glaciers. *Cryosphere Discussions* 6, 2012. Pp. 737–758.

Barsch, Dietrich. *Rock-glaciers: Indicators for the Present and Former Geoecology in High Mountain Environments*. Springer. 1996.

Benn, Douglas I., and D. Evans. *Glaciers and Glaciation*. Arnold: Hodder Headline Group. 1998.

BGC Engineering. *Pascua Lama Permafrost Characterization Study*. BGC Engineering. 2009.

Bianchini, Flaviano. *Impactos de los Emprendimientos Veladero y Pascua Lama sobre los Recursos Hídricos de la Provincia de San Juan*. CEDHA. 2011.

Bonasso, Miguel. *El Mal: El Modelo K y la Barrick Gold: Amos y Servidores en el Saqueo de la Argentina*. Espejo de la Argentina. Planeta. 2011.

Bórquez, Roxana, S. Larraín, R. Polanco, and J. C. Urquidi. *Glaciares Chilenos: Reservas Estratégicas de Agua Dulce para la sociedad, los ecosistemas y la economía*. Chile Sustentable. 2006.

Brenning, Alexander, and Guillermo Azócar. Minería y glaciares rocosos: Impactos ambientales, antecedentes políticos y legales, y perspectivas futuras. 2010. In: *Revista de Geografía Norte Grande* 47, 2010. Pp. 143–158.

Cabrera, Gabriel, and Juan Carlos Leiva. *Monitoreo de Glaciares del Paso Conconta, Iglesia, San Juan Argentina*. Conicet. 2008.

Carey, Mark. *In the Shadow of Melting Glaciers: Climate Change and Andean Society.* Oxford University Press. 2010.

CEDHA. *Derechos Humanos y Ambiente en la República Argentina: Propuestas para una Agenda Nacional.* Advocatus. 2005

Cortázar, Julio. *Rayuela.* Editorial Sudamericana. 1963.

Corte, Arturo E. *Geocriología: El Frío en la Tierra.* Ediciones Culturales de Mendoza. 1983.

Croce, Flavia, and Milana Juan Pablo. Desarrollo de Sistemas Geocriogénicos en la Zona del Paso Agua Negra y su Importancia en Geología Aplicada. In *Actas del XV Congreso Geológico Argentino.* El Calafate. 2002.

Cruikshank, Julio. Glaciers and Climate Change: Perspectives from Oral Tradition. *Arctic* 54(4), December 2001. Pp. 377–393.

Espizua, Lydia. *Ambiente y Procesos Glaciares y Periglaciales en Lama-Veladero, San Juan Argentina.* 2006.

Francou, Bernarad. Montaña y Glaciares. In *Montaña-América Natural,* edited by Antonio Vizcaino and Ximena de la Macorra. Mexico. 2011. Pp. 32–37.

Fauqué, L., and D. Azcurra. Condiciones periglaciales en la vertiente occidental de los Nevados del Aconquija, Catamarca. Argentina. In XII Congreso Geológico Chileno. 2009.

French, Hugh M. *The Periglacial Environment.* Third Edition. Wiley. 2008.

Gascoin, S. et al. Glacier Contribution to Streamflow in Two Headwaters of the Huasco River, Dry Andes of Chile. *Cryosphere* 5, 2011. Pp. 1099–1113.

Gosnell, Mariana. *Ice: The Nature, the History, and the Uses of an Astonishing Substance.* Alfred A. Knopf. 2005.

Gruber, S. Derivation and Analysis of High-Resolution Estimate of Global Permafrost Zonation. *Cryosphere* 6, 2012. Pp. 221–233.

Haemmig, Christopher et al. Hazard Assessment of Glacier Lake Outburst Floods from Kyagar Glacier, Karakoram Mountains, China. *Annals of Glaciology* (55)66, 2014. Pp. 34–44.

Humlum, Ole. *The Climatic and Palaeoclimatic Significance of Rock Glaciers.* UNIS Department of Geology. 2010.

ICIMOD. *Glacier Lakes and Glacier Lake Outburst Floods in Nepal.* ICIMOD. 2011.

IPCC. *Fifth Assessment Report.* IPCC. March 2014.

Iza, Alejandro, and Marta Brunilda Rovere. *Apectos Jurídicos de la Conservación de los Glaciares.* IUCN. 2006.

Jiro, Komori et al. *Glacier Lake Outburst Events in the Bhutan Himalayas. Global Environmental Research* 16, 2012. Pp. 59–70.

Johansson, Emma. *The Melting Himalayas: Examples of Water Harvesting Techniques.* Bachelor's thesis. Lund University. 2012.

Kronenberg, Jakub. Linking Ecological Economics and Political Ecology to Study Mining, Glaciers and Global Warming. *Environmental Policy and Governance* 23, 2013. Pp. 75–90.

Kurter, Ajun. *Glaciers of Turkey: Glaciers of the Middle East and Africa.* US Geological Survey Professional Paper 1386-G-1. 1988.

Lliboutry, Luis. *Nieves y Glaciares de Chile: Fundamentos de Glaciología.* Ediciones de la Universidad de Chile. 1956.

Macdougall, Doug. *Frozen Earth: The Once and Future Story of Ice Ages.* University of California Press. 1994.

MIM Argentina Exploraciones. *Informe de Impacto Ambiental: Etapa de Exploración Proyecto Filo Colorado, para Xstrata Copper.* June 2005.

Mook, Pradeep et al. *Glacial Lakes and Glacial Lake Outburst Floods in Nepal.* ICIMOD. 2011.

Owen, Lewis, and John England. Observations on Rock Glaciers in the Himalayas and Karakoram Mountains of Northern Pakistan and India. *Geomorphology* 26, 1998. Pp. 199–213.

Rasul, G. et al. Glaciers and Glacial Lakes under Changing Climate in Pakistan. *Pakistan Journal of Meteorology* 8(15). Pp. 2–8. July 2011.

Raub, William et al. *Inventory of Glaciers in the Sierra Nevada, California.* USGS. 2006.

Robinson, Charles, and Peter DeaPeter. Quaternary Glacial and Slope Failure Deposits of the Crested Butte Area, Gunnison County Colorado. In *New Mexico Geological Society Guidebook*, 32nd Conference, Western Slope Colorado. Pp. 155–164. 1981.

Romanovsky, Vladimir et al. Frozen Ground. *Global Outlook for Ice and Snow 7.* UNEP. Pp. 181–200. 2007.

Scanu, Marcelo. *Leyendas de los Andes Argentinos.* Cruzpampa Editores. 2012.

Taillant, Jorge Daniel. *Impactos en Glaciares de Roca y en Ambiente Periglacial de los Proyectos Mineros Filo Colorado (Xstrata) y Agua Rica (Yamana Gold).* CEDHA. 2011.

Taillant, Jorge Daniel. *Impacts to Rock Glaciers and Periglacial Environments by El Pachón (Xstrata).* CEDHA. 2011.

Taillant, Jorge Daniel. *Glaciers and Periglacial Environments in Diaguita-Huascoaltino Indigenous Territory, Chile.* CEDHA. 2012.

Taillant, Jorge Daniel. La Democratización de los Glaciares. *Hydria* 41, 2012. Pp. 17–19.

Taillant, Jorge Daniel. *Glaciers and Mining in the Province of La Rioja, Argentina.* CEDHA. 2012.

Taillant, Jorge Daniel. *Barrick's Glaciers: Technical Report on the Impacts by Barrick Gold on Glaciers and Periglacial Environment at Pascua Lama and Veladero.* CEDHA. 2013.

Taillant, Jorge Daniel. The Human Right … to Glaciers? *Journal of Environmental Law and Litigation. University of Oregon School of Law* 28(1), 2013. Pp. 59–78.

Teckle, Nescha, and Krishan Vatsa. *GLOF Risk Reduction through Community-Based Approaches.* UNDP.

Teiji Watanbe, and Daniel Rothacher. Mountain Chronicles: The 1994 Lugge Tsho Glacial Lake Outburst Flood, Bhutan Himalaya. *Mountain Research and Development* 16(1), 1996. Pp. 77–81.

Tapscott, Don, and Anthony Williams. *Wikinomics: How Mass Collaboration Changes Everything.* Portfolio. 2007.

Trombotto, Dario. Survey of Cryogenic Processes, Periglacial Forms and Permafrost Conditions in South America. *Revista do Instituto Geológico, Sao Paulo* 21(1/2), 2000. Pp. 33–55.

Trombotto, Dario. *Mapping of Permafrost and Periglacial Environments, Cordón del Plata, Argentina.* IANIGLA. 2003.

Trombotto, Dario, and E. Borzotta. Indicators of Present Global Warming Through

Changes in Active Layer-Thickness, Estimation of Thermal Diffusivity and Geomorphological Observations in the Morenas Coloradas Rockglacier, Central Andes of Mendoza, Argentina. *Cold Regions Science and Technology* 55, 2009. Pp. 321–330.

Vick, Steven G. Morphology and the Role of Landsliding in the Formation of Some Rock Glaciers in the Mosquito Range, Colorado, Part 1. *Geological Society of America Bulletin* 92, 1981. Pp. 75–84.

Whalley, W. Brian, and Fethi Azizi. Rock Glaciers and Protalus Land Forms: Analogous Forms and Ice Sources on Earth and Mars. *Journal of Geophysical Research* 108(E4), 2003. P. 8032.

White, Christopher. *The Melting World: A Journey Across America's Vanishing Glaciers*. St. Martin's Press. 2013.

Xstrata Copper. *Proyecto El Pachón: Reporte de Sustentabilidad 2010*. 2010.

Yafeng, Shi et al. *Glaciers of Asia: Glaciers of China*. Edited by Richard Williams Jr. and Jane Ferrigno. U.S. Geological Survey Professional Paper 1386-F-2. 2005.

Young, James, and Stefan Hastenrath. *Glaciers of Africa: Glaciers of the Middle East and Africa*. U.S. Geological Survey Professional Paper 1386-G-3. 1987.

{ ABOUT THE AUTHOR }

Jorge Daniel Taillant was born in Buenos Aires, Argentina. His family immigrated to San Francisco, in 1968, where he grew up. He studied political science at the University of California at Berkeley and at the Institute d'Etudes Politiques in Lyon France, with a brief interlude at the Universidad Católica in Chile. He completed his graduate work in Washington DC, at Georgetown University, in Political Economics of Latin America. He has worked for numerous international agencies and organizations, including the US National Oceanic Atmospheric Administration (NOAA), the World Bank, the United Nations, the Organization of American States (OAS), and as a policy advisor to governments.

Daniel's professional career initially focused on development policy and later turned specifically to environmental public policy and human rights. In 1999, he returned to Argentina with his wife to found the Center for Human Rights and Environment (CEDHA), in Cordoba City, later opening an office in Patagonia. He directed CEDHA from 1999 to 2006, and returned again as Director in 2012. CEDHA focuses on the promotion of environmental legislation, public policy, human rights protection, and defending victims of environmental degradation. At CEDHA, he has designed and implemented numerous programs, including CEDHA's globally acknowledged advocacy work on corporate accountability and human rights (receiving the Sierra Club's International Earth Care Award 2007, where he shared the stage that year with both Al Gore and Thomas Friedman), as well as programs centered on international development finance, extractive industries, and hydraulic fracturing (tracking).

During 2006–2008, much of CEDHA's team moved to Argentina's federal environment agency, when CEDHA's co-founder, Romina Picolotti, was named Environment Secretary of Argentina. During that time, Daniel stepped down as CEDHA's director and served as strategic advisor to the Environment Secretariat, helping design numerous national programs, including a Sustainable Cities Program.

In 2008, Daniel launched CEDHA's Democratizing Glaciers Initiative, initially in an effort to bring back Argentina's National Glacier Protection Law, vetoed by President Cristina Fernandez de Kirchner under pressure from the mining industry. He spent much of 2008–2010 studying glaciology and channeling this newly acquired knowledge to legislative representatives during the congressional glacier debates taking place between 2009 and 2010.

Following the enactment of the world's first glacier protection law in 2010, Daniel shifted his work to what he calls "cryoactivism," educating society and policy makers on the role glaciers play in our ecosystems. He has developed glacier training material for children, and he regularly visits communities to teach about glacier relevance and vulnerability.

Daniel is the author of numerous publications on the impacts of mining on glaciers.

Contact: jdtaillant@gmail.com

{ INDEX }